MW01105281

Biomarker Methods in Drug Discovery and Development

METHODS IN PHARMACOLOGY AND TOXICOLOGY™

Y. James Kang, SERIES EDITOR

Biomarker Methods in Drug Discovery and Development
edited by **Feng Wang**, 2008

Pharmacogenomics and Personalized Medicine
edited by **Nadine Cohen**, 2008

Cytokines in Human Health: Immunotoxicology, Pathology, and Therapeutic Applications
edited by **Robert V. House and Jacques Descotes**, 2007

Drug Metabolism and Transport: Molecular Methods and Mechanisms
edited by **Lawrence Lash**, 2005

Optimization in Drug Discovery: In Vitro Methods
edited by **Zhengyin Yan and Gary W. Caldwell**, 2004

In Vitro Neurotoxicology: Principles and Challenges
edited by **Evelyn Tiffany-Castiglioni**, 2004

Cardiac Drug Development Guide
edited by **Michael K. Pugsley**, 2003

Methods in Biological Oxidative Stress
edited by **Kenneth Hensley and Robert A. Floyd**, 2003

Apoptosis Methods in Pharmacology and Toxicology: Approaches to Measurement and Quantification
edited by **Myrtle A. Davis**, 2002

Ion Channel Localization: Methods and Protocols
edited by **Anatoli N. Lopatin and Colin G. Nichols**, 2001

METHODS IN PHARMACOLOGY AND TOXICOLOGY™

Biomarker Methods in Drug Discovery and Development

Edited by

Feng Wang, PhD

Procter & Gamble Pharmaceutical, Inc., Mason, Ohio

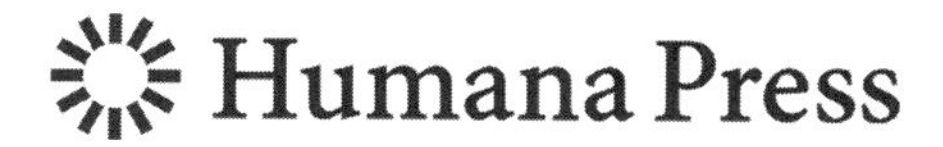

Editor
Feng Wang
Procter & Gamble Pharmaceutical, Inc.
8700 Mason-Montgomery Road
Mason OH, USA

Series Editor
Y. James Kang
Department of Medicine, Pharmacology & Toxicology
School of Medicine
University of Louisville
Louisville, KY, USA

ISBN: 978-1-934115-23-7 e-ISBN: 978-1-59745-463-6

Library of Congress Control Number: 2007943120

Printed on acid-free paper

9 8 7 6 5 4 3 2 1

springer.com

Preface

The pharmaceutical industry has faced many significant challenges since the early 1990s. The fundamental issue that needs to be addressed is how to improve the efficiency of drug discovery and development. The current research and development (R&D) cost for developing a new therapeutic drug is greater than $800 million. Additionally, it takes an average of 12 years to get a new drug to market with an attrition rate greater than 90%. Reviewing the overall pharmaceutical R&D process, it has become clear that many of the drug failures are due to our lack of knowledge in population diversity, which is responsible for differences in drug efficacy and toxicity. In fact, not a single approved drug is 100% safe and efficacious for all patients. For researchers, the key question is: How can we discover biomarkers that can be used to distinguish patients who will respond to the drug without adverse effects from those who will not respond and/or will have adverse effects? There is tremendous urgency to address this question. Biomarkers also fit perfectly with the vision of *personalized medicine*, the new expectation of medical practice. This is why biomarker research has been a central focus in many research labs across academia, government agencies, and the pharmaceutical industry.

There are many different ways to define biomarkers based on molecular properties, applications, and methods. The National Institutes of Health (NIH) suggested an inclusive definition for biomarker as "a characteristic that is objectively measured and evaluated as an indicator of normal biologic processes, pathogenic processes, or pharmacologic responses to a therapeutic intervention." A biomarker can be DNA, protein, metabolite, mRNA, or lipid. This certainly increases the complexity of biomarker research. An array of technologies is needed for biomarker research to increase the success rate. Fortunately, many existing biological and analytical technologies can be, and have already been, directly applied to biomarker research. The "omics" technologies, including genomics, proteomics, and metabonomics, have been developed and can be used for identifying potential biomarkers at several different molecular levels. The unbiased nature of these "omics" technologies is well suited for biomarker research. The goal of *Biomarker Methods in Drug Discovery and Development* is to provide a tool box for those who have a general interest in biomarker research and also for those who are currently specializing in certain technologies but want to gain an understanding of other available methodologies. Many technologies covered in this book are well validated and mature methods, whereas others are rather new but with huge promise. This book also covers some specific issues related to clinical biomarker research, such as clinical sample handling. A total of 17 chapters contributed by many experts in their research areas provide detailed descriptions of biomarker methodologies. This book is intended to be used as a guideline and a protocol reference for biomarker researchers.

Clinical biomarker research often uses patient samples and specimens. This creates significant challenges in sample collection and handling compared with tissue and animal experiments. There are many aspects of sample collection and preservation that need to be considered in order to control and reduce experimental variability. Chapter 1 provides critical insights into careful study planning to ensure that robust data can be generated from clinical samples. The examples given in this chapter cover sample collection and handling for DNA, RNA, and protein analyses from peripheral blood. This chapter also highlights many parts of the assay under development that need to be examined so that the performance characteristics can be well understood. It is important to ensure that the biomarker data generated from the assay will be solid so that accurate conclusions can be reached.

One significant outcome of the Human Genome Project was the development of numerous genomic technologies over the past decade. Many areas of research such as drug target identification, target validation, pharmacogenomics, pharmacogenetics, as well as biomarkers have benefited greatly from these new technologies. In fact, genomics has become a major tool used in biomarker research. This book highlights six genomic methodologies including gene expression, single nucleotide polymorphism (SNP), DNA methylation, and laser capture microdissection, a very useful tissue sample retrieval technique. Adverse effects and toxicity are still being predominately identified at clinical stages of drug development using lengthy and costly approaches. Good toxicity biomarkers with high predictive value are highly desired in drug R&D. Because microarray platforms offer unique advantages in identifying novel mechanism-based biomarkers, it is a powerful method to interrogate perturbations induced by experimental drugs and to pinpoint individual genes or gene sets regulated in parallel with a toxic reaction. Chapter 2 uses three examples of gene expression–based biomarkers for hepatotoxicity, nephrotoxicity, and general toxicity signatures in blood to describe applications for microarray platform technology in toxicity biomarkers. The detailed methodology of the fluorescent microspheres (microbeads) gene expression platform is described in Chapter 3. Using a panel of the "signature" gene expression pattern, the microsphere approach offers advantages in flexibility over the traditional whole-genome microarrays. Real-time PCR is another valuable and widely used gene expression methodology. High-throughput whole-genome microarrays enable screening of large numbers of genes to identify potential biomarkers whose expression levels are correlated with disease state, clinical outcome, and treatment regimens. These candidate biomarkers need to be validated with different sets of samples and, preferably, different methods. Real-time PCR technology fits extremely well for this purpose with excellent design flexibility and fast turnaround. Chapter 4 provides an overview of real-time PCR and practical assay protocols.

Gene expression analysis platforms are also useful in SNP identification. Chapter 5 describes the serial analysis of gene expression (SAGE) methodology and the bioinformatics approach for the applications of SNP analysis. The uniqueness of SAGE is that any molecular biology lab can easily perform the protocol without relying on specialized, expensive equipment. This chapter provides detailed methods and notes so one can readily follow the experimental procedures. Another important genomic biomarker tool is Pyrosequencing, which is covered in Chapter 6. Pyrosequencing is a genotyping method based on sequencing by synthesis. This technique offers accurate and quantitative analysis of DNA sequences without the presence of a restriction enzyme site. It can also be used to identify triallelic, indel, and short repeat polymorphisms, as well as to determine allele percentages for DNA methylation. Chapter 6 provides an overview for the Pyrosequencing method and assay details for commonly analyzed and clinically relevant polymorphisms such as SNPs in the cytochrome P450, as well as assay protocols for DNA methylation measurement.

Biological tissues have high degrees of cell heterogeneity. For some studies aimed at identifying specific biological pathways, analysis is preferably done using the targeted cell type. In Chapter 7, a relatively new tissue selection and retrieval method, laser capture microdissection (LCM), is described. This chapter demonstrates the utility of LCM when it is coupled with gene expression analyses using primate endometrium tissue. The LCM-collected samples certainly can be used for analysis utilizing other biomarker platform technologies.

Six protein biomarker analysis methodologies are described in this book (Chapters 8 to 13) in order to represent a variety of commonly used technologies. As in genomics, there are numerous protein analysis methods that have been used in biomarker research. Traditional two-dimensional gel electrophoresis (2-DE) is a powerful protein separation method that has been constantly improved in reproducibility and ease of use. Chapter 8 uses a clinical biomarker study as an example to illustrate the application of 2-DE/mass spectrometry (MS) method. A specific drug adverse effect was investigated using 2-DE/MS with patient plasma samples. Human plasma is a convenient sample source for clinical biomarker research. However, the large dynamic range of plasma protein concentration poses a significant challenge for analysis of medium- or low-abundance proteins, which are most likely more biologically relevant. Prefractionation step(s) are often necessary before any type of proteomic analysis. Chapter 8 also describes a detailed immunochemistry method that is commonly used in many labs to deplete the most abundant plasma proteins. It demonstrates that the 2-DE/MS approach can be used for complex samples to reveal potential protein biomarkers. In an attempt to simplify the 2-DE process, the difference in-gel electrophoresis (DIGE) method was developed. As an adaptation of conventional 2-DE, DIGE uses novel fluorescent labels so that two to three samples can be resolved on a single gel under identical electrophoretic conditions. It offers advantages in simplifying image analysis, increasing sample throughput, and reducing 2-DE experimental variation. Chapter 9 provides an overview and easy-to-follow protocols for the DIGE technology.

The rapid advancement in MS technologies has resulted in many significant developments in proteomics and protein biomarker research. Currently, there are two MS-based strategies commonly used in quantitative global proteomics. Chapter 10 covers the bottom-up strategy with the shotgun tryptic peptide liquid chromatography–mass spectrometry (LC-MS) approach, and Chapter 11 describes the top-down strategy using a set of novel protein labels to identify and quantify the differences between multiple protein samples. In the bottom-up approach, proteins are first digested by an enzyme into peptide fragments that are analyzed by LC–tandem mass spectrometry (MS/MS) and then identified by database searching. Protein quantification is achieved by measuring chromatographic peak intensity. Chapter 10 provides a comprehensive review of the technique along with the author's insights pertaining to data analysis and bioinformatics. This chapter also provides a case study as a practical example to highlight the experiment design details and data interpretation. This approach is highly sensitive and can be easily automated, offering great feasibility for large-scale protein biomarker analysis. As the top-down proteomics approach involves the analysis of intact proteins, it promises the ability to characterize posttranslational modifications and to reduce false-positive identifications because of its multiplexing capabilities. The top-down approach is well illustrated in Chapter 11 with a newly developed isobaric mass tagging technology, ExacTag Labeling system. The concept of this technology is very similar to isotope code affinity tag (ICAT). But this method labels whole proteins rather than proteolytic peptide fragments, and the subsequent protein mixtures can be enriched under the identical conditions. Experimental variation can be significantly reduced as multiple samples are tagged and mixed before any processing steps.

Other than the traditional 2-DE/MS and extensive proteomic profiling technologies introduced previously, *Biomarker Methods in Drug Discovery and Development* also covers two unique methods that are well suited for specific protein biomarker analysis needs. The surface-enhanced laser desorption ionization mass spectrometry (SELDI-MS) described in Chapter 12 is a high-throughput screening platform especially applicable for large numbers of solution samples such as plasma, urine, and conditioned media. It is a good method to quickly identify the differences among samples based on MS patterns. Further identification of the molecular entities responsible for the differences requires extra effort and most likely other technologies. When a protein biomarker is identified through any approach, validation of the biomarker with a large number of samples is required. The enzyme-linked immunosorbent assay (ELISA) array is the method of choice for validation as well as a diagnostic platform. Chapter 13 provides excellent technical details of how to generate high-quality ELISA microarrays with easy-to-follow directions and notes.

Metabolic biomarkers are also attractive in pharmaceutical biomarker research. This is because metabolite levels can be regarded as the final process readout of biological systems combining both internal factors (genetic) and external factors (disease or drug treatment). Many minor changes at the transcript and protein levels are significant enough to be biologically meaningful. These minor changes themselves may not be detectable. However, after the biological and metabolic process, they can lead to major changes at the metabolite level, which makes analysis more feasible. As an important component of systems biology and biomarker discovery, metabonomics (or metabolomics) technologies have been continually improved. MS and nuclear magnetic resonance (NMR) spectroscopy are the two core technologies used in metabolite analysis. Coupled with liquid chromatographic separation, the MS technique offers advantages in automation and high resolution. Chapter 14 discusses LC-MS methodology and its biomarker applications. It provides detailed descriptions on each component of the LC-MS platform and the multivariate statistical data analysis methods. NMR methodology is covered in Chapter 15 where a variety of NMR analysis approaches and applications are demonstrated. Chapter 16 addresses methodology for a specific group of metabolites, nonpolar analytes, using gas chromatography–MS (GC-MS). Traditional LC-MS and NMR metabolite analyses do not cover molecules with low polarities because of the differences in separation and ionization mechanisms. Cellular lipids play important roles in membrane biology and metabolic dysfunctions. Quantitative lipid analysis, as a biomarker tool, is of interest in the pharmaceutical industry. Chapter 17 introduces MS methodologies in analyzing glycerophospholipids, the main constituents of cellular membranes.

Biomarkers offer tremendous potential and promises for transforming pharmaceutical research and development processes. Biomarker research will positively impact not only new therapeutics but also diagnostics, as well as our overall understanding in the general life sciences. We need to apply multiple technologies at different molecular levels to work on this monumental and complex task. I hope this book will be used as a biomarker technical guideline and reference to stimulate more exciting biomarker research and more technology development.

I sincerely thank all the chapter authors who tirelessly took the extra time beyond their busy daily research activities to contribute to this book, and I express my genuine gratitude to all the authors for their expert knowledge and for their efforts.

Feng Wang, PhD

Contents

Preface .... *v*
Contributors .... *xiii*
Color Plates .... *xvii*

1 Biomarker Sample Collection and Handling in the Clinical Setting to Support Early-Phase Drug Development
Chris B. Russell, Sid Suggs, Kristina M. Robson, Keith Kerkof, Lisa D. Kivman, Kimberly H. Notari, William A. Rees, Natalia Leshinsky, and Scott D. Patterson *1*

2 Gene Expression-Based Biomarkers of Drug Safety
Eric A.G. Blomme and Scott E. Warder *27*

3 Profiling Gene Expression Signatures Using Fluorescent Microspheres
Suzanne M. Torontali, Jorge M. Naciff, Kenton D. Juhlin, and Jay P. Tiesman *51*

4 Real-Time Polymerase Chain Reaction Gene Expression Assays in Biomarker Discovery and Validation
Yulei Wang, Catalin Barbacioru, David Keys, Pius Brzoska, Caifu Chen, Kelly Li, and Raymond R. Samaha *63*

5 SAGE Analysis in Identifying Phenotype Single Nucleotide Polymorphisms
Sandro J. de Souza, I-Mei Siu, Janete M. Cerutti, and Gregory J. Riggins *87*

6 Single Nucleotide Polymorphisms and DNA Methylation Analysis Using Pyrosequencing Methods
Jinsheng Yu and Sharon Marsh *119*

7 The Application of Laser Capture Microdissection for the Analysis of Cell-Type-Specific Gene Expression in a Complex Tissue: *The Primate Endometrium*
William C. Okulicz *141*

8 The Use of Two-Dimensional Gel Electrophoresis for Plasma Biomarker Discovery
Brad Jarrold, Alex Varbanov, and Feng Wang ***171***

9 Difference In-Gel Electrophoresis: *A High-Resolution Protein Biomarker Research Tool*
David S. Gibson, David Bramwell, and Caitriona Scaife ... ***189***

10 Label-Free Mass Spectrometry-Based Protein Quantification Technologies in Protein Biomarker Discovery
Mu Wang, Jin-Sam You, Kerry G. Bemis, and Dawn P.G. Fitzpatrick ***211***

11 Top-Down Quantitative Proteomic Analysis Using a Highly Multiplexed Isobaric Mass Tagging Strategy
Bing Xie, Wayne F. Patton, and Craig E. Parman ***231***

12 SELDI Technology for Identification of Protein Biomarkers
Prasad Devarajan and Gary F. Ross ***251***

13 Sandwich ELISA Microarrays: *Generating Reliable and Reproducible Assays for High-Throughput Screens*
Rachel M. Gonzalez, Susan M. Varnum, and Richard C. Zangar ***273***

14 LC-MS Metabonomics Methodology in Biomarker Discovery
Xin Lu and Guowang Xu ***291***

15 GC-MS-Based Metabolomics
Sally-Ann Fancy and Klaus Rumpel ***317***

16 NMR-Based Metabolomics for Biomarker Discovery
Narasimhamurthy Shanaiah, Shucha Zhang, M. Aruni Desilva, and Daniel Raftery ***341***

17 Unraveling Glycerophospholipidomes by Lipidomics
Kim Ekroos ***369***

Index ***385***

Contributors

CATALIN BARBACIORU, PhD • *Molecular Biology Division, Applied Biosystems, Foster City, California*

KERRY G. BEMIS, PhD • *Indiana Centers for Applied Protein Sciences (INCAPS), Indianapolis, Indiana*

ERIC A.G. BLOMME, DVM, PhD • *Abbott Laboratories, Global Pharmaceutical Research and Development, Department of Cellular, Molecular and Exploratory Toxicology, Abbott Park, Illinois*

DAVID BRAMWELL, PhD • *Nonlinear Dynamics Limited, Newcastle upon Tyne, United Kingdom*

PIUS BRZOSKA, PhD • *Molecular Biology Division, Applied Biosystems, Foster City, California*

JANETE M. CERUTTI, PhD • *Laboratory of Molecular Endocrinology, Department of Medicine, Federal University of São Paulo, São Paulo, Brazil*

CAIFU CHEN, PhD • *Molecular Biology Division, Applied Biosystems, Foster City, California*

M. ARUNI DESILVA, MS • *Department of Chemistry, Purdue University, West Lafayette, Indiana*

PRASAD DEVARAJAN, MD • *Nephrology and Hypertension, Cincinnati Children's Hospital Medical Center, University of Cincinnati, Cincinnati, Ohio*

KIM EKROOS, PhD • *Department of Molecular Pharmacology, AstraZeneca R&D, Mölndal, Sweden*

SALLY-ANN FANCY, PhD • *Pfizer Global Research and Development, Department of Exploratory Medicinal Sciences, Sandwich, Kent, United Kingdom*

DAWN P.G. FITZPATRICK, BS • *Department of Biochemistry and Molecular Biology, Indiana University School of Medicine, Indianapolis, Indiana*

DAVID S. GIBSON, PhD • *Queen's University Belfast, Arthritis Research Group, Musculoskeletal Education and Research Unit, Belfast, United Kingdom*

RACHEL M. GONZALEZ, PhD • *Pacific Northwest National Laboratory, Richland, Washington*
BRAD JARROLD, MS • *Procter & Gamble Pharmaceutical, Inc., Mason Business Center, Mason, Ohio*
KENTON D. JUHLIN, PhD • *The Procter & Gamble Company, Miami Valley Innovation Center, Cincinnati, Ohio*
KEITH KERKOF, BS • *Amgen Inc., Department of Molecular Sciences, Seattle, Washington*
DAVID KEYS, PhD • *Molecular Biology Division, Applied Biosystems, Foster City, California*
LISA D. KIVMAN, BS • *Amgen Inc., Department of Molecular Sciences, Thousand Oaks, California*
NATALIA LESHINSKY, MS • *Amgen Inc., Department of Molecular Sciences, Seattle, Washington*
KELLY LI, PhD • *Molecular Biology Division, Applied Biosystems, Foster City, California*
XIN LU, PhD • *National Chromatographic R&A Center, Dalian Institute of Chemical Physics, Chinese Academy of Sciences, Dalian, China*
SHARON MARSH, PhD • *Internal Medicine, Washington University School of Medicine, St. Louis, Missouri*
JORGE M. NACIFF, PhD • *The Procter & Gamble Company, Miami Valley Innovation Center, Cincinnati, Ohio*
KIMBERLY H. NOTARI, MS • *Amgen Inc., Department of Molecular Sciences, Thousand Oaks, California*
WILLIAM C. OKULICZ, MPH, PhD • *Department of Physiology, University of Massachusetts Medical School, Worcester, Massachusetts*
CRAIG E. PARMAN, PhD • *PerkinElmer Life and Analytical Sciences, Waltham, Massachusetts*
SCOTT D. PATTERSON, PhD • *Amgen Inc., Department of Molecular Sciences, Thousand Oaks, California*
WAYNE F. PATTON, PhD • *PerkinElmer Life and Analytical Sciences, Waltham, Massachusetts*
DANIEL RAFTERY, PhD • *Department of Chemistry, Purdue University, West Lafayette, Indiana*
WILLIAM A. REES, PhD • *Amgen Inc., Department of Molecular Sciences, Seattle, Washington*
GREGORY J. RIGGINS, PhD • *Department of Neurosurgery, Johns Hopkins University Medical School, Baltimore, Maryland*
KRISTINA M. ROBSON, PhD • *Amgen Inc., Department of Molecular Sciences, Thousand Oaks, California*
GARY F. ROSS, PhD • *Bio-Rad Laboratories, Hercules, California*

KLAUS RUMPEL, PhD • *Pfizer Global Research and Development, Department of Exploratory Medicinal Sciences, Sandwich, Kent, United Kingdom*

CHRIS B. RUSSELL, PhD • *Amgen Inc., Department of Molecular Sciences, Seattle, Washington*

RAYMOND R. SAMAHA, PhD • *Molecular Biology Division, Applied Biosystems, Foster City, California*

CAITRIONA SCAIFE, MSc • *University College Dublin, Conway Institute of Biomolecular and Biomedical Research, Proteome Research Centre, Belfield, Dublin, Ireland*

NARASIMHAMURTHY SHANAIAH, PhD • *Department of Chemistry, Purdue University, West Lafayette, Indiana*

I-MEI SIU, PhD • *Department of Neurosurgery, Johns Hopkins University Medical School, Baltimore, Maryland*

SANDRO J. DE SOUZA, PhD • *Ludwig Institute for Cancer Research, Sao Paulo Branch, Hospital Alemão Oswaldo Cruz, São Paulo, Brazil*

SID SUGGS, PhD • *Amgen Inc., Department of Molecular Sciences, Thousand Oaks, California*

JAY P. TIESMAN, PhD • *The Procter & Gamble Company, Miami Valley Innovation Center, Cincinnati, Ohio*

SUZANNE M. TORONTALI, BS • *The Procter & Gamble Company, Miami Valley Innovation Center, Cincinnati, Ohio*

ALEX VARBANOV, PhD • *Procter & Gamble Pharmaceutical, Inc., Mason Business Center, Mason, Ohio*

SUSAN M. VARNUM, PhD • *Pacific Northwest National Laboratory, Richland, Washington*

FENG WANG, PhD • *Procter & Gamble Pharmaceutical, Inc., Mason Business Center, Mason, Ohio*

MU WANG, PhD • *Department of Biochemistry and Molecular Biology, Indiana University School of Medicine, Indianapolis, Indiana*

YULEI WANG, PhD • *Molecular Biology Division, Applied Biosystems, Foster City, California*

SCOTT E. WARDER, PhD • *Abbott Laboratories, Global Pharmaceutical Research and Development, Department of Cellular, Molecular and Exploratory Toxicology, Abbott Park, Illinois*

BING XIE, PhD • *PerkinElmer Life and Analytical Sciences, Waltham, Massachusetts*

GUOWANG XU, PhD • *National Chromatographic R&A Center, Dalian Institute of Chemical Physics, Chinese Academy of Sciences, Dalian, China*

JIN-SAM YOU, PhD • *Indiana Centers for Applied Protein Sciences (INCAPS), Indianapolis, Indiana*
JINSHENG YU, MD, PhD • *Departments of Pathology & Immunology, Washington University School of Medicine, St. Louis, Missouri*
RICHARD C. ZANGAR, PhD • *Pacific Northwest National Laboratory, Richland, Washington*
SHUCHA ZHANG, MS • *Department of Chemistry, Purdue University, West Lafayette, Indiana*

Color Plates

Color plates follow p. 230.

Color Plate 1 Fig. 1, Chapter 2: Comparison of serum alanine aminotransferase (ALT activity) and liver gene expression profiles induced by a hepatotoxicant. Rats were treated for 3 days with either the vehicle or the test article at daily doses causing moderate hepatotoxicity. Serum was collected and ALT activity levels quantified. Livers were sampled, and gene expression profiles were generated with microarrays. Whereas significant interindividual variability can be seen in serum ALT levels, the expression profiles are strikingly consistent among the eight test article–treated rats. Hierarchical cluster analysis was performed using Rosetta Resolver version 6.0 (Rosetta Inpharmatics, Seattle, WA). Genes shown include genes that were upregulated or downregulated by at least twofold with a p value less than 0.01. Green indicates downregulation, and red indicates upregulation (*see* discussion on p. 31).

Color Plate 2 Fig. 6a, b, Chapter 4: Validation of microarray results using TaqMan® Gene Expression Assay data set as reference (*see* complete caption on p. 76 and discussion on p. 74).

Color Plate 3 Fig. 7, Chapter 4: Validation of potential prognostic markers by TaqMan® assay–based real-time PCR. Eighty-five marker genes including the minimal set of 54 genes identified to best distinguish the luminal A and the basal-like subtypes were validated by TaqMan® Gene Expression assays (*see* complete caption on p. 78, 79 and discussion on p. 77).

Color Plate 4 Fig. 3, Chapter 9: Schematic representation of the processes involved in DIGE image analysis (*see* discussion on p. 204).

Color Plate 5 Fig. 1a, b, Chapter 11: Composition of isobaric mass tags. **(A)** Schematic structure of the isobaric mass tag described in this chapter. The tag contains four elements: ***(1)*** a reactive group with specificity toward a thiol (cysteine) or an amine (lysine, α-amino) group; ***(2)*** a cleavage enhancement moiety, which is an aspartic acid (Asp, D) proline (Pro, P) scissile bond group. Distributed around the DP sequence are the ***(3)*** low mass signal reporter and its ***(4)*** corresponding balancer sequence. **(B)** Detailed composition of seven isobaric mass tags. The tag contains a common amino acid composition. G represents glycine. The red bold "G" is heavy glycine and plain "G" is normal glycine. In MS/MS, the tags generate two sets of signals: low mass signals and high mass signals (*see* discussion on p. 234).

Color Plate 6 Fig. 5, Chapter 11: Hierarchical clustering of 440 proteins. The protein list was clustered by dynamic changes in expression levels (the most downregulated in dark-green and the most upregulated in dark-red) (*see* discussion on p. 242).

Color Plate 7 Fig. 6a, b, Chapter 11: Nucleolar protein profiling by the described isobaric mass tagging technology. **(A)** Changes in the abundance of various ribosomal protein subunits after actinomycin D treatment. Relative ratios of quantified ribosomal proteins are shown. **(B)** Changes in the abundances of DEAD box domain containing proteins after actinomycin D treatment. Relative fold changes of DEAD/H domain containing proteins are shown. HLA-B, HLA-B associated transcript (*see* discussion on p. 243).

1

Biomarker Sample Collection and Handling in the Clinical Setting to Support Early-Phase Drug Development

Chris B. Russell, Sid Suggs, Kristina M. Robson, Keith Kerkof, Lisa D. Kivman, Kimberly H. Notari, William A. Rees, Natalia Leshinsky, and Scott D. Patterson

Summary

The era of increased application of pharmacodynamic biomarker measures for early-phase drug development is upon us in response to the introduction to the clinic of targeted therapeutics. This unprecedented opportunity to perform measures of human physiology responding to pharmacologic intervention requires an understanding of sample collection, handling (preanalytical variables), and assay qualification for the results to be informative and allow decisions to be made on the development of the therapeutics under investigation. The literature is replete with biomarkers for many different applications from diagnostic to prognostic to pharmacodynamic. However, few studies have either described fully or even undertaken the necessary studies into each stage of the measurement of a biomarker from collection through to assay. In this chapter, we present insight into some of these issues with the aim of increasing the general understanding of the use of biomarkers in clinical development.

Key Words: assay qualification; biomarker; pharmacodynamic; sample collection

1. INTRODUCTION

The concept of applying pharmacodynamic (PD) biomarkers into early clinical development is gaining in acceptance *(1)*. In the past, early-phase studies (phase Ia and Ib) were typically designed to evaluate safety, tolerability, and pharmacokinetics (PK) of new therapeutics. This approach was

From: *Methods in Pharmacology and Toxicology: Biomarker Methods in Drug Discovery and Development*
Edited by: F. Wang © Humana Press, Totowa, NJ

understandable when the target of the therapeutic was not well defined. However, now that the target(s) of most new therapeutics is known, the addition of PD measures into early-phase clinical studies to determine whether the therapeutic is modulating the desired pathway has become more important. It should be noted that observation of target pathway modulation does not necessarily reflect ultimate clinical benefit; however, lack of target pathway modulation would support early termination of the clinical program. By demonstrating measurable PD effects in early clinical trials, meaningful exposure:response (PK:PD) relationships can be determined that will help guide the dose ranges to be explored in later trials and support continued development of the therapeutic agent.

Providing early evaluation of the PD effects of a new therapeutic could positively impact the high cost of drug development. A large portion of the cost of drug development is due to the expense of late-stage trials because of the large number of patients enrolled and the large number of clinical sites that are typically required to accrue patients to these studies. Currently, greater than 50% of phase III trials fail to meet their efficacy end points ***(2)***. If unambiguous no-go decisions can be made based on biomarkers showing failed mechanism of action or exposure:response relationships that are not efficacious within a predicted safety margin, the execution of more expensive later-phase trials could be eliminated thus resulting in reduced overall drug development costs.

An increasing number of phase I studies incorporate biomarkers, some of which are well understood to be direct measures of therapeutic impact on a specific pathway, whereas others are more general measures of physiologic changes ***(3–6)***. Some of these measures can be conducted on samples that are collected as part of routine practice (such as plasma or serum). However, for the promise of PD biomarkers to be fully realized, tissue types that are not routinely collected need to be obtained, and all types of samples need to be collected in a manner that is compatible with subsequent analyses. In some cases, the assays to be conducted will be well established, whereas others will be cutting-edge assays, new even to the research field. Performing these types of PD assays poses a number of issues that need to be addressed for the strategy to be successful.

This chapter will deal with these issues, from the evaluation of preanalytical variables, to sample collection, to fit-for-purpose assay qualification. All the aspects dealt with in this chapter are critical to the success of early-stage biomarker programs, particularly as an increasing number of novel assays (with respect to the clinic) are being employed. However, diligence in these issues can provide the degree of confidence necessary for the execution of robust assays, whose results can provide evidence for clear go/no-go decisions. Our overall aim in this chapter is to enhance the discussion of issues that affect biomarker programs by providing a background of some of our experiences in this setting. With the range of biomarkers that can be analyzed, the ideas presented here should not be considered all-encompassing but rather provide a basis for discussion of how individual biomarkers may be measured in the clinical setting.

2. SAMPLE TYPES AND THEIR PREANALYTICAL PROPERTIES

2.1. *Sample Selection Considerations*

The choice of samples to obtain is dictated by the markers to be tested, the accessibility of the preferred tissue, and the capabilities of the clinical site. Fixed samples on slides, flash-frozen tissue pieces, or fresh (often refrigerated) tissue can be collected in cases where preserving structural relationships is important. Whereas the fresh tissue will have the least damage directly caused by preservation methods, it will undergo the most spontaneous change, and those changes should be evaluated before using fresh tissue as a source for measuring critical biomarkers. One should not forget that the cells/tissue will continue to respond to their perturbed environment until they are rendered physiologically incapable of doing so. Whole blood can be used as a source of viable cells for measuring a biological response that does not require a larger physiologic structure (i.e., cell-cell interactions of differing cell types such as skin inflammation responses). Consistent staining of different cell types can be achieved more readily in the homogenous environment of blood (in which the concentration of drug in the blood is also maintained within the window of *ex vivo* metabolism of the drug) and can enhance measurements of biomarkers in cellular subsets. For molecular measurements where homogenization is acceptable, other tissue collections are possible, including immediate lysis in guanidinium *(7)*.

For PD assessments in early-phase clinical trials, primary considerations in tissue sample choice are data quality, ease of acquisition, value of the biomarker in informing clinical decisions, and the subjects' willingness to donate tissue. A subject's decision to enroll in a trial with PD sampling may be influenced by a range of factors including frequency and invasiveness of specimen collection. For phase I studies performed in patient populations, specimen collection from the site of active disease is preferred because these tissues provide the best PD readout; however, collection of these tissue samples may not be feasible. Thus, if the target tissue is not easily accessible, consideration should be given to surrogate tissues that can provide information on drug activity.

When the preferred tissue is not readily accessible or plausible, such as for multiple tumor biopsies from patients in a cancer trial, surrogate tissues such as skin may be considered ***(8,9)***. Less-invasive surrogates can also sometimes be found in urine (e.g., collagen breakdown products as measures of bone turnover), saliva ***(10)***, buccal punch biopsies ***(11)***, or hair follicles ***(12)***. Ideally, the surrogate tissue should process signal through the same pathways as is being targeted in the disease tissue; however, resistance mechanisms can arise in the target tissue that are not present in the surrogate, and compensatory mechanisms may be present in the surrogate that are not present in the tumor.

Histologic examination of slides is a standard practice, with a large set of qualitative assessments available such as tumor staging, psoriasis plaque staging, or tumor cell presence in lymph nodes. For more objective quantitative measurements, laser scanning cytometry ***(13)*** can also be used on stained slides, and in recent years mass spectrometry has been used to examine the physical distribution of different lower-molecular-mass proteins ***(14)***. A limitation for use of slides can be loss of molecular signals due to the chemical fixation process. Traditionally, proteins and nucleic acids from fixed cells are cross-linked or degraded and thereby less accessible or are qualitatively different in uncontrolled ways. There are now increasingly better methods for dealing with molecular assays on formalin-fixed cells ***(15,16)***, and the use of molecular biomarkers from these samples may increase.

In some cases, opportunistic sampling may be appropriate. When tissue samples are obtained, they may be under the control of the pathologist for diagnosis, and until their needs are complete, the sample will not be released for biomarker studies. Although sampling lymph nodes as a general practice may present risks to a patient, if a sample from the lymph node has been collected as part of standard patient management or monitoring, subsequent use of a portion of the collected sample to measure PD biomarkers is possible. Likewise, the major portion of prostate biopsy samples must go to the pathologist for diagnostic evaluation; however, removing a small tissue print micropeel ***(17)*** may provide an opportunity for biomarker analysis without compromising the required pathology evaluation.

Although many different individual tissues may be of value for a particular biomarker program, blood is one of the easiest to sample. Almost all clinical sites are capable of handling blood draws and have established methods for processing and shipping samples. The circulating hematopoietic cells in blood are actively involved in many aspects of inflammation, and even if the core inflammatory reactions in a disease are isolated in an inaccessible tissue, the circulating cells may carry information from when they transited from or through an inflammatory site. Furthermore, blood can accumulate proteins or other analytes from remote sites, thereby providing surveillance of multiple tissues, such as with the classic liver enzymes used to assess hepatitis and other diseases or toxicities that affect the liver. Assays on blood can be either a snapshot of the analytes present in the draw sample or a measure of the response of the blood cells to some stimulus applied *ex vivo*.

Maximizing the information to be gleaned from biomarkers in the blood requires selecting appropriate methods and minimizing intrinsic variations in collection, storage, and downstream processing. The removal of blood from circulation activates many cellular processes, the most notable of which is the initiation of coagulation. These processes can be exploited (for instance, in the deliberate use of coagulation in serum separation tubes) but must be properly controlled for to have interpretable data. An example of where this has not been rigorously considered is in the mass spectrometric analysis of the so-called low-molecular-weight serum peptidome in which the clotting process generates peptides measured by matrix-assisted laser desorption ionization mass spectrometry (MALDI-MS) or surface-enhanced laser desorption ionization mass spectrometry (SELDI-MS). The complications that arise during this process have been described ***(18)***. Several blood collection products are on the market, including tubes containing anticoagulant suitable for maintaining whole blood and specialized tubes for isolating subsets of cells or that can be used in processes that yield biochemical products such as serum, RNA, or DNA. A clinical trial may require blood collection into several different tube types for each patient because of the multiple analyte types to be measured. The use of multiple different tube types may present challenges for some clinical sites. The local clinician may not be actively supervising the phlebotomy, and the phlebotomist is often focused on filling the order for the tubes he or she is provided. Although the different tube types come with different colored caps, incorrect tubes can inadvertently be used. It is best to have the tubes prepackaged for the clinical site, with all the tubes, coded labels, and summary instructions in a single kit for each patient. Finally, it is recommended to provide a means for the clinical site to document sample collection, storage time, and temperature, which helps provide information about the preanalytical variables.

2.2. *RNA or DNA Collection from Blood*

Nucleic acid biomarkers include both cellular and free nucleic acids, where the free nucleic acids generally are assumed to represent products from damaged cells, and the cellular RNA and DNA represent the current transcriptional program and genetic and epigenetic components of the cell.

Free nucleic acids can be measured in circulation and may serve as biomarkers of systemic lupus erythematosus ***(19)***, other autoimmune diseases ***(20)***, and cancer ***(21)***. Surveillance of specific sequences in circulation can provide fetal ***(22)*** or tumor ***(21)*** genotypes. Free DNA can be collected from either serum or plasma, with serum having a higher yield ***(23–25)***. The higher yield of DNA in serum may be due to differences in the centrifugation (as discussed below for some protein analytes) or to DNA release from hematopoietic cells during serum preparation ***(26)***.

Cellular RNA has also been used as a biomarker for autoimmune disorders such as lupus ***(27)*** and leukemias ***(28)***. RNA quality is essential for generating reproducible measurements for transcriptionally based biomarkers. Traditional methods of assessing RNA quality were based on 28S to 18S rRNA ratio as determined by agarose gel electrophoresis, as well as the 260 nm/280 nm UV absorption ratio ***(29)***. The premier method for assessing quality of RNA preparations is currently an electrophoretic profile on the Agilent BioAnalyzer (Agilent, Santa Clara, CA), which provides both measurements of 28S to 18S rRNA ratios, as well as concentration estimates on samples of 25 to 500 ng. Agilent has also recently developed RIN (RNA Integrity Number) as a single metric of RNA quality and comparability that is predictive of downstream consistency in gene expression measurements ***(30,31)***. This can be combined with a UV absorption measurement of concentration and purity with the NanoDrop instrument (NanoDrop Technologies, Wilmington, DE).

We have found that PAXgene tubes from PreAnalytix/Qiagen (Valencia, CA) have consistently generated good-quality RNA from both fresh and fresh-frozen blood samples and provide a means for convenient sample collection. It is however, necessary to evaluate and stipulate the appropriate sample collection and storage when using samples collected in this tube type. In the ideal case, one could draw blood into PAXgene tubes, leave the tubes for hours to days at ambient temperature, and/or freeze the tubes and leave them for weeks to years, after which the expression profile for all transcripts would be unchanged. As described below, we have tested the effect of extended time at ambient temperature after draw and time in the freezer on the consistency of PAXgene RNA, finding that weeks in the freezer have little effect but that a difference of 24 hours at room temperature can strongly perturb the RNA profile. We use the ratio of donor-specific gene expression differences as a measure of reproducible signal. Changes seen in gene expression between PAXgene tubes collected from the same donor at a single draw but handled differently are a measure of the process-specific noise introduced.

There can be remarkable reproducibility in expression profiling from two different preparations from the same draw left for different times in the freezer. Multiple PAXgene tubes were collected from three individuals, incubated for 2 to 3 h at room temperature, transferred to a –70ºC freezer, and stored upright in a rack in a secondary sealed plastic box. Tubes were removed from the freezer at different times, thawed, and RNA prepared. **Figure 1** compares the results of expression profiling for two of the donors, with RNA prepared from PAXgene samples after 1 week or 26 weeks of freezing. After RNA preparation, the samples were profiled on Affymetrix U133A2 chips (Affymetrix, Santa Clara, CA), and the normalized intensities of the approximately 13,000 sequences with the highest intensity were compared. Some random fluctuation in intensity measurements can be expected from the microarray process, and this can be statistically evaluated by various means. Using the Rosetta Resolver Error Model (Rosetta Biosoftware, Seattle, WA), set with a p value detection cut-off of 0.01, we would expect about 130 differences between two identical preparations from this gene set. For each of the three donors, comparing the two paired preparations (from each donor, one at 1 week and 26 weeks) only 69, 104, or 243 interdonor differences can be detected, which indicates very little process-specific differences. In fact, more than 90% of the 13,000 sequences have expression levels measured within a factor of 1.5 of each other when comparing the preparations from the same individual processed 25 weeks apart. The remaining sequences have generally lower intensity and noisier measurements and can be statistically evaluated as having lower probability of being a true change. These results are similar to the intradonor differences in side-by-side preparations suggesting that moderate-term storage at –70ºC does not compromise RNA quality or change the relative levels of the different transcripts. Comparatively, between any two individuals, more than 2000 differences are detected at either time point or across time points. The magnitude of these interdonor differences are remarkably consistent, and the correlation coefficient is greater than 0.93 for all the sequence expression levels that are detected as changed, in either ratio (**Fig. 1A, B**). When all 13,000 sequences are included, the correlation coefficients are still a minimum of 0.73, indicating broad reproducibility. This high correlation continues when RNA from tubes frozen for up to a year are compared with RNA processed within the first week.

A different story emerges when PAXgene tubes are held at ambient temperature for variable lengths of time prior to processing. Although the PAXgene literature from PreAnalytix suggests that the sample is stable at room temperature for days after blood draw and shows consistency of transcript stability for a few sequences, it is recommended that this be

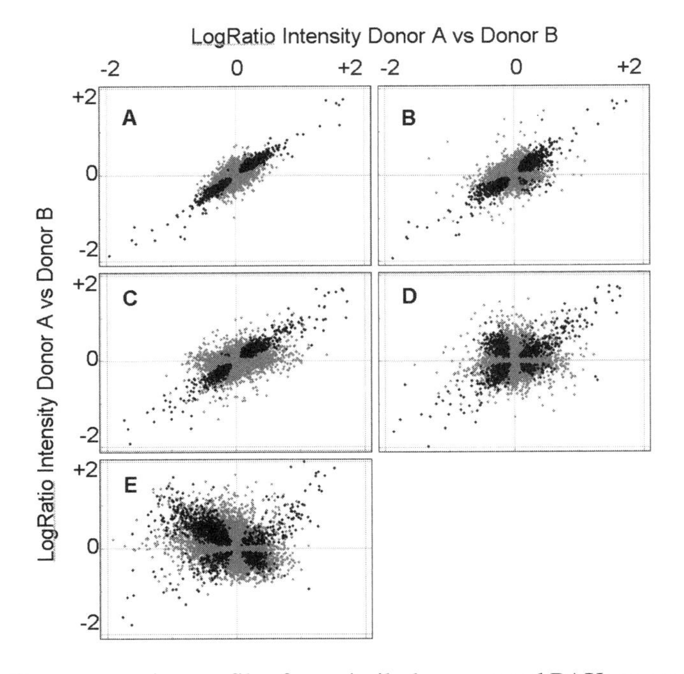

Fig. 1. Gene expression profiles from similarly processed PAXgene preparations show a high degree of similarity, whereas those with different ambient incubation times can show strong differences. Each panel is a comparison of the log ratio of intensities of two samples from different donors, such that a value of +2 indicates 100-fold higher expression in donor B, and –2 indicates 100-fold lower expression compared with donor A. Gene sequences that show increased expression in donor B lie in the upper-right quadrant, and those with decreased expression lie in the lower-left quadrant. If the two comparisons were perfectly identical, all points would lie on a 45-degree line from lower left to upper right The black points indicate sequences that were called "significantly changed" in the two ratios by the software. In all cases, the approximately 13,000 most highly expressed sequences are shown. **(A)** The comparison between profiles from two donors prepped after either 1 week (x-axis) or 26 weeks (y-axis) frozen at –70°C. The ratios are very similar with most points close to the 45-degree diagonal and high correlations (0.91 for changed genes, 0.90 for all). **(B)** A comparison of the same preps, but now the x-axis is the log ratio between the 26-week frozen sample of donor A and the 1-week sample of donor B (the reciprocal of time/samples). Even with preps at very different times, the results still identify the differences between donor A and donor B, with high correlations (0.93 for genes called "changed," and 0.73 for all). Incubation of the PAXgene tubes for different times at ambient temperatures resulted in less-consistent profiles. **(C)** Comparisons for two different donors where on the x-axis is the ratio between samples incubated for 2 h at room temperature before prepping, and the y-axis compares identical tubes left for 24 h. Within either time point, the comparison is very similar as shown by the strong correlation

tested for the transcripts of interest. We conducted a similar experiment to the one above; however, in this experiment, samples were collected and then left at room temperature for 2, 24, or 48 h before processing. When gene expression profiles are compared between donors at either 2 or 24 h, a consistent set of interdonor differences are still seen, with correlation coefficients greater than 0.95 (**Fig. 1C**), though the differences somewhat diminish by 48 hours. The problem is that comparisons across different incubation times are now vastly different. In fact, intradonor differences between 2 and 24 h or 24 and 48 h are of the same order as the interdonor differences observed at a single time point When an interdonor comparison is confounded with an inter–time-point comparison, the true donor-to-donor differences can be masked. For instance, comparing interdonor cross–time-point ratios for the donor A sample that sat at ambient for 2 h versus the 24-h sample for donor B, and the ratio for the donor A 24-h preparation versus the donor B 2-h preparation, the different ambient incubation times have a large confounding effect and the correlation coefficients drop to less than 0.4 for the identified changes or less than 0.2 for all sequences (**Fig. 1D**). The differences between 2 and 48 h are even greater, vastly exceeding the interindividual variation and generating a negative correlation coefficient (**Fig. 1E**). It should be noted that whereas 70% of the RNA transcripts do not show more than a twofold intradonor change over the first 48 h after blood collection, close to 30% of the transcripts do change. This exceeds the number that would change in many treatment situations, and the confounding noise could swamp out the true signal. It would be potentially disastrous if the transcripts of interest in a particular PD assessment would be included in the set of process-sensitive transcripts. Complex profiles such as those for lupus involve many transcripts and thus increase the odds that the transcript profile would change because of variable incubation time of the PAXgene tubes at ambient temperature.

Fig. 1. (Continued) (0.97 for changed sequences, 0.78 for all). In contrast with extended freezing, the deviations in room temperature incubations can result in a loss of cross-comparability. **(D)** The results when ratios are built between the 2-h donor A and 24-h donor B sample (x-axis) and the 24-h donor A and 2-h donor B sample (y-axis). Now the correlation has dropped considerably (0.36 for changed sequences, 0.17 for all), and in fact 40% of the sequences called "changed" have the opposite change in the other ratio. Comparing samples prepped with 24 h difference in the incubation could lead to loss of signal and actually misidentifying noise as signal. The situation becomes even more extreme when samples prepped after 2 h and after 48 h are compared in **(E)** where the correlation coefficients become negative and most identified changes are artifacts.

As such, ensuring that samples are collected and handled in the manner specified in the protocol and defined by experiments such as those described above is critical to obtaining robust data from any samples collected from clinical studies. Just as important is the recording of any deviations from the sample collection protocol so that those samples can be flagged for destruction and not included in any analysis.

2.3. *Protein Analyte Measurement from Blood*

For analytes measured from blood, different results will be detected from the same sample collected in different blood collection tubes. A broad look at protein analytes from normal serum or plasma shows a number of specific differences generated based on the tube type used for collection. Rules Based Medicine (RBM, Austin, TX), a company that provides multi-analyte profiling testing, looked at the differences in results from the same sample collected as serum, and in three different plasma collection tubes, with Ethylenediaminetetraacetic acid (EDTA), citrate, or heparin as anticoagulant ***(32)***. As expected, they found that circulating fibrinogen was easily measured in the plasma tubes and absent in serum tubes, presumably consumed in the clotting reaction. However, they also observed that other analytes were more easily detected in serum, such as growth factors EGF (17× to 44× higher in serum than plasma), and brain-derived neurotrophic factor (BDNF) (8× to 17×). Some analytes measured were below the detectable range in EDTA plasma, such as creatine kinase (CK-MB) and matrix metalloproteinase-3 (MMP-3), whereas others such as monocyte chemoattractant protein-1 (MCP-1) and vascular endothelial growth factor (VEGF) show of the order twofold reduced detection compared with serum. Conversely, serum glutamic oxaloacetic transaminase (SGOT) was only detectable when the sample is collected in EDTA-plasma. In independent enzyme-linked immunosorbent assays (ELISAs), we have also seen MCP-1 concentration reduced even more drastically in tumor necrosis factor-α (TNF-α) whole-blood stimulations performed from samples collected in EDTA plasma versus heparin plasma.

It is not clear what drives the differences seen between the various collection tubes in all cases, but it is notable that differences exist in the measurements of analytes collected in either EDTA-plasma or citrate-plasma collection tubes. As sample stability often seems similar, and the anticlotting mechanism is presumed for both to be based on chelation of divalent cations, this would suggest that other interactions of the analytes with the tube buffer are responsible. One candidate to explain the differences in analyte concentrations observed between the different collection tubes could also be differential sedimentation or lysis rates of platelets or other particles when the tubes are centrifuged as is discussed below. With the time constraints normally present when qualifying an assay for an upcoming clinical trial, it may be impossible to fully understand all the factors driving differential results for any particular analyte. Therefore, given that individual analytes perform differently in different settings, one should approach an assay for a particular analyte with the knowledge that sample collection in different tube types is a likely source of variability. A complete readout for multiple analytes may require more than one blood collection tube, which may or may not be feasible. Even then, complete characterization of the process from blood collection to sample measurement is necessary to have confidence in the results. As an example of this, we review our experience with a multianalyte measurement of angiogenic cytokines.

3. FIT-FOR-PURPOSE ASSAY QUALIFICATION

3.1. Components of the Qualification of a Multiplexed Assay for Angiogenic Cytokines

Angiogenesis, the formation of new blood vessels, has been a focus for the development of anticancer therapies. Several antiangiogenic therapeutics have been approved or are in clinical development including bevacizumab (anti-VEGF antibody), sunitinib, and PTK787/ZK 222584 [small-molecule antagonists of KDR (VEGFR-2) and other tyrosine kinases]. Amgen is developing new antiangiogenic therapeutics, and as part of the development of these molecules, we have investigated the use of several angiogenic cytokines and soluble receptors as PD biomarkers of drug efficacy *(6)*. Angiogenic cytokines have been shown to be elevated in the serum and plasma of cancer patients *(33)* and have previously been shown to be modulated upon dosing of antiangiogenic therapeutics *(34)*. The following describes a few of the issues that need to be evaluated for the measurement of these analytes including sample freeze-thaw stability, analyte interference, and identification of conditions to diminish analyte interference. What are not presented are the other aspects of fit-for-purpose assay qualification that we employ, which include parameters such as assay linearity, assay range (upper and lower limits of quantitation), accuracy, precision (inter-/intra-assay reproducibility), parallelism, interference (drug, matrix, abnormal samples [hemolyzed, icteric, and lipemic]), and sample stability.

To measure levels of angiogenic biomarkers, we used two multiplexed assays developed by Meso Scale Discovery (MSD; Gaithersburg, MD). We selected the MSD assay format because multiplexing conserves sample and allows multiple analytes to be assessed in the same assay. Another beneficial feature of using MSD assays is that they have wide dynamic ranges (in some cases 3 to 4 logs), which can reduce the need to reanalyze samples, and this in turn can reduce the need for additional freeze-thaw cycles and minimize volume requirements of precious samples. One of the MSD assays used to measure levels of angiogenic cytokines is a 4-plex that includes basic fibroblast growth factor (bFGF), placental growth factor (PlGF), VEGF, and soluble Flt-1 (fms-like tyrosine kinase 1; VEGFR-1). The second MSD assay is a 2-plex that includes soluble KDR (kinase insert domain receptor; VEGFR2) and soluble cKit.

During assay qualification, bFGF, soluble Flt-1, and VEGF were shown to be unstable over multiple freeze-thaw cycles. Previous studies indicated that sample preparation is important for the evaluation of VEGF and bFGF, as these analytes are known to be sequestered by platelets ***(35–37)***. Because VEGF is released from platelets during the preparation of serum, plasma is the matrix of choice for measuring this analyte ***(38)***. Several different centrifugation conditions can be used in the preparation of plasma, resulting in plasma containing varying amounts of platelets. We investigated the impact of different centrifugation conditions used in preparing plasma on the measurement of levels of the analytes in our angiogenic panel.

Plasma was prepared from blood of a normal donor using three centrifugation conditions: $760 \times g$ for 15 min, $1500 \times g$ for 10 min, or $2800 \times g$ for 30 min. The latter preparation conditions result in plasma from which most of the platelets have been removed and is referred to as platelet-poor plasma ***(38)***. The plasma samples were subjected to multiple freeze-thaw cycles, and the levels of bFGF, VEGF, PlGF, and soluble Flt-1 were measured using the MSD 4-plex assays. As shown in **Table 1**, when plasma samples were prepared from a normal donor using centrifugation at either $760 \times g$ for 15 min or $1500 \times g$ for 10 min, levels of bFGF, soluble Flt-1, and VEGF increased after a single cycle of freeze-thaw. Levels of these three analytes continue to increase with increasing numbers of freeze-thaw cycles, with the increase in VEGF being particularly notable. In contrast, the level of PlGF did not change appreciably when subjected to three cycles of freeze-thaw regardless of the centrifugation conditions used to prepare plasma. Similar results were obtained with plasma prepared from three other normal donors (data not shown). We also investigated the effect of centrifugation conditions used for plasma preparation on the recovery of soluble cKit and soluble KDR after multiple cycles of freeze-thaw. As shown in **Table 2**, recovery of these analytes is stable through three freeze-thaw cycles, regardless of centrifugation conditions.

The increases in measured levels of VEGF and bFGF observed upon freeze-thaw of plasma prepared by centrifugation at $760 \times g$ for 15 min or $1500 \times g$ for 10 min are consistent with release of these cytokines from platelets ***(38)***. However, the sequestration of soluble Flt-1 by platelets has

Table 1
Plasma Prepared Under Different Conditions and Subjected to Three Freeze-Thaw Cycles Shows Altered Analyte Recovery for bFGF, Soluble Flt-1, and VEGF but Not PlGF*

		Centrifugation speed (× *g*)					
		760 (15 min)		1500 (10 min)		2800 (30 min)	
	Freeze-thaw cycle	pg/mL	Recovery	pg/mL	Recovery	pg/mL	Recovery
bFGF	0	116	100	76	100	3	
	1	177	153	124	163	2	
	2	185	159	143	189	3	
	3	202	174	142	188	2	
Soluble Flt-1	0	90	100	79	100	100	100
	1	111	124	91	115	97	97
	2	121	134	110	139	101	101
	3	124	138	101	127	116	116
PlGF	0	15	100	14	100	13	100
	1	15	103	15	106	13	102
	2	16	111	16	112	14	102
	3	16	111	15	109	15	113
VEGF	0	242	100	163	100	139	100
	1	362	150	239	146	143	103
	2	473	196	322	197	144	103
	3	504	208	313	191	151	108

*The concentrations shown are the mean value of three replicates. Recovery is calculated as percent of cycle 0 (fresh). Recovery was not calculated for bFGF measured in plasma prepared at 2800 × *g* for 30 min due to the low amounts recovered.

not been previously reported. VEGF is a ligand for the Flt-1 receptor, and soluble Flt-1 may bind to VEGF in circulation. One possible explanation for the increase in Flt-1 concentration upon multiple freeze-thaw cycles is that soluble Flt-1 binds to VEGF in circulation and in turn is sequestered with VEGF by platelets.

VEGF and PlGF are ligands for the Flt-1 receptor. Maynard et al. ***(39)*** reported that soluble Flt-1 interferes with the measurement of VEGF and PlGF by ELISA. We decided to test the effect of soluble Flt-1 on the measurement of VEGF and PlGF in our MSD immunoassays. Using Flt-1 standard obtained from both MSD and R&D Systems (Minneapolis, MN), we tested the impact of spiking increasing amounts of standard into plasma

Table 2
Plasma Collected Under Different Conditions and Subjected to Four Freeze-Thaw Cycles Showed Similar Recovery of cKit and KDR*

		Centrifugation speed (× *g*)					
		760 (15 min)		1500 (10 min)		2800 (30 min)	
	Freeze-thaw cycle	pg/mL	Recovery	pg/mL	Recovery	pg/mL	Recovery
cKit	0	2691	100	2564	100	2447	100
	1	2850	106	2512	98	2423	99
	2	2644	98	2686	105	2493	102
	3	2444	91	2551	99	2519	103
KDR	0	536	100	532	100	508	100
	1	581	108	527	99	542	107
	2	526	98	552	104	514	101
	3	499	93	541	102	547	108

*The concentrations shown are the mean value of three replicates. Recovery is calculated as percent of cycle 0 (fresh).

samples on the levels of analytes as determined in our MSD 4-plex assay. As shown in **Table 3**, the detectable levels of VEGF decreased with the addition of increasing concentrations of Flt-1, with the highest spike level resulting in 60% recovery of VEGF.

Spiking Flt-1 into the sample did not effect the recovery of PlGF or bFGF (**Table 3**). In a separate experiment, the recovery of PlGF was not affected by a Flt-1 spike in plasma samples with higher intrinsic levels of PlGF under conditions in which VEGF recovery was reduced as before (data not shown). The lack of Flt-1 interference in the measurement of PlGF contradicts a previous study in which spiked Flt-1 interfered with the measurement of PlGF *(39)*. The authors of the previous report used an ELISA from R&D Systems to measure levels of PlGF. We obtained PlGF ELISA kits from R&D Systems and evaluated the impact of spiked Flt-1 on PlGF recovery, and again we did not observe interference by Flt-1 in the measurement of PlGF (data not shown).

To further evaluate interference in the VEGF assay, we investigated the impact of KDR on the measurement of VEGF in plasma. As shown in **Table 4**, the recovery of VEGF is reduced in the presence of spiked KDR, and addition of spiked Flt-1 exacerbates the reduction in the quantitation of VEGF. To mitigate the interference of Flt-1 and KDR in the measurement

Table 3
The Concentrations of VEGF, PlGF, Soluble Flt-1, and bFGF Measured in Presence of Increasing Levels of Spiked Flt-1*

		Concentration (pg/mL)			
	Spike	VEGF	PlGF	Soluble Flt-1	bFGF
Meso Scale Discovery Flt-1	None	2274 (100%)	25	206	28
	Calibrator diluent	2441 (107%)	25	192	26
	0.5 ng/mL Flt-1	1877 (83%)	20	705	21
	5 ng/mL Flt-1	1799 (79%)	23	4783	24
	15 ng/mL Flt-1	1356 (60%)	24	> ULOQ	26
R&D Systems Flt-1	None	2675 (100%)	30	247	31
	Calibrator diluent	2650 (99%)	26	219	25
	0.5 ng/mL Flt-1	2169 (81%)	24	698	27
	5 ng/mL Flt-1	2192 (82%)	26	5503	28
	15 ng/mL Flt-1	1590 (60%)	28	> ULOQ	31

ULOQ, upper limit of quantitation.

*The concentrations shown are the mean value of three replicates. Values in parentheses are percent recovery of VEGF.

Table 4
The Concentrations of VEGF and PlGF Measured in the Presence of Spiked Flt-1 and/or KDR*

	VEGF			PlGF		
Spike	Concentration (pg/mL)	CV%	% Mock spike	Concentration (pg/mL)	CV%	% Mock spike
None	1943	2		22	3	
Calibrator diluent	1827	3		20	3	
KDR	1577	0	86	21	1	106
Flt-1	1588	5	87	23	2	117
KDR + Flt	1384	0	76	22	4	109

*The concentrations (pg/mL) are the mean value of three replicates. The percent mock spike is relative to the concentration measured in calibrator diluent. CV% is the Coefficient of Variation expressed as a percentage.

Table 5
The Effect of Pretreatments on Recovery of VEGF and PlGF in Plasma*

	% Mock spike	
Treatment	VEGF	PlGF
80°C for 10 min	64	174
30 mM acetic acid	81	86
0.005% SDS	81	104
0.1% Tween	81	110
0.1% Triton	81	109
100 mM Urea	76	109
200 mM NaCl	91	113
400 mM sodium deoxycholate	30	48

*Flt-1 was spiked into plasma from a normal donor at a level of 15 ng/mL. Recovery is the percentage of untreated plasma. The percent mock spike is relative to the concentration measured in calibrator diluent.

of VEGF, several pretreatment methods were evaluated. As shown in **Table 5**, pretreatment with acetic acid, SDS, Tween, Triton X-100, or urea did not improve recovery of VEGF, whereas pretreatment with heat (80°C for 10 min) or sodium deoxycholate resulted in poorer recovery of VEGF. Of the pretreatment conditions tested, 200 mM NaCl resulted in the best recovery of VEGF. To further optimize pretreatment conditions, we evaluated pretreatment with different concentrations of NaCl on the recovery of VEGF and PlGF in plasma from a normal donor spiked with Flt-1 and/or KDR (**Table 6**). Interference in VEGF quantitation was observed by spiking each VEGFR alone and was increased when both VEGF receptors were spiked together. Treatment with NaCl increased recovery of VEGF when each VEGFR was spiked alone, but was not able to completely overcome the inhibition by both Flt-1 and KDR spiked together. These spiking experiments are meant to explore the endogenous analyte interactions occurring in the samples, which are likely to impact assay results. Readout of a single analyte may be affected by levels of other analytes, due to expected physiologic binding. However, to obtain accurate measurement of an analyte, these parameters should be considered and ideally the assay developed to eliminate these assay variables.

As can be seen from this discussion, there were several aspects to sample handling and cross-analyte interference that were found to affect the results of the assays. In some cases, the process could be controlled to reduce the

Table 6
The Effect of NaCl Pretreatment on Recovery of VEGF and PlGF in Plasma*

	% Mock spike	
Pretreatment	VEGF	PlGF
KDR spike	86	106
Flt spike	87	117
KDR and Flt spike	76	109
Flt spike + 400 mM NaCl	97	126
Flt spike + 300 mM NaCl	93	121
Flt spike + 200 mM NaCl	93	122
KDR spike + 400 mM NaCl	106	127
KDR spike + 300 mM NaCl	99	130
KDR spike + 200 mM NaCl	98	117
KDR and Flt spike + 400 mM NaCl	87	129
KDR and Flt spike + 300 mM NaCl	87	125
KDR and Flt spike + 200 mM NaCl	82	123

*Normal donor plasma was spiked with Flt-1 at 15 ng/mL and/or KDR at 50 ng/mL.

interfering noise that was generated (i.e., altering centrifugation conditions). In the case of VEGF-VEGFR interference, we were unable to completely control the effect but were able to minimize it and collect a body of data that defined the nature of the interference, such that it could be considered in analysis. This highlights the need to understand the performance characteristics of the assay under development. Other issues not touched upon in this section that need to be considered in multiplex assays include the placement of samples and standard in the assay plate (due to the close proximity of the analytes in the well) and identification of low, medium, and high controls. All of these parameters must be addressed if the performance characteristics of the assay are to be understood and the assay qualified to the appropriate level (fit-for-purpose).

3.2. Ex Vivo *Stimulation of Blood: Preanalytical and Assay Qualification Considerations*

Ex vivo stimulation of whole blood generally encounters all of the same variables observed with direct plasma collection, with additional variables that are introduced from continuing cellular activity and the reaction of the blood with the anticoagulant or other components of the tube. *Ex vivo* stimulation assays can be especially valuable in PD assays of biochemical coverage where the therapeutic is expected to modulate a well-expressed stimulus-response pathway in a common circulating cell type. The stimulus is often the addition of a natural cellular signal such as a cytokine or lipopolysaccharide (LPS). The response can be measured as changes in protein phosphorylation, cytokine release, gene expression, or cell surface marker expression. Such measurements are employed in drug development programs where the aim is to determine the effect of therapeutic intervention on a specific response (represented in whole blood) prior to and at various times after dosing.

Informative comparison of the response in the presence or absence of therapeutic requires a minimal level of robustness of the response in the range of clinical settings in which it will be measured, in spite of an underlying inherent dynamic response of the blood to its new environment. Almost immediately, the blood temperature is lowered in the phlebotomy tubing even before the blood gets to the receiving tube where it encounters the anticoagulant, usually in a dried matrix that dissolves in the blood. The environment within the collection tube is also much different than that of the body (no carbon dioxide, oxygen, nutrient, or byproduct exchange) and is missing all the homeostatic influences of the rest of the body.

Unfortunately, the program the blood executes *ex vivo* may seriously influence the readout of an assay in ways that can translate poorly between different laboratory and clinical settings. To get an overview of the magnitude of the collection tube response and differences between anticoagulants, we examined the changes in the transcriptional program over the first 6 h after collection using DNA microarrays. Blood was collected from three healthy donors into tubes with EDTA, citrate, or heparin as anticoagulant, as well as directly into PAXgene tubes. The anticoagulant samples were left at room temperature, with individual samples removed from their collection tubes and transferred to PAXgene at 1-h intervals, for up to 6 h. As observed in microarray experiments using Affymetrix U133A2 chips, there was induction of several early-response genes, including *FOSB, JUN, EGR1, EGR2*, and *EGR3*, by up to 100-fold in 2 h compared with sample transferred to PAXgene immediately after draw (data not shown). There is also a consistent induction of some genes involved in inflammatory reactions, including *IL8, DUSP2*, and *CD83* as well as *STK17b*, the latter of which is associated with induction of apoptosis ***(40)***. **Table 7** lists some of the modulated genes and their altered expression levels in different anticoagulants to illustrate three (immediate early gene, inflammatory and apoptotic induction responses) of the many processes that appear to be taking place in any plasma blood draw at ambient temperature. These common responses

Table 7
The Effect of Anticoagulant on the Expression of Specific Genes*

	Citrate	EDTA	Heparin
JUN	33.2	4.9	5.7
FOSB	100.0	50.4	21.0
EGR1	75.0	49.9	39.8
EGR2	18.9	48.2	36.3
EGR3	35.6	10.5	4.7
IL8	91.0	42.3	14.8
CD83	19.6	2.5	1.3
DUSP2	33.0	6.2	5.3
STK17B	2.7	3.4	2.7

*Blood was collected into citrate, EDTA, or heparin as an anticoagulant and incubated for 2 h. RNA was prepped after transfer to PAXgene tubes and compared with a time zero collection. The values in the table represent RNA quantity, measured as a ratio, from Affymetrix intensity measurements analyzed in Rosetta Resolver. The citrate tubes had a measurable endotoxin effect, which is probably responsible for some of the induction of *IL8, CD83*, and *DUSP2*.

were seen in all three donors, as well as two other donors in a separate experiment (data not shown). During the first 6 h, more than 2000 detectable transcripts showed changes relative to the baseline, with greater than 80% of these changes similar (to different extents) for all three donors and in all three anticoagulants.

In addition to the common time-dependent transcriptional changes observed across donors, there were also some transcript changes that were specific to individual donors or anticoagulants. However, by comparing within a single lot of a particular collection tube, almost all responses observed were consistent between donors, suggesting that keeping good control on collection conditions will result in a consistent background activity in a whole-blood stimulation assay in plasma.

Unsurprisingly, an even stronger inflammatory signal can be seen when the collection tubes contain endotoxin contamination. Whereas the manufacturers strive to produce collection tubes with minimal endotoxin, the products themselves are not often tested for this. We have observed significant lot-to-lot variation in our laboratory, and we and others ***(41)*** have occasionally seen extreme endotoxin effects in certain lots of tubes. In fact, retrospective analysis of the tubes in the previous time-course experiment showed detectable endotoxin contamination in the citrate tubes. When

Table 8
The Effect of Endotoxin Contamination on Gene Expression in Whole Blood*

	Citrate	EDTA	Heparin
CCL4	18.3	2.2	0.6
IL1B	10.4	0.8	0.4
CCL3	40.6	5.3	1.5
CCL20	58.9	1.0	0.7
TNF	44.2	1.8	0.7
CXCL2	100.0	3.6	1.0

*The citrate tubes that were known to be endotoxin contaminated exhibited broad induction of inflammation genes by 2 h, which was not seen in the other tubes without endotoxin contamination (or in other lots of citrate tubes without endotoxin). The values in the table represent RNA quantity, measured as a ratio compared with time zero as in Table 7.

present at high levels, the endotoxin induces expression of a wide range of cytokine and chemokine genes, observable within an hour, and predominate the expression profile by 2 h. Expression of effected cytokine and chemokine genes can be increased more than 100-fold even in only moderate levels of endotoxin. Listed in **Table 8** are six genes with consistently increased expression as a result of contaminating endotoxin. The table shows the change in gene expression at 2 h relative to immediately after draw for the contaminated citrate tubes and the endotoxin-free EDTA and heparin tubes. This lot of citrate tubes tested could fail well-designed quality-control criteria, such as consistent response in a known sample, or low baseline in a 24-h whole-blood stimulation assay, and implementation of this type of quality-control criteria is recommended to mitigate confounding effects due to endotoxin contamination. Undetected endotoxin contamination could generate the internal production of TNF-α, IL-1β, or other signal could independently stimulate the cells and possibly overwhelm the experimental effect.

Endotoxin contamination can vary dramatically between different lots of tubes. For example, we have tested several different lots of heparin tubes from one manufacturer and found variable endotoxin signals from all of them. Laboratory safety considerations have led many people to consider plastic collection tubes as a standard. Unfortunately, we have yet to find a lot of plastic tubes with low levels of endotoxin, whereas most lots of glass tubes do have sufficiently low levels and therefore allow reproducible measurements from *ex vivo* stimulation assays. The danger of endotoxin contamination makes it essential to perform up-front quality assessments for blood collection tubes to be used for a whole-blood assay and, if possible, to perform an entire study with one lot of tubes.

The complex reaction of blood in the collection tube can play out in the background of a whole-blood stimulation assay. The reaction can produce interfering signals and also sensitize the blood to other stimuli. Assay qualification of a stimulus release assay should always include testing the time between blood draw and initial stimulation. **Figure 2** shows the result of assay qualification tests on two donors for an 80% whole-blood TNF-α stimulation. The blood was collected into heparin and kept sealed at ambient temperature until being transferred to 96-well plates and diluted to 80%

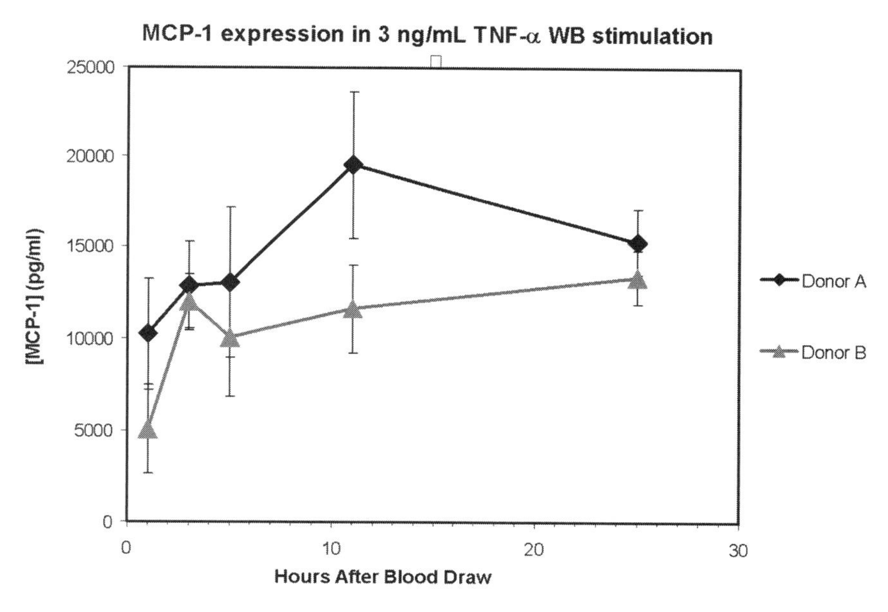

Fig. 2. Monocyte chemoattractant protein expression in TNF stimulation of whole blood varies with time after blood draw. Blood from two different donors was drawn in heparin and incubated at room temperature for various times until being transferred into the wells of a 96-well plate where it was diluted to 80% whole blood with media containing TNF-α at a final concentration of 3 ng/mL and placed in a 37°C incubator at 0.5% CO . The plates were spun 24 h later to separate the plasma, which was assayed for MCP-1 by ELISA. The x-axis is the time at ambient temperature from blood draw until stimulation, and the y-axis is the measured concentration of MCP-1 in the plasma. All points are the result of ELISA measurements on triplicate stimulations. In both donors, the amount of MCP-1 produced dramatically increased as a result of a few extra hours at ambient temperatures in sealed tubes.

with media containing the TNF-α stimulus. The samples were immediately transferred to a 37°C incubator at 0.5% CO_2. All the samples had a strong response to TNF-α, but the size of the response noticeably increased as the heparin tubes were incubated prior to being transferred to the 96-well plate. In some PD assays, this change could be very misleading if the time dependence was not well characterized. One can imagine a TNF-α stimulation assay for an anti-inflammatory agent that would require a blood draw from a patient prior to administration of the therapeutic and then an additional blood draw 4 h after the first sample. If the predose and postdose samples from a placebo-treated individual were then simultaneously stimulated an hour after the second draw, they would effectively represent the third and first points of the graph, respectively. One would conclude that donor A had a predose induction of about 13,000 units and a postdose induction of about 10,000, whereas donor B dropped from 10,000 to 5000. The variation introduced by sample collection will outweigh that observed effect within the study and negatively impact the ability to interpret the data.

The major conclusion to be drawn from these examples is the absolute necessity of qualifying an assay to cover the variables that could be encountered in a clinical trial. The assay range, together with how expected fluctuations in assay conditions can affect the results, must be characterized and considered as a package in deciding if the assay will satisfy the fit-for-purpose requirements. Again, not all these effects can be eliminated, but characterization of the effects during assay qualification can give one the confidence that the assay is fit-for-purpose or can identify its insufficiency.

It is also worth considering the transfer of the assay from its laboratory setting into the clinical setting. In a whole-blood stimulation assay, where stimulus, incubation, and subsequent harvest are to be performed at the site, the nature of the patient visit and standard practices at distant contract sites should be considered. A protocol that specifies sample processing at 6 or 8 h after blood collection may not be feasible, and a protocol with a 4-h or 24-h incubation may give more consistent results. Even a 4-h incubation will work better if patient visits can be restricted to mornings and the harvest step streamlined.

Thus, careful planning of the actual protocol to be executed at the clinical site (including how much an operator can perform in a given time period) is paramount to a successful program. We have discussed how particular details such as protocol timing, centrifugation speed, and blood collection products can affect the results. There can be other unsuspected variables that will affect the results. A pilot study using untreated volunteer donors at each clinical site can test how well these complicated assays can be transferred to the site. The pilot study should use all the same production lots of reagents and sample collection tubes and all the same equipment as will be used in the actual trial, as well as include all sample collection and all biomarker analyses that will be part of the main study to ensure that the entire protocol is properly tested. For complex assays, it may even be useful for individuals from both the sponsor and the clinical site to visit each other's labs. Whether the final analyte measurements are taken at the clinical site, at some other contractor site, or at the sponsor's laboratory, the complete analysis of the pilot study should be compared with the development results to confirm successful transfer. This requires that the assay be fully qualified in the sponsor's laboratory in time for such a pilot study to be accomplished and evaluated before the start of the main trial but can be valuable in identifying and resolving minor issues, maximizing the ability to provide the highest quality information from the biomarker program.

4. CONCLUSION

In this chapter, we provided some examples of the various aspects of biomarker analysis that need to be considered so that robust data can be generated in support of early-phase clinical drug development. This is by no means a comprehensive treatise on this topic but rather meant to point to some issues that one needs to consider when embarking on such studies. As described, even one of the simplest of samples to be collected, whole blood and its component parts (e.g., cells, plasma/serum), can change after collection. This emphasizes the requirement for careful planning of the clinical site capabilities and ensuring that the needs of the biomarker program can be met before the clinical study and, therefore, sampling begins. It also highlights how many parts of the assay under development need to be examined so that the performance characteristics are well understood. If these aspects of biomarker analysis are not examined and understood, the data generated may lead to inaccurate conclusions being drawn.

The hope for biomarkers to positively impact drug development costs will only succeed if the data generated are sufficiently robust, and for that to occur, all aspects of the biomarker program need to be carefully considered and tested. The generation of robust data sets can enable decisions on dosing and program progression to be made.

ACKNOWLEDGMENT

We would like to thank Rules Based Medicine (Austin, TX) for allowing us to use the data posted on their Web site.

REFERENCES

1. Severino ME, DuBose RF, Patterson SD. A strategic view on the use of pharmacodynamic biomarkers in early clinical drug development. IDrugs 2006;9: 754–755.
2. Williams SA, Slavin DE, Wagner JA, et al. A cost-effectiveness approach to the qualification and acceptance of biomarkers. Nat Rev Drug Discov 2006;5: 897–902.
3. Burris, HA III, Hurxitz, HI, Dees, EC, et al. Phase I safety, pharmacokinetics, and clinical activity study of Lapatinib (GW572016), a reversible dual inhibitor of epidermal growth factor receptor tyrosine kinases, in heavily pretreated patients with metastatic carcinomas, J Clin Oncol 2005;23:5305–5313.
4. Richly H, Henning BF, Kupsch P, et al. Results of a phase I trial of sorafenib (BAY 43–9006) in combination with doxorubicin in patients with refractory solid tumors. Ann Oncol 2006;17:866–873.
5. Faivre S, Delbaldo C, Vera K, et al. Safety, pharmacokinetic, and antitumor activity of SU11248, a novel oral multitarget tyrosine kinase inhibitor, in patients with cancer. J Clin Oncol 2006;24:1–11.
6. Rosen L, Kurzrock R, Mulay M, et al. Safety, pharmacokinetics, and efficacy of AMG 706, an investigational, oral, multikinase inhibitor in patients with advanced solid tumors. J Clin Oncol 2007;25:2369–76.
7. Chirgwin JM, Przybyla AE, MacDonald RJ, et al. Isolation of biologically active ribonucleic acid from sources enriched in ribonuclease. Biochemistry 1979;18:5294–5299.
8. Malik SN, Siu LL, Rowinsky EK et al. Pharmacodynamic evaluation of the epidermal growth factor receptor inhibitor OSI-774 in human epidermis of cancer patients. Clin Cancer Res 2003;9:2478–2486.
9. Tan AR, Yang X, Hewitt SM, et al. Evaluation of biologic end points and pharmacokinetics in patients with metastatic breast cancer after treatment with erlotinib, an epidermal growth factor receptor tyrosine kinase inhibitor. J Clin Oncol 2004;22:3080–3090.
10. Park NJ, Li Y,Yu T, et al. Characterization of RNA in saliva. Clin Chem 2006;52:988–994.
11. Camidge DR, Pemberton MN, Growcott JW, et al. Assessing proliferation, cell-cycle arrest and apoptotic end points in human buccal punch biopsies for use as pharmacodynamic biomarkers in drug development. Br J Cancer 2005;93: 208–215.
12. Camidge DR, Randall KR, Foster JR, et al. Plucked human hair as a tissue in which to assess pharmacodynamic end points during drug development studies. Br J Cancer 2005;92:1837–1841.
13. Juan G, Darzynkiewicz Z. Detection of cyclins in individual cells by flow and laser scanning cytometry. Methods Mol Biol 1998;91:67–75.
14. Cornett DS, Mobley JA, Dias EC, et al. A novel histology directed strategy for MALDI-MS tissue profiling that improves throughput and cellular specificity in human breast cancer. Mol Cell Proteomics 2006;5:1975–1983.

15. Paik S, Shak S, Tang G, et al. A multigene assay to predict recurrence of tamoxifen-treated, node-negative breast cancer. N Engl J Med 2004;351: 2817–2826.
16. Masuda N, Ohnishi T, Kawamoto S, et al. Analysis of chemical modification of RNA from formalin-fixed samples and optimization of molecular biology applications for such samples. Nucleic Acids Res 1999;27:4436–4443.
17. Gaston S, Soares MA, Siddiqui, MM, et al. Tissue-print and print-phoresis as platform technologies for the molecular analysis of human surgical specimens: mapping tumor invasion of the prostate capsule. Nat Med 2004;11:95–101.
18. Davis MT, Patterson SD. Does the serum peptidome reveal hemostatic dysregulation? In: Bringmann P, Butcher EC, Parry G, Weiss B, eds. Ernst Schering Research Foundation Workshop 61. Systems Biology: Applications and Perspectives. Heidelberg: Springer-Verlag; 2007:23–44.
19. Tan EM, Schur PH, Carr RI, et al. Deoxyribonucleic acid (DNA) and antibodies to DNA in the serum of patients with systemic lupus erythematosus. J Clin Invest 1966;45:1732–1740.
20. Koffler D, Agnello V, Winchester R, et al. The occurrence of singe-stranded DNA in the serum of patients with systemic lupus erythematous and other diseases. J Clin Invest 1973;52:198–204.
21. Nawroz H, Koch W, Anker P, et al. Microsatellite alterations in serum DNA of head and neck cancer patients. Nat Med 1996;2:1035–1037.
22. Lo YM, Leung TN, Tein MS, et al. Quantitative abnormalities of fetal DNA in maternal serum in preeclampsia. Clin Chem 1999;45:184–188.
23. Fujimoto A, O'Day SJ, Taback B, et al. Allelic imbalance on 12q22–23 in serum circulating DNA of melanoma patients predicts disease outcome. Cancer Res 2004;64:4085–4088.
24. Holdenrieder S, Stieber P, von Pawel J, et al. Circulating nucleosomes predict the response to chemotherapy in patients with advanced non-small cell lung cancer. Clin Cancer Res 2004;10:5981–5987.
25. Umetani N, Hiramatsu S, Hoon DSB. Higher amount of free circulating DNA in serum than in plasma is not mainly caused by contaminated extraneous DNA during separation. Ann N Y Acad Sci 2006;1075:299–307.
26. Lui YY, Chik KW, Chiu RW, et al. Predominant hematopoietic origin of cell-free DNA in plasma and serum after sex-mismatched bone marrow transplantation. Clin Chem 2002;48:421–427.
27. Baechler EC, Batliwalla FM, Karypis G, et al. Interferon-inducible gene expression signature in peripheral blood cells of patients with severe lupus Proc Natl Acad Sci U S A 2003;100:2610–2615.
28. Bullinger L, Dohner K, Bair E, et al. Use of gene-expression profiling to identify prognostic subclasses in adult acute myeloid leukemia. N Engl J Med 2004;350:1605–1616.
29. Sambrook J, Russell D. Molecular Cloning: A Laboratory Manual. Woodbury, NY: Cold Spring Harbor Laboratory Press; 2001.

30. Schroeder A, Mueller O, Stocker S, et al. The RIN: an RNA integrity number for assigning integrity values to RNA measurements. BMC Mol Biol 2006;7:3.
31. Jones L, Goldstein DR, Hughes G, et al. Assessment of the relationship between pre-chip and post-chip quality measures for Affymetrix GeneChip expression data, *BMC Bioinformatics* 2006;7:211.
32. Rules Based Medicine. White paper Comparison of serum to three plasma anticoagulant samples (EDTA, Citrate and Heparin) using the human MAP available at http://www.rulesbasedmedicine.com/serum_or_plasma.asp.
33. Fuhrman-Benzakein E, Ma MN, Rubbia-Brandt L, et al. Elevated levels of angiogenic cytokines in the plasma of cancer patients. Int J Cancer 2000;85: 40–45.
34. Drevs J, Zirrgieble U, Schmidt-Gersbach CIM, et al. Soluble markers for the assessment of biological activity with PTK787/ZK 222584 (PTK/ZK), a vascular endothelial growth factor receptor (VEGFR) tyrosine kinase inhibitor in patients with advanced colorectal cancer from two phase I trials. Ann Oncol 2005;16:558–565.
35. Brunner G, Nguyen H, Gabrilove J, et al. Basic fibroblast growth factor expression in human bone marrow and peripheral blood cells. Blood 1993;81:631–638.
36. Martyré M-C, Le Bousse-Kerdiles M-C, Romquin N, et al. Elevated levels of basic fibroblast growth factor in megakaryocytes and platelets from patients with idiopathic myelofibrosis. Br J Haemotol 1997;97:441–448.
37. Möhle R, Green D, Moore MAS, et al. Constitutive production and thrombin-induced release of vascular endothelial growth factor by human megakaryocytes and platelets, Proc Natl Acad Sci U S A 1997;94:663–668.
38. Wynendaele W, Derua R, Hoylaerts MF, et al. Vascular endothelial growth factor measured in platelet poor plasma allows optimal separation between cancer patients and volunteers: a key to study an angiogenic marker *in vivo?* Ann Oncol 1999;10:965–971.
39. Maynard SE, Min J-Y, Merchan J, et al. Excess placental soluble fms-like tyrosine kinase 1 (SFlt-1) may contribute to endothelial dysfunction, hypertension, and proteinuria in preeclampsia. J Clin Invest 2003;111:649–658.
40. Sanjo H, Kawai T, Akira S. DRAKs, novel serine/threonine kinases related to death-associated protein kinase that trigger apoptosis. J Biol Chem 1998;273:29066–29071.
41. Aziz N, Irwin MR, Dickerson SS, et al. Spurious tumor necrosis factor-α and interleukin-6 production by human monocytes from blood collected in endotoxin-contaminated vacutainer blood collection tubes. Clin Chem 2004;50:2215–2216.

2

Gene Expression-Based Biomarkers of Drug Safety

Eric A.G. Blomme and Scott E. Warder

Summary

Large-scale gene expression profiling with microarray platforms represents a new approach to identify much needed, novel mechanism-based biomarkers of toxicity for use in preclinical and clinical studies. These biomarkers may have diagnostic and/or predictive values and may consist of single gene products or of gene sets or gene expression signatures. Derivation and validation of these molecular markers involves supervised classification methods of reference data with sophisticated statistical methodologies. In this chapter, we review the methods for the identification and development of toxicity biomarkers with gene expression profiling. This chapter also describes how these novel multigene molecular markers can be integrated in a discovery pipeline, using the examples of hepatotoxicity, nephrotoxicity, and blood-based markers to illustrate successful or promising applications in toxicology.

Key Words: drug discovery; microarrays; signature; toxicity; toxicogenomics

1. INTRODUCTION

The cost of discovering and developing a new drug is estimated to range from $800 million to $1.1 billion *(1,2)*. This low research and development (R&D) productivity is not sustainable and is being aggressively addressed by the pharmaceutical industry. One of the main reasons behind these drug discovery and development economics is the high failure rate of experimental molecules at the costly stages of development. Providing an earlier triage of compounds that are ultimately destined to fail would significantly reduce R&D costs. To decrease clinical attrition, an understanding of the causes of termination is necessary. In the late 1980s, poor pharmacokinetic properties and clinical safety were the major causes of attrition, each accounting for 40% and 30% of the failures, respectively *(3)*. In contrast, by early 2000, termination was mostly driven by clinical

From: *Methods in Pharmacology and Toxicology: Biomarker Methods in Drug Discovery and Development*
Edited by: F. Wang © Humana Press, Totowa, NJ

safety and lack of efficacy, each accounting for approximately 30% of the failures *(4)*. The remarkable improvement in the pharmacokinetic profiles of compounds is largely due to the development of tools and models with improved predictive value for humans and animals. This is in sharp contrast with the limited advances in the methods for toxicity prediction. Hence, toxicity is still being predominately identified at advanced stages using lengthy and costly approaches. In particular, there is a lack of markers that can detect toxicity at the early stages of drug discovery. For decades, the only reliable biomarkers of toxicity have included panels of hematology and serum chemistry parameters. Although these indicators are often sufficient to rapidly eliminate toxic compounds, the addition of sensitive and complementary biomonitoring methods would theoretically contribute to an even earlier detection. Furthermore, these novel markers would enable a more efficient use of preclinical and clinical resources by focusing on compounds with improved likelihood for success and, in turn, limit the exposure of humans to potentially harmful compounds.

Large-scale gene expression profiling with microarray platforms represents an attractive solution to identify novel mechanism-based biomarkers. Through a global evaluation of cellular and tissue transcriptomes, microarray technology provides a powerful approach to interrogate perturbations induced by compounds and to pinpoint genes or gene sets regulated in parallel to a toxic reaction. This chapter covers the current status in the identification and development of novel biomarkers of toxicity using gene expression profiling. As for any new and relatively untested technologies, the optimal positioning and effective implementation of these biomarkers in an R&D organization can be quite challenging. Consequently, this chapter will also discuss how these novel molecular toxicology approaches can be integrated in a discovery pipeline, using specific examples to illustrate successful or promising applications in toxicology.

2. IDENTIFICATION OF BIOMARKERS OF TOXICITY WITH GENE EXPRESSION PROFILING

2.1. Gene Expression Signatures as Biomarkers

The use of large-scale expression profiling in toxicology has led to the development of the toxicogenomics discipline *(5–7)*. Through a global evaluation of transcriptomes, microarrays allow toxicologists to formulate novel hypotheses about mechanisms of toxicity and to isolate potential genes or gene sets correlated with a particular toxicologic outcome *(5,8)*. Toxicogenomics is based on the finding that toxicants acting through similar mechanisms affect related molecular pathways and lead to common gene expression changes. These shared changes are signatures or molecular multigene markers that can be used as specific and sensitive end points to identify and classify toxicants ***(8–11)***. The signature concept is not unique to toxicology and was initially demonstrated for classification of various types of diseases, in particular for cancer diagnosis ***(12–15)***.

The National Institutes of Health (NIH) Biomarkers Definitions Working Group defined *biomarkers* as measurable and quantifiable characteristics that served as indicators of pathologic or pharmacologically related events ***(16)***. Based on this definition, microarray signatures should be classified as biomarkers, as they can be used as semiquantitative or quantitative indicators of concurrent or developing toxic reactions. The two main applications for toxicity biomarkers include detection of a specific toxic outcome and signaling that a toxic outcome may develop. To clearly distinguish the two categories, it is useful to define them as *diagnostic biomarkers* and *predictive biomarkers*, respectively.

2.2. Development and Use of Diagnostic Biomarkers

Using large-scale gene expression profiling, one can scan for perturbations in transcript levels that are linked to the mechanism by which toxicants injure cells. These perturbations can then be used to isolate specific gene products that may be useful for monitoring the occurrence of a particular toxicity. The diagnostic accuracy of these specific gene products can be interrogated in subsequent, appropriately designed experiments. Two recent studies illustrate the application of gene expression profiling to understand toxic mechanisms and to identify specific biomarkers for a particular toxicity. Functional γ-secretase inhibitors (FGSIs) are being developed as potential therapeutic agents for Alzheimer's disease ***(17,18)***. FGSIs can block the cleavage of several transmembrane proteins, including the cell fate regulator Notch-1, which plays an important role in cellular differentiation in the immune system and gastrointestinal tract. Several FGSIs elicit a unique alteration in the intestinal tract of rats, characterized by a mucoid enteropathy related to ileal goblet cell hyperplasia ***(17,18)***. Microarray analysis of the small intestine of FGSIs-treated rats identified changes in the expression levels of several genes, supporting the notion that perturbation in Notch signaling was the likely mechanism for this unique enteropathy. Furthermore, the gene encoding the serine protease adipsin was significantly upregulated after treatment with FGSIs. This finding was confirmed at the protein level by demonstrating elevated levels of adipsin in the gastrointestinal contents and feces of FGSIs-treated rats and increased numbers of ileal epithelial cells expressing adipsin by immunohistochemistry. These results suggest that adipsin may represent a noninvasive biomarker of FGSIs-induced gastrointestinal toxicity with clinical utility.

Global gene expression changes are quite challenging to interpret, as they reflect a large number of complex, interacting molecular processes that may or may not be related to a toxic reaction ***(19)***. Simply stated, not all gene expression changes induced by toxicants are toxicologically relevant. In addition, changes in transcript level may reflect a general, nonspecific toxic outcome and consequently may not add significant value as a biomarker compared with the battery of current markers. The induction of cytochrome P450 1A1 (CYP1A1) represents such an example. CYP1A1 is a member of the xenobiotic metabolizing enzymes involved in detoxification of polycyclic aromatic compounds but also contributes to the generation of mutagenic metabolites and reactive metabolites ***(20)***. CYP1A1 induction is a hallmark of aryl hydrocarbon receptor (AhR) pathway activation and is involved in the pathogenesis of the toxicity caused by 2,3,7,8-tetrachlorodibenzo-*p*-dioxin ***(21,22)***. Therefore, one may envision that CYP1A1 mRNA levels could be used as a surrogate marker of AhR pathway activation and of dioxin-like toxicity. However, when evaluating reference databases of gene expression profiles, we and others have identified large numbers of non–AhR-activating compounds, including a significant portion of Food and Drug Administration (FDA)-approved drugs from multiple classes, that significantly upregulate CYP1A1 expression levels in rat liver at toxic doses ***(23)***. Therefore, upregulation of CYP1A1 mRNA levels by compounds cannot be considered a specific marker of dioxin-like toxicity or of any other toxicity. This example is a good illustration of the use of reference expression profiles for the facile *in silico* validation of proposed biomarkers.

A pivotal step in developing novel diagnostic biomarkers entails the identification of gene expression changes that are specific to the toxic end point of interest. This requires access to a repository of reference expression profiles induced by a variety of toxicants, as well as by compounds without toxic effects or with unrelated toxic effects. This contextual information can then be used to develop robust statistical inferences between changes in expression of a specific gene set and a particular toxic phenotype, a process sometimes referred to as phenotypic anchoring or relating gene expression changes to a particular phenotype ***(24–26)***. There are many different methods to recognize the deregulated biological pathways and genes associated with the development of a toxic phenotype, and these will be further covered in our discussion on predictive biomarkers.

Several toxicogenomics companies have reported diagnostic signatures of toxicity for several tissues ***(24,27–29)***. However, because of the commercial value of these signatures, detailed published reports are rare, and therefore it is difficult to fully interrogate their performance ***(26)***. Our laboratory has acquired access to DrugMatrix (Iconix Biosciences, Mountain View, CA) signatures and has successfully applied them for toxicology ***(29)***. In addition, we have generated internal, fit-for-purpose diagnostic signatures using DrugMatrix reference profiles. These gene expression–based tools are used as a complement to confirm changes detected by other methods or to more specifically define the nature of a toxic change. For instance, compound-induced increases in liver weight are frequent in rats ***(30)***. Those liver enlargements can be caused by various mechanisms, such as by induction of metabolizing or conjugating enzymes or by peroxisome proliferation, and ultimately influence decisions on the fate of compounds ***(31)***. Whereas current biomarkers cannot provide a rapid mechanistic explanation for liver weight increase, gene expression profiles by adding mechanistic clarity can address this issue. Gene signatures are also quite useful to the toxicologist or pathologist to interpret changes of questionable toxicologic significance. For instance, in short-term toxicology studies, histopathologic changes can be subtle or not easily distinguishable from spontaneous changes. Furthermore, clinical pathologic changes can show striking interindividual variability that complicates the interpretation. In contrast, in our experience, the interindividual variability of compound-induced gene expression profiles is rather limited, increasing the confidence in data interpretation (**Fig. 1** and Color Plate 1, following p. 230).

2.3. Development and Use of Predictive Gene Expression–Based Biomarkers

Predictive toxicology refers to the use of *in vitro* assays or of short-term *in vivo* studies to predict toxic changes that will occur after longer exposure. The objective is to provide an assessment of the toxic profiles of compounds at an early stage before committing additional resources to further evaluate compounds in lengthy and costly studies. After exposure to compounds at toxic doses, transcriptional changes typically occur before the development of the functional and morphologic changes detected by clinical observations, clinical pathology, or histopathology. It is important to clarify that, although toxicologically relevant gene expression changes occur before the development of a toxic phenotype, they only occur at exposures that result in phenotypic changes. For this reason, toxicogenomics represents a unique alternative to characterize toxicity early in the drug discovery process and has great potential to significantly improve R&D productivity ***(5,6, 32)***. To be fully predictive, these signatures must be anchored to toxic changes occurring after longer exposure and not to concurrent toxic changes. However, it is interesting to point out that some signatures designed to be diagnostic can also have excellent predictive value.

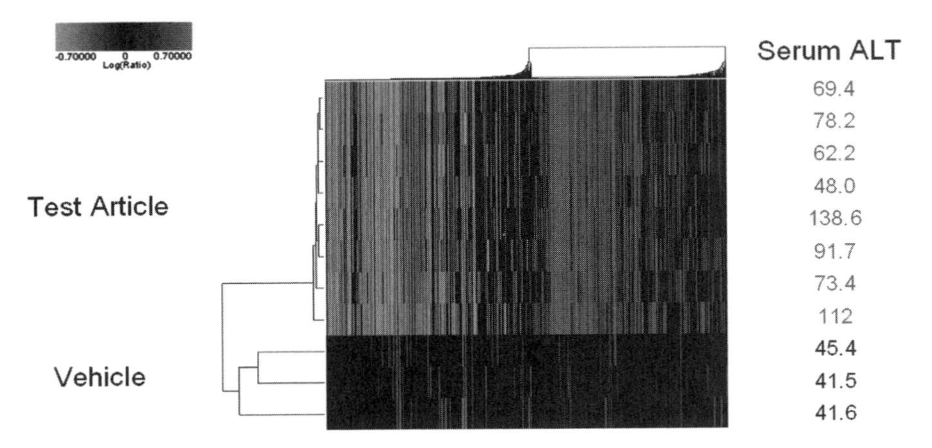

Fig. 1. Comparison of serum alanine aminotransferase (ALT activity) and liver gene expression profiles induced by a hepatotoxicant. Rats were treated for 3 days with either the vehicle or the test article at daily doses causing moderate hepatotoxicity. Serum was collected and ALT activity levels quantified. Livers were sampled, and gene expression profiles were generated with microarrays. Whereas significant interindividual variability can be seen in serum ALT levels, the expression profiles are strikingly consistent among the eight test article–treated rats. Hierarchical cluster analysis was performed using Rosetta Resolver version 6.0 (Rosetta Inpharmatics, Seattle, WA). Genes shown include genes that were upregulated or downregulated by at least twofold with a p value less than 0.01. Green indicates downregulation, and red indicates upregulation. (*see* Color Plate 1, following p. 230)

It is beyond the scope of this chapter to provide an in-depth review of the methods used to develop predictive gene expression signatures. Several reviews are available on this topic *(5,24,33–35)*. Briefly, the development of predictive models to classify compounds as toxic or nontoxic requires the use of a variety of sophisticated statistical methodologies and a strong biostatistical expertise. The first step consists of selecting appropriate training sets composed of a large repository of gene expression profiles encompassing both toxic and nontoxic compounds covering multiple chemical classes and various mechanisms of toxicity. Computational algorithms, such as linear discriminant analysis, artificial neural networks, and support vector machines, are then trained to classify compounds as toxic or nontoxic (**Fig. 2**) *(15,35–41)*. The algorithms have unique strengths and weaknesses, and the selection of the appropriate machine learning methodology should be made based on the predetermined expectations for the signature. The prediction accuracy of the model is then typically estimated using a testing set containing gene expression profiles independent from those in the

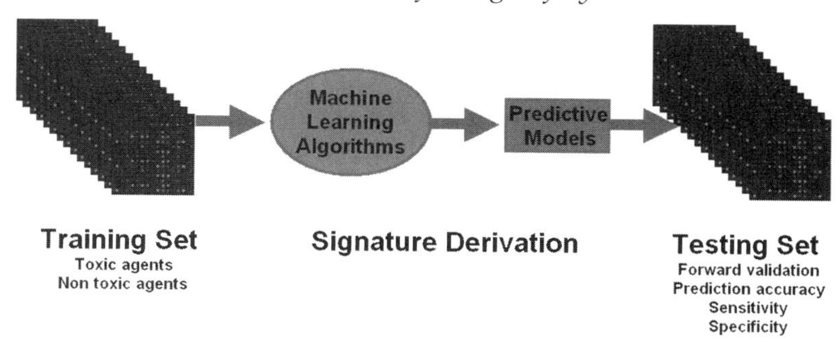

Fig. 2. Development of predictive gene expression signatures. Predictive signatures are derived using a training set composed of gene expression profiles induced by a wide range of structurally and pharmacologically diverse compounds that are toxic or nontoxic in the tissue studied. This training set is analyzed using various machine learning algorithms that generate a model to classify compounds as toxic or nontoxic in the tissue of interest. This classification may be based on nonspecific toxic changes or may be refined to more specific toxic changes, such as bile duct hyperplasia or lipidosis for the liver. These predictive models are then evaluated in a forward validation step by using a testing set of expression profiles distinct from those of the training set. The objective of this validation step is to determine in an unbiased manner the prediction accuracy of the model, as well as its sensitivity and specificity. (Reprinted from American Drug Discovery, Eric A.G. Blomme, Use of toxicogenomics in drug discovery, February-March 2007, Copyright © 2007, with permission from Russell Publishing.)

training set during a forward validation step *(5,35,42)*. This step is critical and the best approach to correctly estimate the performance of a predictive model. Also, it is highly recommended to use compounds related to the chemical space under investigation.

2.4. *Development of* In Vitro *Gene Expression–Based Biomarkers*

Toxicologic assessment is ultimately conducted in animal and clinical studies. Although there are well-publicized cases of human toxicities that were not predicted in animal studies, preclinical toxicology assessments are relatively efficient at detecting molecules toxic to humans. In fact, according to an industry report, 94% of human toxicities are detected in preclinical toxicologic evaluations *(43)*. And this number probably underrepresents the predictive value of animal studies, as a vast majority of exploratory compounds are eliminated after animal testing and are never pursued in the clinic. However, assessments in animal models require large amounts of compound (in the gram range), are expensive, time-consuming, and do not allow for a rapid toxicologic characterization of many compounds. Therefore, during the lead optimization stage of drug discovery, compounds are mostly selected based on pharmacologic, physicochemical, and pharmacokinetic criteria rather than on any robust toxicologic evaluation. This biased approach leads to a high attrition rate driven by toxicity in later stages. *In vitro* systems with their high throughput, low cost, and minimal compound requirements (in the milligram range) are optimal for the toxicologist during lead optimization.

Several studies have demonstrated that toxic compounds can be classified using gene expression profiles and signatures generated from *in vitro* systems ***(11,44–49)***. A gene expression–based approach offers some non-negligible advantages compared with the traditional *in vitro* toxicologic profiling used at various institutions. Indeed, the traditional paradigm typically involves the use of several cell types and multiple concentrations to evaluate the dose response, thereby generating an overwhelming amount of data that may be challenging to interpret in the context of their *in vivo* significance. In contrast, the gene expression–based approach can simultaneously interrogate several toxicologically relevant end points using one cell type and a single concentration, thereby resulting in a significant increase in throughput and easier interpretation of the data (**Fig. 3**). The derivation and validation of *in vitro* signatures involves supervised classification methods similar to those covered before. In particular, an appropriate repository of gene expression profiles in the model of interest needs to be established. The choice of an *in vitro* model, therefore, represents a pivotal first step.

There is no consensus on what represents the optimal concentration for an *in vitro* toxicogenomics assessment. In our experience, relatively high concentrations are required, at least concentrations sufficient to cause some low degree of cytotoxicity. Lower concentrations are typically associated with inconsistent gene expression responses, low sensitivity, and large interassay variability. We typically characterize compounds at concentrations sufficient to cause 20% cell death, and this has led to consistent and robust gene expression changes. It is probably not feasible to detect all toxicities in one cell type, and therefore selecting a cell system that can cover the broader range of toxic end points is recommended. To identify markers of general toxicity (such as DNA damage, apoptosis, or oxidative stress), any cell type should be appropriate, whereas to evaluate more tissue-specific end points, a cell type derived from or that mimics the targeted tissue may be more relevant. Most published *in vitro* toxicogenomics studies have evaluated liver-derived *in vitro* systems, as the liver is a common target

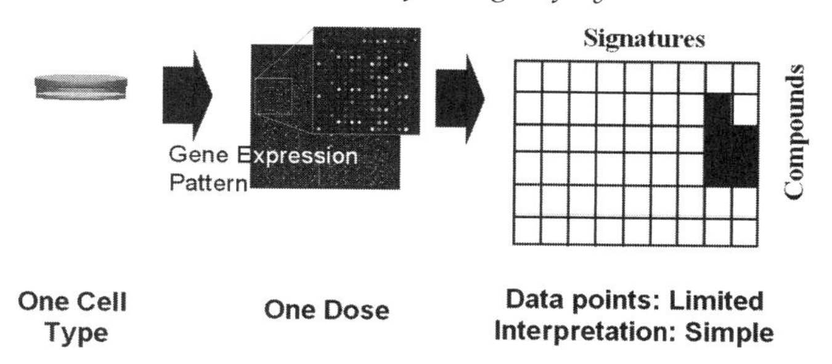

Fig. 3. *In vitro* toxicogenomics paradigm. Whereas traditional *in vitro* screening systems typically employ multiple assays and concentrations and generate large amounts of complex data, several *in vitro* gene expression signatures can be interrogated in a single cell type treated at a single concentration and used as binary markers for the presence of absence of selected toxicologically relevant end points. This streamlined approach increased the throughput of testing and facilitates the decision-making process. In the heatmap, black squares indicate a positive score for a signature, and white squares indicate a negative score for a signature.

organ of toxicity, and liver cell preparations are routinely used for *in vitro* toxicology in the industry.

Our laboratory has selected primary rat hepatocytes to characterize compounds by gene expression profiling. We have generated a database containing more than 1500 gene expression profiles derived from primary hepatocytes treated at toxic concentrations with a wide variety of reference toxicants. This database was used to derive predictive signatures for several toxicologic end points, such as phospholipidosis, DNA damage, peroxisome proliferation, mitochondrial dysfunction, and glutathione depletion. In general, these *in vitro* gene expression–based signatures have acceptable accuracy to be useful in a discovery screening setting when they are designed to detect specific mechanism-based toxic effects, such as activation of a nuclear receptor or damage to a particular organelle, such as mitochondria. In contrast, identifying more complex toxic end points, such as necrosis, is a more challenging task, as it is difficult to cover all mechanisms leading to tissue necrosis in the positive reference class of the training set.

We have recently released the identity of two predictive signatures for *in vitro* use ***(11)***. These signatures were built to detect two prototypical toxicologic classes: aryl hydrocarbon receptor (AhR) agonists and peroxisome proliferator activated receptor alpha (PPAR-α) agonists. To achieve this, primary rat hepatocytes were exposed to a number of AhR agonists, PPAR-α agonists, and compounds not associated with these two mechanisms. Global gene expression profiles were generated with microarrays for each class of compound. Using linear discriminant analysis coupled with permutation-based *t*-test, gene signatures were established to classify compounds according to a discriminant score. The final gene signatures consist of eight genes for AhR agonism and 11 genes for PPAR-α agonism. These signatures were further validated using additional compounds naïve to the training set and with different chemical structures and different potencies for activating the AhR or PPAR-α pathway. The overall accuracy was excellent, indicating that these types of *in vitro* biomarkers represent a feasible approach to screen compounds early in the discovery process.

Studies in other laboratories have confirmed our findings. In particular, one report used gene expression analysis to identify biomarkers of phospholipidosis that could be used in an *in vitro* screen for drug-induced phospholipidosis ***(46)***. Specifically, gene expression profiling of HepG2 cells exposed to 12 compounds known to induce phospholipidosis led to the identification of 17 candidate marker genes. Real-time polymerase chain reaction (PCR) analysis confirmed that 12 of those 17 gene markers showed significant concordance with the phenotype of phospholipidosis, and the magnitude of the changes for these markers correlated well with the phenotypic change. In a separate study, this set of genes was independently validated by another group using the same cell line ***(50)***. Based on these genes, 11 of 12 compounds known to induce phospholipidosis were identified as positive, and all the negative compounds (5 of 5) were confirmed as negative. In a second experiment using 26 compounds, 14 of 15 compounds known to induce phospholipidosis and all the negative controls (6 compounds) were correctly classified. These predictive accuracy values clearly demonstrate the utility of gene expression–based biomarkers of toxicity using *in vitro* systems.

Gene expression–based biomarkers have also been investigated for genotoxicity. The current *in vitro* genotoxicity assays using mammalian cells (mammalian mutation and/or chromosomal damage assays) have low specificity with questionable relevance to the *in vivo* situation ***(24,51,52)***. Assays that could differentiate DNA-reactive from non–DNA-reactive mechanisms of genotoxicity would improve the risk assessment of positive findings in the *in vitro* mammalian cell–based assays. Several gene expression profiling studies have demonstrated differences in gene expression profiles between DNA-damaging and non–DNA-damaging compounds ***(52–54)***. This suggests that gene expression signatures could be developed to more specifically evaluate compounds with positive findings in the standard *in vitro* genotoxicity assays.

In vitro toxicogenomics is still not an integral component of the battery of routine assays used at the lead selection and lead optimization stages of drug discovery. Gene expression profiling platforms with appropriate throughput and costs are needed for practical applications of toxicogenomics in drug discovery. We and others have shown that microarray-generated signatures can easily be transferred to more cost-effective platforms with better throughput, such as the Taqman Low Density Array (Applied Biosystems, Foster City, CA) or ArrayPlate (High Throughput Genomics, Inc., Tucson, AZ) platforms ***(11,55)***. Clearly, this area is rapidly evolving, and it is likely that gene expression–based *in vitro* assays will become part of the battery of assays for compound characterization and prioritization in discovery.

3. GENE EXPRESSION–BASED BIOMARKERS OF HEPATOTOXICITY

Like most pioneering toxicogenomics studies, our laboratory had focused its early efforts on hepatotoxicity. The liver is a common target organ of toxicity and is very homogenous, making it an attractive choice for genome-scale expression profiling. Our initial goal was to generate a quantitative gene expression–based model predictive of hepatotoxicity in rats ***(5)***. The first step entailed the construction of a database of microarray-generated expression profiles from the livers of rats treated daily with a variety of prototypical compounds toxic or nontoxic to the rat liver. A set of 40 marker genes was then selected using analysis of variance (ANOVA) analysis based on the ability to discriminate the hepatotoxicants from the non-hepatotoxicants. A quantitative predictive model was then developed using an artificial neural network algorithm. This model classified compounds according to a composite score that indicated the probability that the test compound would induce hepatotoxicity in rats upon dosing for additional days. Using a limited testing set as part of a forward validation step, the accuracy of this assay was estimated to be approximately 89%. However, it was unclear whether this performance would translate when evaluating proprietary compounds. Therefore, we recently conducted a reevaluation of the performance of this model using an extensive set of data from internal projects. For 52 compounds, data were available from the short-term (1 to 5 days) and the 2-week rat studies. This allowed us to estimate the predictive accuracy of this multigene marker by comparing the results generated by the model in the short-term studies to the histopathologic and serum chemistry changes in the 2-week studies (**Fig. 4**). The assay correctly identified eight compounds as hepatotoxic and 42 compounds as non-hepatotoxic. Only two compounds were incorrectly classified, indicating a sensitivity of 89%, a specificity of 98%, and a predictive accuracy of 96%. For eight compounds, the model was also used in long-term (>2 weeks) studies, allowing us to evaluate its diagnostic accuracy. In these studies, the assay

a

Hepatotoxicity in 2-week Study

Model Prediction Short-term Studies

	Positive	Negative
Positive	8	1
Negative	1	42
Accuracy (96.2%)	Sensitivity 88.9%	Specificity 97.7%

b

Hepatotoxicity in 2-week Study

Model Prediction 2-week studies

	Positive	Negative
Positive	2	0
Negative	0	6
Accuracy (100%)	Sensitivity 100%	Specificity 100%

Fig. 4. Evaluation of the performance of the predictive hepatotoxicity model. Hepatotoxicity was predicted with the model using liver expression profiles from **(a)** short-term or **(b)** 2-week studies, and the predictions (vertical axis) were compared with the gold standards, histopathology and serum chemistry in the 2-week studies (horizontal axis). Overall, this neural network–based model assay had **(a)** a predictive accuracy of 96% and **(b)** an excellent diagnostic accuracy.

correctly identified two compounds as hepatotoxic and six compounds as non-hepatotoxic (**Fig. 4**). Although this represents a sensitivity, a specificity, and a diagnostic accuracy of 100%, the number of test compounds is obviously too small to confirm the actual predictive value. However, these preliminary data suggest that this model represents an excellent diagnostic tool that could be used as a complement to the current, well-established biomarkers of hepatoxicity.

Gene expression–based biomarkers can also be developed for more specific hepatic toxic end points. For instance, bile duct hyperplasia is a hepatic change that occurs after days or weeks of exposure to various chemicals. This pathologic change is typically identified late in drug discovery, and no reliable biomarker is available for rapid detection in preclinical species. For this reason, we were interested in finding biomarkers of bile duct hyperplasia for rat. In collaboration with Iconix Biosciences, we derived two gene expression signatures for bile duct hyperplasia using liver gene expression profiles from the commercial DrugMatrix toxicogenomic database. The predictive signature was designed to predict, within 1 to 5 days of dosing, the occurrence of bile duct hyperplasia after weeks of continued dosing. The diagnostic signature was designed to correlate with the presence of bile duct hyperplasia after 5 days of dosing. The performance characteristics (sensitivity and specificity) of these two signatures were further characterized in a forward validation. Male rats were treated for 1, 5, or 28 days with eight compounds not used to generate the two signatures. Liver gene expression profiles from the 1- and 5-day time points were generated, and liver histopathology was assessed for all time points.

Table 1
Predictive Accuracy of a Gene Expression–Based Signature for Bile Duct Hyperplasia in Rats*

Compound	Dose(mg/kg)	Duration (day)	Diagnostic signature score	Histology at day 28
Aflatoxin B1	0.3	1	+	+
Aflatoxin B1	0.3	5	+	+
Allyl alcohol	16	1	–	–
Allyl alcohol	16	5	–	–
Carbon tetrachloride	3178	1	–	+
Carbon tetrachloride	3178	5	+	+
Ethanol	3000	1	–	–
Ethanol	3000	5	–	–
Isoniazid	50	1	–	–
Isoniazid	50	5	–	–
N-Nitrosodiethylamine	100	1	+	+
N-Nitrosodiethylamine	100	5	+	+
N-Nitrosodimethylamine	10	1	+	+
N-Nitrosodimethylamine	10	5	+	+
Thioacetamide	200	1	–	+
Thioacetamide	200	5	+	+

*Rats were treated for 1, 5, and 28 days with eight compounds. Liver gene expression profiles were generated for the day 1 and day 5 time points, and histopathology was evaluated for the day 28 time point. Note the concordance between the signature's prediction (+ or –) and the histopathology (+ and – indicate the presence or absence of bile duct hyperplasia, respectively).

Using these data, the diagnostic signature had an accuracy of 69% with a sensitivity of 100% and a specificity of 62%, and the predictive signature had an accuracy of 88% with a sensitivity of 90% and a specificity of 83%. Interestingly, the diagnostic signature also had excellent predictive properties, correctly identifying the five compounds that induced bile duct hyperplasia after 28 days (**Table 1**).

4. GENE EXPRESSION–BASED BIOMARKERS OF NEPHROTOXICITY

After an initial emphasis on liver, the field of toxicogenomics has expanded to other tissues. Although the heterogeneity of certain tissues such as brain or intestine poses a technical challenge, recent data suggest that multigene biomarkers of toxicity are feasible for many tissues. This is best exemplified by a current publication illustrating the use of gene expression profiles from the kidneys of rats to derive a predictive signature of nephrotoxicity ***(10)***. In this study, the signature of renal tubular toxicity correctly predicted renal tubular injury in 76% of the compound treatments. This predictive value represents a significant improvement over the current biomarkers typically used, such as blood urea nitrogen or serum creatinine. Another study used a similar, yet slightly different approach to generate a predictive gene expression–based classifier of proximal tubular injury in the rat ***(56)***. In this study, rats were treated with a wide variety of classic nephrotoxicants, and renal gene expression profiles were evaluated 1, 3, and 7 days after initiation of dosing. A predictive classifier was developed using a support vector machine–based algorithm and a training set of 120 gene expression profiles. The signature was able to predict the type of pathology of a testing set composed of 28 gene expression profiles with 100% selectivity and 82% sensitivity.

Our laboratory was also interested in generating a predictive model of nephrotoxicity for rat. We used an approach similar to the one used in our predictive hepatotoxicity model. After careful curation of the DrugMatrix database, we identified 173 reference gene expression profiles that met our inclusion criteria. These profiles covered 62 compounds (33 compounds toxic and 29 compounds nontoxic to the kidney) dosed for 3 or 5 days in male rats. Discriminative probe sets between the classes of compounds were selected with an ANOVA analysis, and a quantitative model was developed with an artificial neural network algorithm. In the forward validation step, this predictive nephrotoxicity model correctly classified 10 of 11 compounds (six nephrotoxicants and five non-nephrotoxicants), further supporting the concept of gene expression–based biomarkers for various tissues.

Ultimately, having similar predictive models for all major organs and tissues would represent a significant advantage and should lead to a more precise evaluation of the toxic characteristics of compounds. Typically, these predictive models are built upon 20 to 150 genes and therefore, it is quite realistic to think that these models could be interrogated using smaller, customized, and cost-efficient platforms. With these customized tools, the cost and resources necessary to generate the mRNA profiles should not prohibit an extensive survey of tissues, and it is reasonable to envision that these predictive models could become part of the toxicologist's toolbox, similar to serum chemistry panels.

5. GENE EXPRESSION–BASED BIOMARKERS OF TOXICITY IN BLOOD

Signatures derived from tissue expression profiles represent a definite scientific advancement in the field of biomarkers of toxicity, but their use is mostly limited to preclinical samples. Indeed, gene expression analysis can only be performed on human tissues in rare circumstances, such as tumor samples, and therefore an alternative is needed to obtain signatures that can be used to monitor toxic events in humans. Blood represents the ideal tissue for this approach, as it is easily accessible without the need for invasive collection procedures and allows for repeated measurements. There have been several attempts to develop single-analyte assays by evaluating the abundance of certain transcripts in blood samples ***(57)***. These transcript measurements using standardized RT-PCR (StaRT-PCR; Gene Express, Inc., Toledo, OH) are a new source of molecular markers for evaluation of disease risk, diagnosis of pathologic or toxicologic conditions, or predictors of drug response ***(58,59)***. Furthermore, blood could serve as a surrogate for other tissues by monitoring perturbations throughout the organism ***(60,61)***. The assumption is that circulating blood cells survey the state of the various components of an organism and that their transcriptome is partly regulated by this surveillance function. Blood-based signatures could also present major advantages for preclinical studies. First, they would limit the generation of expression profiles to one, rather than multiple tissues. Second, they could help to bridge the gap between preclinical and clinical toxicity data, ultimately dramatically improving our ability to monitor patient health in clinical trials. Several studies have shown that various insults or pathologic conditions induce characteristic gene expression changes in blood in rats and humans ***(62–67)***. Likewise, peripheral blood has been shown to be an appropriate matrix for molecular markers of diseases such as Parkinson's disease or Huntington's disease ***(68,69)***. Finally, a recent study has shown that diagnostic molecular profiles of radiation exposure could be generated in mice and transferred to humans ***(70)***.

Using blood as a sentinel tissue for toxicogenomics analysis presents both technical and scientific challenges ***(61)***. First, unlike tissues such as liver, blood contains a heterogeneous cell population subject to changes related or unrelated to the toxicity. Therefore, these variations make it difficult to determine what type of cell is actually driving any gene expression changes and whether gene expression profiles simply reflect alterations in cell population, changes in the transcriptome of a specific cell subpopulation, or simply intraindividual sample variation ***(71)***. Second, globin mRNA transcripts make up as much as 70% of whole-blood mRNA transcripts, causing technical issues exemplified by decreased percent presence calls on microarray chips and increased replicate variability ***(72)***. Globin reduction protocols or white blood cell isolation procedures (such as using buffy coats) are used to address these technical issues (**Fig. 5**) ***(61)***. However, these protocols are considerably more time-consuming, difficult, and, in our experience, subject to interassay variability ***(72)***. Third, the field

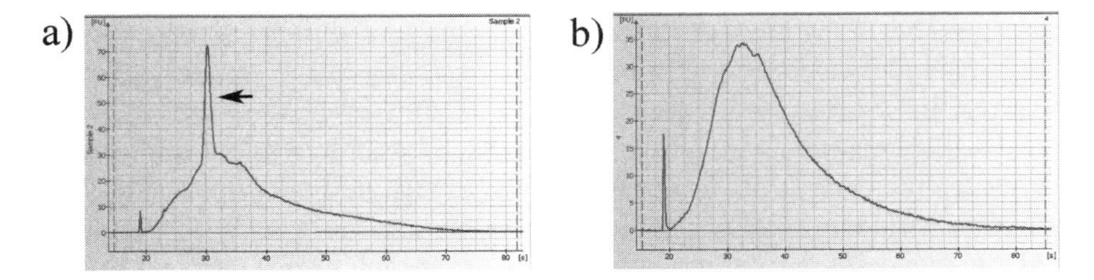

Fig. 5. Globin reduction in peripheral blood samples. RNA was isolated from whole rat blood (**a**) without or (**b**) after two rounds of globin reduction. Note the globin peak (arrow) on the Agilent Bioanalyzer–generated electrophoretic image (Santa Clara, CA).

of blood genomics has still not matured enough for standardization of practical collection procedures, which may compromise detection of some transcripts ***(61)***.

Currently, it is unclear if hemogenomics can be used to predict or diagnose toxic events occurring in other parts of the organism. It is likely that blood-derived signatures can be generated to detect toxic changes in other tissues, but a critical question is whether these signatures provide any additional value compared with our current battery of blood-based biomarkers. In our laboratory, we have started to ask these critical questions by building a database of blood expression profiles from rats treated with various compounds causing known toxicities. Our goal is very similar to what has been accomplished with other tissues. Briefly, this database of profiles will be used to ensure that the technical variation in the process and biological reproducibility are sufficiently limited for this tissue to be used as a reliable matrix for the proposed signatures. If an adequate reproducibility is demonstrated, then molecular markers for several distinct toxic changes occurring in various tissues will be derived and validated. Although it is too early to know if hemogenomics has practical use in toxicology, the objective would be to use it as a complement to the current serum chemistry and hematology panels to scan the presence of toxic changes in an organism with improved sensitivity and mechanistic clarity. Signals from these signatures could then be further investigated by evaluating the transcriptome of the tissues identified as targets of toxicity.

We are also evaluating the possibility of using human peripheral blood mononuclear cells (PBMCs) in an *ex vivo* setting to interrogate efficacy and toxicity end points. PBMCs represent an attractive system, as they constitute a circulating organ with high and monitorable exposure to drugs and with a potential to serve as a surrogate target cell. In contrast with other tissues where local exposure to drugs may be challenging to estimate, exposure of PBMCs can be easily monitored, such that one could better predict therapeutic indices for experimental compounds. Our early experience with human PBMCs suggests that they represent a viable *in vitro* toxicology system. More specifically, we have shown that PBMCs collected from different donors respond similarly to the same compound, indicating limited concerns with donor-to-donor variability. Furthermore, in preliminary experiments, gene responses from multiple donors correlated well with compound structure-activity classes, suggesting that specific signatures could be derived. Finally, gene expression changes induced by compounds, such as the glucocorticoid receptor agonists, were very consistent with their known *in vivo* pharmacologic effects. Others have also reported results consistent with our experience. For instance, using PBMCs, the pharmacologic properties of an engineered form of IL-2 was successfully interrogated and compared with that of recombinant IL-2 *(73)*. Likewise, exposure of PBMCs to cigarette smoke condensate led to the discovery of several potential biomarkers for population studies on the early biological effects caused by cigarette smoke *(74)*. These early results are encouraging and suggest that PBMCs could represent a robust system to interrogate early toxicities in a format more relevant to humans. In addition, this cell system would allow for a facile transfer of *in vitro*–generated, gene expression–based signatures to the clinic. We currently are evaluating whether this system truly has the potential to become a bridge between *in vitro* and *in vivo*.

6. CONCLUSION

The microarray technology offers a novel approach to discover biomarkers capable of predicting or monitoring pathologic processes including toxic reactions. This chapter provided a flavor of how this technology is affecting and will continue to enable the discipline of toxicology and the drug discovery and development process. Through supervised classification of repository data and with the help of advanced computational tools, numerous multigene markers of various toxic states are likely to be publicly available in the near future. The reliability and precision of microarray platforms in the context of biomarkers was recently confirmed by a consortium led by the FDA *(75)*. However, these platforms still have not reached a stage at which standardized, reproducible, and high-throughput applications of diagnostic assays is feasible. In addition, these novel markers need to be further validated in prospective, well-controlled studies and across multiple institutions with well-established standards for all steps in the process *(76)*. The examples used in this chapter are a good indication of the scale involved in the generation of robust signatures with preclinical or clinical utility. This requires access to large numbers of diverse reference data generated according to consistent experimental procedures. In other words, such endeavors are only achievable with significant resources, and it is likely that the generation and validation of multigene signatures will entail a combined effort of the broad scientific community *(26)*. These efforts have already been initiated in the form of industry consortia. For instance, the Predictive Safety Testing Consortium between the C-Path Institute (www.c-path.org) and several large pharmaceutical companies has been formed to share proprietary markers or laboratory methods that can help predict and monitor the safety of experimental compounds. Members contribute proprietary data, including gene expression profiles, with the aim of validating novel molecular markers of adverse events. These collaborative efforts will allow institutions to enable and accelerate the acceptance and use of novel biomarkers for use in preclinical and/or clinical settings.

REFERENCES

1. Rawlings MD. Cutting the cost of drug development? Nat Rev Drug Discov 2004;3:360–364.
2. Service RF. Surviving the blockbuster syndrome. Science 2004;303:1796–1799.
3. Prentis RA, Lis Y, Walker SR. Pharmaceutical innovation by the seven UK-owned pharmaceutical companies (1964–1985). Br J Clin Pharmacol 1988;25:387–396.
4. Kola I, Landis J. Can the pharmaceutical industry reduce attrition rates? Nat Rev Drug Discov 2004;3:711–715.
5. Yang Y, Blomme EA, Waring JF. Toxicogenomics in drug discovery: from preclinical studies to clinical trials. Chem Biol Interact 2004;150:71–85.
6. Searfoss GH, Ryan TP, Jolly RA. The role of transcriptome analysis in preclinical toxicology. Curr Mol Med 2005;5:53–64.
7. Luhe A, Suter L, Ruepp S, Singer T, Weiser T, Albertini S. Toxicogenomics in the pharmaceutical industry: hollow promises or real benefit? Mutat Res 2005; 575:102–115.
8. Segal E, Friedman N, Kaminski N, Regev A, Koller D. From signatures to models: understanding cancer using microarrays. Nat Genet 2005;37(Suppl): S38–S45.
9. Hamadeh HK, Bushel PR, Jayadev S, Martin K, DiSorbo O, Sieber S, et al. Gene expression analysis reveals chemical-specific profiles. Toxicol Sci 2002; 67:219–231.
10. Fielden MR, Eynon BP, Natsoulis G, Jarnagin K, Banas D, Kolaja KL. A gene expression signature that predicts the future onset of drug-induced renal tubular toxicity. Toxicol Pathol 2005;33:675–683.
11. Yang Y, Abel SJ, Ciurlionis R, Waring JF. Development of a toxicogenomics in vitro assay for the efficient characterization of compounds. Pharmacogenomics 2006;7:177–186.
12. Khan J, Wei JS, Ringner M, Saal LH, Ladanyi M, Westermann F et al. Classification and diagnostic prediction of cancers using gene expression profiling and artificial neural networks. Nat Med 2001;7:673–679.

13. Ringner M, Peterson C. Microarray-based cancer diagnosis with artificial neural networks. Biotechniques 2003;Suppl:30–35.
14. Xu Y, Selaru FM, Yin J, Zou TT, Shustova V, Mori Y, et al. Artificial neural networks and gene filtering distinguish between global gene expression profiles of Barrett's esophagus and esophageal cancer. Cancer Res 2002;62:3493–3497.
15. Gant TW. Classifying toxicity and pathology by gene-expression profile—taking a lead from studies in neoplasia. Trends Pharmacol Sci 2002;23:388–393.
16. Biomarkers Definitions Working Group. Biomarkers and surrogate endpoints: preferred definitions and conceptual framework. Clin Pharmacol Ther 2001;69: 89–95.
17. Searfoss GH, Jordan WH, Calligaro DO, Galbreath EJ, Schirtzinger LM, Berridge BR, et al. Adipsin, a biomarker of gastrointestinal toxicity mediated by a functional gamma-secretase inhibitor. J Biol Chem 2003;278:46107–46116.
18. Milano J, McKay J, Dagenais C, Foster-Brown L, Pognan F, Gadient R, et al. Modulation of notch processing by gamma-secretase inhibitors causes intestinal goblet cell metaplasia and induction of genes known to specify gut secretory lineage differentiation. Toxicol Sci 2004;82:341–358.
19. Guerreiro N, Staedtler F, Grenet O, Kehren J, Chibout SD. Toxicogenomics in drug development. Toxicol Pathol 2003;31:471–479.
20. Whitlock JP Jr. Induction of cytochrome P4501A1. Annu Rev Pharmacol Toxicol 1999;39:103–125.
21. Mimura J, Fujii-Kuriyama Y. Functional role of AhR in the expression of toxic effects by TCDD. Biochim Biophys Acta 2003;1619:263–268.
22. Nebert DW, Dalton TP, Okey AB, Gonzalez FJ. Role of aryl hydrocarbon receptor-mediated induction of the CYP1 enzymes in environmental toxicity and cancer. J Biol Chem 2004;279:23847–23850.
23. Hu W, Sorrentino C, Denison MS, Kolaja K, Fielden MR. Induction of cyp1a1 is a nonspecific biomarker of aryl hydrocarbon receptor activation: results of large scale screening of pharmaceuticals and toxicants in vivo and in vitro. Mol Pharmacol 2007;71:1475–1486.
24. Ganter B, Tugendreich S, Pearson CI, Ayanoglu E, Baumhueter S, Bostian KA, et al. Development of a large-scale chemogenomics database to improve drug candidate selection and to understand mechanisms of chemical toxicity and action. J Biotechnol 2005;119:219–244.
25. Kolaja K, Fielden M. The impact of toxicogenomics on preclinical development: from promises to realized value to regulatory implications. Preclinica 2004;2:122–129.
26. Fostel JM. Future of toxicogenomics and safety signatures: balancing public access to data with proprietary drug discovery. Pharmacogenomics 2007;8: 425–430.
27. Fielden MR, Halbert DN. Iconix Biosciences, Inc. Pharmacogenomics 2007;8: 401–405.
28. Castle AL, Carver MP, Mendrick DL. Toxicogenomics: a new revolution in drug safety. Drug Discov Today 2002;7:728–736.

29. Ganter B, Snyder RD, Halbert DN, Lee MD. Toxicogenomics in drug discovery and development: mechanistic analysis of compound/class-dependent effects using the DrugMatrix database. Pharmacogenomics 2006;7:1025–1044.
30. Schulte-Hermann R. Induction of liver growth by xenobiotic compounds and other stimuli. CRC Crit Rev Toxicol 1974;3:97–158.
31. Amacher DE, Schomaker SJ, Burkhardt JE. The relationship among microsomal enzyme induction, liver weight and histological change in rat toxicology studies. Food Chem Toxicol 1998;36:831–839.
32. Suter L, Babiss LE, Wheeldon EB. Toxicogenomics in predictive toxicology in drug development. Chem Biol 2004;11:161–171.
33. Nikolsky Y, Ekins S, Nikolskaya T, Bugrim A. A novel method for generation of signature networks as biomarkers from complex high throughput data. Toxicol Lett 2005;158:20–29.
34. Natsoulis G, El Ghaoui L, Lanckriet GR, Tolley AM, Leroy F, Dunlea S, et al. Classification of a large microarray data set: algorithm comparison and analysis of drug signatures. Genome Res 2005;15:724–736.
35. Maggioli J, Hoover A, Weng L. Toxicogenomic analysis methods for predictive toxicology. J Pharmacol Toxicol Methods 2006;53:31–37.
36. Bushel PR, Hamadeh HK, Bennett L, Green J, Ableson A, Misener S, et al. Computational selection of distinct class- and subclass-specific gene expression signatures. J Biomed Inform 2002;35:160–170.
37. Brown MP, Grundy WN, Lin D, Cristianini N, Sugnet CW, Furey TS, et al. Knowledge-based analysis of microarray gene expression data by using support vector machines. Proc Natl Acad Sci U S A 2000;97:262–267.
38. Bishop CM. Neural networks for pattern recognition. Oxford: Clarendon; 1995.
39. Thomas RS, Rank DR, Penn SG, Zastrow GM, Hayes KR, Pande K, et al. Identification of toxicologically predictive gene sets using cDNA microarrays. Mol Pharmacol 2001;60:1189–1194.
40. Cristianini N, Shawe-Taylor J. An introduction to support vector machines. Cambridge, UK: Cambridge University Press; 2000.
41. Furey TS, Cristianini N, Duffy N, Bednarski DW, Schummer M, Haussler D. Support vector machine classification and validation of cancer tissue samples using microarray expression data. Bioinformatics 2000;16:906–914.
42. Martin R, Yu K. Assessing performance of prediction rules in machine learning. Pharmacogenomics 2006;7:543–550.
43. Olson H, Betton G, Robinson D, Thomas K, Monro A, Kolaja G, et al. Concordance of the toxicity of pharmaceuticals in humans and in animals. Regul Toxicol Pharmacol 2000;32:56–67.
44. Waring JF, Ciurlionis R, Jolly RA, Heindel M, Ulrich RG. Microarray analysis of hepatotoxins in vitro reveals a correlation between gene expression profiles and mechanisms of toxicity. Toxicol Lett 2001;120:359–368.
45. Hamadeh HK, Bushel PR, Jayadev S, DiSorbo O, Bennett L, Li L, et al. Prediction of compound signature using high density gene expression profiling. Toxicol Sci 2002;67:232–240.

46. Sawada H, Takami K, Asahi S. A toxicogenomic approach to drug-induced phospholipidosis: analysis of its induction mechanism and establishment of a novel in vitro screening system. Toxicol Sci 2005;83:282–292.
47. Hong Y, Muller UR, Lai F. Discriminating two classes of toxicants through expression analysis of HepG2 cells with DNA arrays. Toxicol in vitro 2003;17: 85–92.
48. Martin R, Rose D, Yu K, Barros S. Toxicogenomics strategies for predicting drug toxicity. Pharmacogenomics 2006;7:1003–1016.
49. Burczynski ME, McMillian M, Ciervo J, Li L, Parker JB, Dunn RT, et al. Toxicogenomics-based discrimination of toxic mechanism in HepG2 human hepatoma cells. Toxicol Sci 2000;58:399–415.
50. Atienzar F, Gerets H, Dufrane S, Tilmant K, Cornet M, Dhalluin S, et al. Determination of phospholipidosis potential based on gene expression analysis in HepG2 cells. Toxicol Sci 2007;96:101–114.
51. Snyder RD, Green JW. A review of the genotoxicity of marketed pharmaceuticals. Mutat Res 2001;488:151–169.
52. Newton RK, Aardema M, Aubrecht J. The utility of DNA microarrays for characterizing genotoxicity. Environ Health Perspect 2004;112:420–422.
53. Aubrecht J, Caba E. Gene expression profile analysis: An emerging approach to investigate mechanisms of genotoxicity. Pharmacogenomics 2005;6: 419–428.
54. Dickinson DA, Warnes GR, Quievryn G, Messer J, Zhitkovich A, Rubitski E, et al. Differentiation of DNA reactive and non-reactive genotoxic mechanisms using gene expression profile analysis. Mutat Res 2004;549:29–41.
55. Sawada H, Taniguchi K, Takami K. Improved toxicogenomic screening for drug-induced phospholipidosis using a multiplexed quantitative gene expression ArrayPlate assay. Toxicol in vitro 2006;20:1506–1513.
56. Thukral SK, Nordone PJ, Hu R, Sullivan L, Galambos E, Fitzpatrick VD, et al. Prediction of nephrotoxicant action and identification of candidate toxicity-related biomarkers. Toxicol Pathol 2005;33:343–355.
57. Peters EH, Rojas-Caro S, Brigell MG, Zahorchak RJ, des Etages SA, Ruppel PL, et al. Quality-controlled measurement methods for quantification of variations in transcript abundance in whole blood samples from healthy volunteers. Clin Chem 2007;53:1030–1037.
58. Willey JC, Crawford EL, Knight CR, Warner KA, Motten CA, Herness EA, et al. Standardized RT-PCR and the standardized expression measurement center. Methods Mol Biol 2004;258:13–41.
59. Willey JC, Knight CR, Crawford EL, Olson DE, Hammersly J, Yoon Y, et al. Use of standardized reverse transcription-polymerase chain reaction and the standardized expression measurement center in multi-institutional trials to develop meaningful lung cancer classification based on molecular genetic criteria. Chest 2004;125:155S–156S.
60. Rockett JC, Burczynski ME, Fornace AJ, Herrmann PC, Krawetz SA, Dix DJ. Surrogate tissue analysis: monitoring toxicant exposure and health status of

inaccessible tissues through the analysis of accessible tissues and cells. Toxicol Appl Pharmacol 2004;194:189–199.
61. Burczynski ME, Dorner AJ. Transcriptional profiling of peripheral blood cells in clinical pharmacogenomic studies. Pharmacogenomics 2006;7:187–202.
62. Burczynski ME, Peterson RL, Twine NC, Zuberek KA, Brodeur BJ, Casciotti L, et al. Molecular classification of Crohn's disease and ulcerative colitis patients using transcriptional profiles in peripheral blood mononuclear cells. J Mol Diagn 2006;8:51–61.
63. Rockett JC, Kavlock RJ, Lambright CR, Parks LG, Schmid JE, Wilson VS, et al. DNA arrays to monitor gene expression in rat blood and uterus following 17beta-estradiol exposure: biomonitoring environmental effects using surrogate tissues. Toxicol Sci 2002;69:49–59.
64. Baechler EC, Batliwalla FM, Karypis G, Gaffney PM, Ortmann WA, Espe KJ, et al. Interferon-inducible gene expression signature in peripheral blood cells of patients with severe lupus. Proc Natl Acad Sci U S A 2003;100:2610–2615.
65. Aune TM, Maas K, Moore JH, Olsen NJ. Gene expression profiles in human autoimmune disease. Curr Pharm Des 2003;9:1905–1917.
66. Olsen NJ, Moore JH, Aune TM. Gene expression signatures for autoimmune disease in peripheral blood mononuclear cells. Arthritis Res Ther 2004;6: 120–128.
67. Tang Y, Gilbert DL, Glauser TA, Hershey AD, Sharp FR. Blood gene expression profiling of neurologic diseases: a pilot microarray study. Arch Neurol 2005;62:210–215.
68. Scherzer CR, Eklund AC, Morse LJ, Liao Z, Locascio JJ, Fefer D, et al. Molecular markers of early Parkinson's disease based on gene expression in blood. Proc Natl Acad Sci U S A 2007;104:955–960.
69. Borovecki F, Lovrecic L, Zhou J, Jeong H, Then F, Rosas HD, et al. Genome-wide expression profiling of human blood reveals biomarkers for Huntington's disease. Proc Natl Acad Sci U S A 2005;102:11023–11028.
70. Dressman HK, Muramoto GG, Chao NJ, Meadows S, Marshall D, Ginsburg GS, et al. Gene expression signatures that predict radiation exposure in mice and humans. PLoS Med 2007;4:e106.
71. Whitney AR, Diehn M, Popper SJ, Alizadeh AA, Boldrick JC, Relman DA, et al. Individuality and variation in gene expression patterns in human blood. Proc Natl Acad Sci U S A 2003;100:1896–1901.
72. Kim SJ, Dix DJ, Thompson KE, Murrell RN, Schmid JE, Gallagher JE, et al. Effects of storage, RNA extraction, genechip type, and donor sex on gene expression profiling of human whole blood. Clin Chem 2007;53:1038–1045.
73. Steppan S, Kupfer K, Mayer A, Evans M, Yamasaki G, Greve JM, et al. Genome wide expression profiling of human peripheral blood mononuclear cells stimulated with BAY 50–4798, a novel T cell selective interleukin-2 analog. J Immunother 2007;30:150–168.
74. van Leeuwen DM, Gottschalk RW, van Herwijnen MH, Moonen EJ, Kleinjans JC, van Delft JH. Differential gene expression in human peripheral blood

mononuclear cells induced by cigarette smoke and its constituents. Toxicol Sci 2005;86:200–210.
75. Shi L, Reid LH, Jones WD, Shippy R, Warrington JA, Baker SC, et al. The MicroArray Quality Control (MAQC) project shows inter- and intraplatform reproducibility of gene expression measurements. Nat Biotechnol 2006;24:1151–1161.
76. Ludwig JA, Weinstein JN. Biomarkers in cancer staging, prognosis and treatment selection. Nat Rev Cancer 2005;5:845–856.

3

Profiling Gene Expression Signatures Using Fluorescent Microspheres

Suzanne M. Torontali, Jorge M. Naciff, Kenton D. Juhlin, and Jay P. Tiesman

Summary

One of the primary goals of genome-wide expression profiling is to identify subsets of gene transcripts that can be associated with specific biological conditions. Once these gene expression "signatures" have been elucidated, they can be used in screening assays to identify compounds that induce similar gene expression changes. To increase the throughput of these screening assays, we have optimized the development of a fluorescent microsphere–based, flow cytometric platform that allows the specific and sensitive screening of up to 100 transcripts per sample in a 96-well plate format. This chapter will outline experimental details of such an experiment using an estrogenic gene expression signature as a specific example.

Key Words: beads; biomarker; expression profiling; fingerprint; gene expression; microspheres; RNA; signature; screening

1. INTRODUCTION

RNA expression profiling using high-density, whole-genome microarrays has revolutionized the field of molecular biology. In addition to providing important insights into biological systems, it is being successfully integrated into the identification of gene expression biomarkers. Gene expression biomarkers are genes whose changes in expression are associated with a specific biological effect (for instance, the response of cells to treatment with a chemical compound). Efforts to identify biomarkers are affecting the fields of diagnostics, drug discovery, and toxicology. Research in these fields has shown that most biological effects are best represented not by a single gene biomarker but by a finite set of biomarkers that have been statistically associated with the biological effect (i.e., a gene *signature* or *fingerprint*) ***(1)***.

From: *Methods in Pharmacology and Toxicology: Biomarker Methods in Drug Discovery and Development*
Edited by: F. Wang © Humana Press, Totowa, NJ

In many cases, once this gene signature has been identified, it would be useful to develop a secondary screening platform to identify compounds capable of inducing this signature. To address this need, we have developed a bead-based secondary screening platform that can be used for high-throughput gene signature screening. Based on the Luminex xMAP platform, we have implemented this technology as secondary screen for gene signatures that have been identified using high-density microarrays *(2)*. For instance, in the field of environmental toxicology, there is a need to develop a screening platform for compounds capable of inducing biological responses similar to estrogen (i.e., *estrogenics*) *(3)*. The identification of such an estrogenic signature will be used to illustrate the methods used to (1) identify a gene expression signature associated with a biological response; (2) develop a set of microspheres representing that gene signature; and (3) assay compounds for that gene signature.

2. MATERIALS

1. Luminex xMAP microspheres (Luminex, Austin, TX)
2. Trizol (Invitrogen, Carlsbad, CA) and chloroform
3. Polytron homogenizer (Polytron, Inc., Norcross, GA)
4. Eppendorf microcentrifuge (Eppendorf AG, Hamburg, Germany)
5. Vortex (VWR Scientific, West Chester, PA)
6. Heavy and Light Phase Lock Gel Tubes (Eppendorf AG)
7. Qiagen RNEasy Mini Columns (Qiagen NV, Venlo, The Netherlands)
8. Invitrogen Superscript II cDNA Synthesis kit (Invitrogen)
9. Enzo BioArray High Yield cRNA Synthesis kit (Enzo Biochem, New York, NY)
10. NanoDrop ND-1000 Spectrophotometer (NanoDrop, Wilmington, DE)
11. Agilent 2100 Bioanalyzer (Agilent Technologies, Santa Clara, CA)
12. MJ Research ThermoCycler Plates with Caps (or equivalent) (Waltham, MA)
13. Eppendorf Thermo-Sealer (Eppendorf AG)
14. Eppendorf Thermo-Mixer (Eppendorf AG)
15. BioRad (Hercules, CA) BioPlex (or equivalent) Flow Cytometer

3. METHODS

3.1. Selection of Biomarker Signature

In multiple studies using high-density oligonucleotide arrays (i.e., Affymetrix GeneChip Arrays; Affymetrix, Santa Clara, CA), we have shown that exposure to various estrogen receptor agonists induced a characteristic gene expression profile in the developing reproductive system of the female rat *(2–4)*. From the global analysis of the gene expression changes induced by estrogen exposure, we determined that the expression of more than 1000 genes showed a significant change with respect to control (ANOVA, $p \leq 0.0001$), with almost equal representation of upregulated (53.7%) and downregulated (46.3%) genes. From these genes, we selected 20 genes that showed a robust and significant response to estrogen exposure in a dose-dependent manner (listed in **Table 1**). Further, this subset of estrogen-responsive genes includes genes identified by others (using various approaches) as being estrogen responsive (see Refs. *3* and *4* for references). Among them are complement component 3, intestinal calcium-binding protein (also known as calbindin 3), steroidogenic acute regulatory protein, cytochrome P450 (family 17, subfamily a, polypeptide 1), insulin-like growth factor–binding protein 3, brain creatine kinase, Kruppel-like factor 4, cathepsin B, progesterone receptor, and eotaxin (also known as chemokine [C-C motif] ligand 11). We also included two internal control

Table 1
Genes in Which Expression Is Modified by Estrogen Exposure in a Dose-Dependent Manner in the Uterus and the Ovaries of the Immature Rat

Gene name	Gene symbol
Steroidogenic acute regulatory protein	*Star*
Uterus-ovary specific putative transmembrane protein or CUB and zona pellucida-like domains 1	*Cuzd1*
Peroxisomal enoyl-CoA-hydrotase-3-hydroxyacyl-CoA	*Ehhadh*
Fibronectin 1	*Fn1*
Cyclin-dependent kinase 4	*Cdk4*
Kruppel-like factor 4	*Klf4*
Osteopontin or secreted phosphoprotein 1	*Spp1*
Complement component 3	*C3*
Insulin-like growth factor–binding protein 3	*Igfbp3*
Creatine kinase, brain	*Ckb*
Heat shock protein 27	*Hspb1*
Cathepsin B	*Ctsb*
Eotaxin or chemokine (C-C motif) ligand 11	*Scya11*
Vascular alpha-actin	*VaACTIN*
Progesterone receptor	*Pgr*
Bacteriophage M13	*M13*
Intestinal calcium-binding protein or calbindin 3	*Calb3*
Cytochrome P450, family 17, subfamily a, polypeptide 1	*Cyp17a1*
Hydroxysteroid 11-beta dehydrogenase 2	*Hsd11b2*
Cyclophilin B or peptidylprolyl isomerase B	*Ppib*

genes, vascular alpha actin and cyclophilin B, that are also expressed in the uterus/ovaries of the immature rat, but whose expression is not regulated by estrogens in these organs (**Table 1**).

3.2. Oligonucleotide Probe Selection

The probes are chosen from among those used on the Affymetrix GeneChip arrays. Each probe set consists on the order of 11 to 20 probe pairs, one "Perfect Match" probe and a corresponding "Mismatch" probe in each pair. The selection process has three components: one is for a gene that varies across experimental conditions, one is for a gene that remains constant across experimental conditions (such as a *housekeeping gene*), and the last component is for genes used in assessing quality of the experiment, such as GAPDH (glyceraldehyde 3-phosphate dehydrogenase: 3′ end, middle, 5′ end).

Application of the selection algorithms requires having one or more suitable microarray experiments. For a gene that varies across conditions, a probe is chosen so that its probe level intensities, taken across all samples, best correlate with the corresponding signal values. In some cases, two or more probes may be averaged to achieve a better correlation. These pairs (or triplets) may be used if the amount of overlap in the sequences is not substantial.

For a gene that does not vary across experimental conditions, the objective is to minimize a measure of variability that captures the signal-to-noise ratio. Specifically, a relative standard deviation (RSD) is used, which is expressed as the ratio of the standard deviation to the mean, and the probe having the lowest RSD is chosen.

If a gene is used for assessing the quality of the experiment, the signal value measure of quality (such as the GAPDH 3′/5′ ratio) will determine the appropriate selection measures at the probe level.

3.3. Linking Probes to Luminex xMAP Beads

All oligonucleotides were ordered from Luminex Corp. Antisense strand oligonucleotides that were bound to the microspheres included an amine substitution to allow the oligonucleotides to bind to the carboxyl group on the surface of the beads. Twenty sets of microspheres of differing fluorescence addresses were ordered from Luminex. The oligonucleotides were coupled to the microspheres using established procedures (Radix Biosolutions, Georgetown, TX). After coupling, the microspheres were washed of unbound oligonucleotides, counted on a hemacytometer, and adjusted to a concentration of 1×10^7 microspheres per milliliter and stored at 4°C until use. Complimentary sense strand oligonucleotides were ordered with a 5'-biotin modification to be used for titration curves and quality control.

3.4. Tissue Treatment

For the sake of specific illustration, uteri from rats treated with estradiol were assayed for gene signature in this experiment. However, it should be clear that a primary goal of developing these assays is to identify signatures in treated cultured cells. Toward this end, Naciff et al. *(5)* have demonstrated similar estrogenic expression signatures using a cultured human endometrial adenocarcinoma cell line.

Fifteen-day-old female Sprague-Dawley rats were obtained (Charles River VAF/Plus; Charles River Laboratories, Wilmington, MA) in groups of 10 pups per surrogate mother. This rat strain was chosen because it is a commonly used strain in reproductive and developmental toxicity studies. The rats were acclimated to the local vivarium conditions (24°C; 12-h light/12-h dark cycle) for 5 days. Starting on postnatal day (PND) 20 and during the experimental phase of the protocol, all rats were singly housed in 20 × 32 × 20 cm plastic cages. The experimental phase was run to test the effects of 17-alpha-ethynyl estradiol (EE) at the relatively high dosages of 0.1, 1.0, and 10.0 μg EE kg^{-1} day^{-1} in animals fed with a standard laboratory rodent diet (Purina 5001; Purina Mills, St. Louis, MO). The Purina 5001 diet contains phytoestrogens, mostly genistein and daidzein derived from soy and alfalfa *(6)*, but is not uterotrophic and does not elicit estrogen-related gene expression changes *(7)*. We chose to use this diet to avoid a potential negative shifting of the baseline data, thereby diminishing the value of historical comparisons of estrogen-dependent gene expression data already published *(3,4)*. Furthermore, using a natural diet makes the work more relevant for extrapolation to humans and other free-living species. Whereas this raised the possibility of a modest additive estrogenic effect, we believe it is negligible as it was diluted at the highly active dose levels of EE used in this study. The experimental protocol was carried out according to Procter and Gamble's animal care approved protocols, and animals were maintained in accordance with the NIH Guide for the Care and Use of Laboratory Animals.

Starting on PND 20, the animals were dosed, by subcutaneous injection, with 0.1, 1.0, or 10.0 μg kg^{-1} day^{-1} 17-α-ethynyl estradiol in peanut oil (10 animals per dose group were used). Animals received 5 mL/kg body weight dose solution once a day for 4 days. A 4-day dosing regime was selected to optimize detection of any effect of EE exposure at the low dose, both at the histologic level as well as at the gene expression level. The dose was administered between 8 and 9 AM each day. Controls received 5 mL/kg peanut oil once a day for 4 days. Dose volumes were adjusted daily for weight changes. Body weight (nearest 1.0 g) and the volume of the dose administered (nearest 0.1 mL) were recorded daily. The animals were sacrificed by CO asphyxiation on PND 24, 24 h after the last dosing. The body of the uterus was cut just above its junction with the cervix and leaving the ovaries attached to it, carefully dissected free of adhering fat and mesentery and weighed, as a whole. Then, the ovaries were dissected free, and the uterine and ovarian wet weight was recorded. The tissues were then placed into RNALater (50 to 100 mg/mL solution; Ambion, Austin, TX), and then placed at 4°C. After 24 h incubation in RNALater solution at 4°C, RNALater was aspirated and tissues were frozen at –80°C indefinitely.

3.5. RNA Isolation

3.5.1. RNA Extraction

1. Tissues are removed from freezer and 1 mL Trizol is added immediately to frozen samples.
2. Tissues are homogenized using a Polytron homogenizer until sample is homogenous. (*Note*: If using cultured cells, lysis may be performed by simply pipetting the solution.)
3. After harvesting tissues, lysates can be stored in 1 mL Trizol at –80°C until processed.
4. Samples are thawed at room temperature. After thawing, samples are incubated an additional 5 min at room temperature and centrifuged briefly at maximum speed if there is any debris.
5. Add 0.2 mL chloroform and vortex. Incubate samples at room temperature for 3 min.
6. Prepare 2 mL Heavy Phase Lock Gel (PLG-Heavy) tubes by centrifuging at maximum speed for 20 to 30 s.
7. Transfer sample to PLG tubes and centrifuge at 12,000 × *g* for 5 min at 4°C.
8. Collect supernatant and transfer to fresh tubes. Add 0.5 mL isopropanol and vortex. Incubate at room temperature for 10 min.
9. Spin samples at 12,000 × *g* for 10 min at 4°C.
10. Decant supernatant and overlay pellets with 1.0 mL –20°C, 70% ethanol. Invert several times to wash pellet and re-centrifuge at 7500 × *g* for 5 min at 4°C.
11. Carefully decant supernatant and air-dry pellets. Resuspend pellets in 100 μL diethylpyrocarbonate-treated water (DEPC-H_2O).

3.5.2. RNA Cleanup Using Qiagen RNEasy Mini Columns

1. To RNA sample in 100 μL DEPC-H_2O, add 350 μL Qiagen Buffer RLT (containing β-mercaptoethanol), cap tubes, and vortex.
2. Add 250 μL absolute ethanol, cap tubes, and vortex. Apply sample to RNEasy Micro column and centrifuge for 30 s at maximum speed.
3. Reapply flow-through to column and re-centrifuge (30 s at maximum speed).
4. Move column to fresh wash collection tube and add 0.5 mL Qiagen Buffer RPE (containing ethanol). Centrifuge for 30 s at maximum speed. Discard flow-through.

5. Wash with another 0.5 mL buffer RPE. Spin 30 s at maximum speed.
6. Transfer columns to labeled sample collection tubes and elute by pipetting a minimumof 30 μL RNase-free water, warmed at least to 65°C, directly onto the column membrane. Incubate at room temperature for 5 min and then centrifuge for 1 min at maximum speed.
7. Quantitate RNA using spectrophotometer.

3.6. Preparation of Biotinylated Targets

The Invitrogen Superscript II cDNA Synthesis kit is used to generate double-stranded cDNA template, and the Enzo BioArray High Yield cRNA Synthesis kit is used to create cRNA targets for hybridization.

3.6.1. First Strand cDNA Synthesis

1. Add 1 μL 100 pmol T7-(dT) primer to 10 μL purified total RNA (5 to 10 μg).
2. Incubate for 10 min at 70°C.
3. Centrifuge briefly at maximum speed and place on ice for 7 to 10 min.
4. Add:

   4μL 5X First Strand buffer
   2μL 0.1 M dithiothreitol (DTT)
   1μL 10 mM deoxynucleotide triphosphate (dNTP) mix
   2μL SuperScript II Reverse Transcriptase

   20μL Total Volume

5. Incubate for 1 h at 42°C.

3.6.2. Second Strand DNA Synthesis

1. Place first strand reaction tubes on ice for 10 min and centrifuge briefly at maximum speed.
2. Add:

   91μL DEPC-H_2O
   30μL 5X Second Strand buffer
   3μL 10 mM dNTP mix
   1μL 10 U/μL *E. coli* DNA ligase
   4μL 10 U/μL *E. coli* DNA polymerase I
   1μL 2 U/μL RNAse H

   162μL Total Volume

3. Mix by pipetting and centrifuge briefly at maximum speed.
4. Incubate for 2 h at 16°C.
5. Store at –20°C or proceed to cDNA cleanup, keeping samples on ice.

3.6.3. cDNA Cleanup

1. To cDNA samples, add an equal volume (162μL) of (25:24:1) phenol: chloroform:isoamyl alcohol and vortex.

2. Prepare Light Phase Lock Gel (PLG-Light) tubes by centrifuging at maximum speed for 20 to 30 s.
3. Transfer sample to prepared Phase Lock Gel tubes. (*Important*: Do not vortex.)
4. Centrifuge for 2 min at maximum speed.
5. Transfer upper (aqueous) phase to fresh 1.5-mL Eppendorf tubes.
6. Add 1 μL NF PelletPaint, 0.5 volumes (80 μL) 7.5 M NH_4OAc, and 2.5 volumes (400 μL) –20ºC, 100% ethanol.
7. Immediately centrifuge for 20 min at room temperature at maximum speed.
8. Aspirate ethanol and overlay pellet with 0.5 mL –20ºC, 70% ethanol.
9. Centrifuge at maximum speed for 2 min at room temperature.
10. Aspirate ethanol and air-dry pellet (a Speed-Vac may be used to accelerate drying).
11. Resuspend pellet in 10.5 μL DEPC-H_2O. Place tubes on ice for 10 min. (*Important*: Do not delete this incubation step.) To ensure that cDNA is completely resuspended, flick tubes, centrifuge briefly, and heat for 5 min at 65ºC. (*Important*: Do not mix by pipetting. Following this procedure precisely ensures complete suspension of cDNA and is critical for high yields of cRNA.)
12. Remove 0.5 μL sample for gel analysis (e.g., using the Agilent 2100 Bioanalyzer Total RNA Nano Assay) and store remainder at –20ºC or proceed to cRNA synthesis.

3.6.4. cRNA Synthesis

The Enzo BioArray High Yield RNA Transcription Labeling Kit is used for cRNA synthesis.

1. Using the mix, the following at room temperature in the listed order (all reagents, except DTT, should be kept on ice before adding):

10.0μL	cDNA
12μL	DEPC-H O
4μL	10X Hyb Rxn buffer
4μL	10X biotin-labeled ribonucleotides
4μL	10X DTT (*bring to room temperature to prevent precipitation*)
4μL	10X RNase inhibitor mix
2μL	20X T7 RNA polymerase
40μL	**Total Volume**

2. Incubate 6 h at 37ºC in thermal cycler.
3. Place tubes on ice for 10 min (or at –20ºC for long-term storage) after incubation.

3.6.5. cRNA Cleanup

cRNA is cleaned up using Qiagen's RNeasy mini cleanup protocol.

1. In an RNAse-free 1.5-mL Eppendorf tube, mix 40 μL cRNA with 60 μL DEPC-H_2O.

2. In a separate tube, mix 2 mL RLT buffer with 20 μL β-mercaptoethanol to make RLT+β-mercaptoethanol stock reagent (this is enough reagent to clean up four samples; adjust as needed).
3. Add 350 μL RLT+β-mercaptoethanol stock reagent to each cRNA tube.
4. Add 250 μL 100% ethanol and vortex. (*Important*: Do not centrifuge.)
5. Apply sample to RNeasy column.
6. Centrifuge for 15 s at maximum speed.
7. Reapply flow-through to column and centrifuge another 15 s at maximum speed.
8. Transfer column to fresh collection tube and dispose of flow-through.
9. Add 500 μL RPE buffer to column and centrifuge for 15 s at maximum speed.
10. Discard flow-through.
11. Add another 500 μL RPE buffer to column and centrifuge for 2 min at maximum speed. It is important to make sure there is no residual RPE left on the column membrane, as this can decrease yield of purified cRNA recovered during elution.
12. Transfer column to a fresh 1.5-mL collection tube.
13. Add 30 μL DEPC-H_2O to column (make sure water is dispensed onto filter within the column), and incubate for 5 min at room temperature.
14. Optional: If yields are low, another elution of 30 to 40 μL RNase-free water over the column can be performed. The second elution should be quantified separately from the first.
15. Centrifuge for 1 min at maximum speed to elute cRNA.
16. Determine concentration of cRNA using a spectrophotometer.
17. If cRNA is not at least 0.6 μg/μL, concentrate sample by vacuum centrifugation (e.g., using a Speed-Vac concentrator).

3.6.6. *cRNA Fragmentation*

The following reaction is performed in 8-well strip or 96-well PCR reaction tubes (e.g., MJ Research).

1. Mix the following:

   X μL (20 μg) cRNA
   8 μL 5X Fragmentation buffer
   X μL DEPC-H_2O

   40 μL Total Volume

2. Incubate 94°C for 35 min.
3. Place on ice.
4. Remove 1 μL fragmented cRNA for gel analysis (e.g., using the Agilent 2100 Bioanalyzer Total RNA Nano Assay).
5. Store remainder of fragmented cRNA at –20°C. Fragmented cRNA may be stored at –20°C indefinitely before hybridization.

3.7. Bead Processing

Hybridization, washing, and staining of beads are performed using reagents and protocols modified from Affymetrix for use in their GeneChip reagent system. Bead assays are performed using the BioRad BioPlex instrument designed for analysis of Luminex xMAP beads.

3.7.1. Dilution of Bead Sets and cRNA Targets

1. Linked beads for each probe (see **Section 3.3**) are diluted to 10^7 beads/mL in 0.5X TMAC buffer (1X TMAC buffer = 3 M tetramethylammonium chloride, 0.1% Sarcosyl, 50 mM Tris-HCl pH 8.0, 4 mM EDTA). Forty microliters of each bead set is added to each well (~1000 beads per well for each probe).
2. Fragmented cRNA target is diluted to 0.1 μg/μL in 0.5X TMAC buffer. Twenty microliters (2 μg) of target is added to each well.

3.7.2. Hybridization, Washing, and Staining of Beads

1. Add 20 μL diluted cRNA target (see **Section 3.7.1, step 2**).
2. Add 40 μL bead mix to each well (see **Section 3.7.1, step 1**).
3. Place strip caps over wells and incubate at 95°C for 2 min.
4. Transfer plate to Eppendorf Thermo Mixer, cover, and hybridize overnight at 45°C with shaking at 500 rpm.
5. Spin samples in centrifuge at 2250 × *g* for 2 min, flick plates, and tap off solution.
6. Wash beads with 100 μL 0.5X TMAC; shake 500 rpm at 25°C for 2 min. (*Important*: Do not mix by pipette; this applies for all wash steps.)
7. Spin samples in centrifuge at 2250 × *g* for 2 min, flick plate, and tap off solution.
8. Wash beads with 100 μL PBS-BSA (Stock = 9.7 mL phosphate-buffered saline pH7.2 + 330 μL 30% bovine serum albumin); shake 500 rpm at 25°C for 2 min.
9. Spin samples in centrifuge at 2250 × *g* for 2 min, flick plate, and tap off solution.
10. Add 100 μL StreptAvidin-PE (StAv-PE) stain mix (stock concentration 1 mg/mL: Affymetrix); shake 500 rpm at 25°C for 10 min. Seal plate with foil using Eppendorf Thermo-Sealer for remainder of protocol.
11. Spin samples in centrifuge at 2250 × *g* for 2 min, flick plate, and tap off solution.
12. Wash beads with 100 μL PBS-BSA; shake 500 rpm at 25°C for 2 min.
13. Spin samples in centrifuge at 2250 × *g* for 2 min, flick plate, and tap off solution.
14. Add 100 μL Second Stain (anti-StAv and nGtIgG; Affymetrix); shake 500 rpm at 25°C for 10 min.
15. Spin samples in centrifuge at 2250 × *g* for 2 min, flick plate, and tap off solution.
16. Wash beads with 100 μL PBS-BSA; shake 500 rpm at 25°C for 2 min.

17. Spin samples in centrifuge at 2250 × *g* for 2 min, flick plate, and tap off solution.
18. Add 100 μL Third Stain (StAv-PE); shake 500 rpm at 25°C for 10 min.
19. Spin samples in centrifuge at 2250 × *g* for 2 min, flick plate, and tap off solution.
20. Wash beads with 100 μL PBS-BSA; shake 500 rpm at 25°C for 10 min.
21. Spin samples in centrifuge at 2250 × *g* for 2 min, flick plate, and tap off solution.
22. Resuspend in 65 μL PBS-BSA and read on BioPlex. Shake 500 rpm at 25°C for 2 min. (*Important*: Shake plate thoroughly before running on BioPlex.)

3.7.3. Bead Assay

1. Calibrate BioPlex for Low PMT reading first then reread plate at High PMT. (*Important*: ***(1)*** Between PMT readings, perform a wash between plates then calibrate; ***(2)*** shake plate at 500 rpm at 25°C for 2 min before High PMT reading.)
2. Read 50 μL, 100 count per probe. (*Important*: Evaluate needle height before reading MJ Research PCR plate using BioPlex instrument.)

3.8. Data Analysis

The BioPlex instrument measures the total fluorescence at the surface of each microsphere to quantify the amount of reporter bound to it. The time for data capture was limited by requiring 100 microsphere measurements from each set. From these readings, the median fluorescence intensity (MFI) for each set was derived. The MFI from five independent samples (biological replicates) from control or treated animals were determined and used to determine the average fold change on the expression for each transcript of interest.

4. NOTES

1. This procedure is ideally used to measure the effects of cells treated in culture. As such, the RNA isolation procedure may be modified accordingly for cultured cells.
2. The goal of this procedure is to move rapidly from high-density oligonucleotide arrays to high throughput screening. As such, efforts were made to develop protocols that were very similar between the two platforms. Many reagents used for identification of the gene signature using a microarray can also be used for signature screening. This includes RNA isolation, target labeling, and hybridization reagents.
3. Beads are processed using a BioPlex instrument capable of reading Luminex xMAP beads. Similar instruments from other manufacturers are available and will likely provide similar results.

4. Pay special attention to notes marked *Important* in the protocol. Deviations from these parts of the procedure have been shown to result in procedural failure or reduced data quality.

ACKNOWLEDGMENTS

The authors wish to thank Lynn Jump, Brian Richardson, Cindy Ryan, and Karen Lammers for their assistance in developing this protocol.

REFERENCES

1. Tiesman JP, Tortontali SM. Screening gene signatures: strategies for secondary screening of gene expression biomarkers. Am Drug Discov 2007;2:6–15.
2. Naciff JM, Richardson BD, Oliver KG, Jump ML, Torontali SM, Juhlin KD, et al. Design of a microspheres-based high-throughput gene expression assay to determine estrogenic potential. Environ Health Perspect 2005;113:1164–1171.
3. Naciff JM, Overmann GJ, Torontali SM, Carr GJ, Tiesman JP, Richardson BD, Daston GP. Gene expression profile induced by 17-α-ethynyl estradiol in the prepubertal female reproductive system of the rat. Toxicol Sci 2003;72: 314–330.
4. Naciff JM, Jump ML, Torontali SM, Carr GJ, Tiesman JP, Overmann GJ, Daston GP. Gene expression profile induced by 17α-ethynyl estradiol, bisphenol A, and genistein in the developing female reproductive system of the rat. Toxicol Sci 2002;68:184–199.
5. Naciff JM, Khambatta ZS, Thomason RG, Carr GJ, Tiesman JP, Daston GP. The genomics response of a human endometrial adenocarcinoma cell line to 17α-ethynyl estradiol and bisphenol A: a potential alternative to animal testing for estrogens. Presented at: 46th Annual Society of Toxicology Meeting; March 25–29, 2007; Charlotte, North Carolina.
6. Thigpen JE, Setchell KDR, Goelz MF, Forsythe DB. The phytoestrogen content of rodent diets. Environ Health Perspect 1999;107:A182–A183.
7. Naciff JM, Overmann GJ, Torontali SM, Carr GJ, Tiesman JP, Daston GP. Impact of the phytoestrogen content of laboratory animal feed on the gene expression profile of the female reproductive system of the immature rat. Environ Health Perspect 2004;112:1519–1526.

4

Real-Time Polymerase Chain Reaction Gene Expression Assays in Biomarker Discovery and Validation

Yulei Wang, Catalin Barbacioru, David Keys, Pius Brzoska, Caifu Chen, Kelly Li, and Raymond R. Samaha

Summary

Biomarkers have shown great potential in molecular classification and targeted treatment of human diseases. Biomarkers are also expected to play an increasingly important role in all phases of drug development as well as regulatory decision making. However, translating biomarker discovery into clinically useful tests in a time- and cost-effective manner remains a significant challenge. Real-time polymerase chain reaction (PCR) technology, in particular TaqMan®-based real-time PCR gene expression assays, provides a simple, robust, and practical tool that has shown great potential in bridging the gaps between biomarker discovery and clinical practice. In this chapter, we will survey the principle of this technology and its wide applications in biomarker discovery and validation.

Key Words: biomarker; gene copy number; gene expression; microRNA; real-time PCR; TaqMan®

1. INTRODUCTION

Clinical medicine is in the midst of a revolution and is continually evolving toward individualized medicine that is based on molecular classification and targeted treatments of diseases. As a result, it has become increasingly desirable to develop new biomarkers with wide clinical applications, including improved resolution of molecular classification of diseases to achieve better precision in diagnosis, monitoring treatment effects or disease progression, and predicting clinical outcomes. Biomarkers are also expected to play an increasingly important role in all phases of drug development and

From: *Methods in Pharmacology and Toxicology: Biomarker Methods in Drug Discovery and Development*
Edited by: F. Wang © Humana Press, Totowa, NJ

regulatory decision making. Panels of relevant biomarkers may be applicable for predicting the efficacy or safety of a drug and used by regulatory agencies as a basis for approval and market access of drugs.

Recent advances in molecular biotechnology have been one of the major driving forces in transforming the process of biomarker development and its adoption in clinical practice (**Fig. 1**). For example, high-throughput genomic technology, in particular DNA microarray technology, has greatly accelerated the discovery phase of gene expression biomarker development. This technology allows screening of large numbers of genes to identify potential biomarkers whose expression levels are correlated with disease status, clinical outcomes, and treatment regimens. However, before these candidate biomarkers can become useful in clinical settings, they must go through rigorous verification, typically involving both retrospective and prospective clinical studies. These verification studies remain a limiting step in translating biomarker discovery to clinically useful tests in a time- and cost-effective manner. In this chapter, we will focus on the real-time polymerase chain reaction (PCR) technology, in particular TaqMan®-based real-time PCR gene expression assays (Applied Biosystems, Foster City, CA), a simple, robust, and practical tool that has shown great potential in bridging the gaps between biomarker discovery and clinical practice. The principle of this technology and its applications in biomarker development will be discussed in detail.

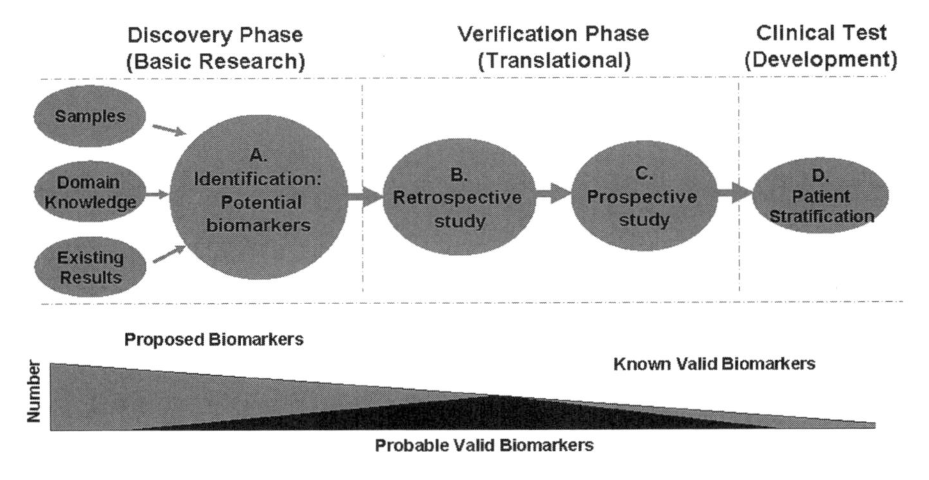

Fig. 1. Critical phases in biomarker development.

2. PRINCIPLES OF TAQMAN® REAL-TIME PCR ASSAYS

2.1. *Overview of Real-Time PCR Technology*

Quantitative, real-time PCR was first developed in the mid-1990s. This method capitalizes on the fact that there is a quantitative relationship between the amounts of specific transcripts and the amount of PCR products at any given PCR cycle number. For characterizing gene expression, a real-time PCR reaction is typically performed in a two-step reverse transcription–polymerase chain reaction (RT-PCR) (**Fig. 2**). In the reverse transcription step, RNA from a biological sample is reverse transcribed into cDNA using a reverse polymerase and random priming, oligo-dT primers or gene-specific primers. The use of gene-specific primers minimizes background priming, whereas the use of random or oligo-dT primers maximize the number of mRNA molecules that can be analyzed from a small sample of RNA. The second step is the polymerase chain reaction. Various methods are available to combine the processes of amplification and detection of the target to permit the monitoring of PCR reactions in real time. The simplest method uses fluorescent dyes (i.e., SYBR Green Applied Biosystems, Foster City, CA) that bind specifically to double-stranded-DNA *(1)*, however, the specificity of this method is determined entirely by PCR primers and therefore

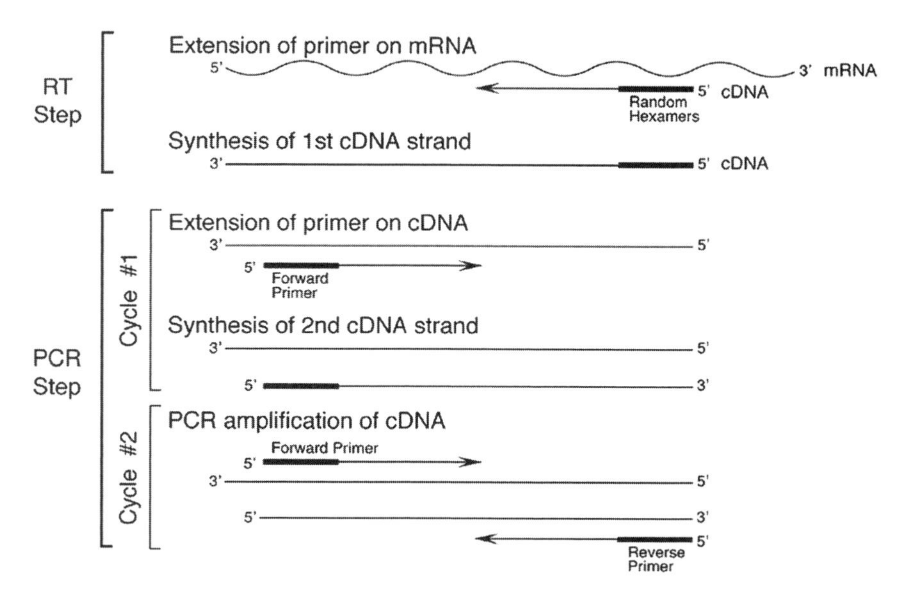

Fig. 2. Scheme of two-step real-time PCR reaction.

less optimal. Other methods provide more specificity by using hybridization of fluorescence-labeled probes to the targeted amplicon. These include molecular beacons *(2,3)*, hybridization probes *(4)*, and TaqMan® probes that utilize the 5' nuclease activity of the DNA polymerase to hydrolyze a hybridization probe bound to its target amplicon. We will take the TaqMan®-based real-time PCR assay as an example and discuss its chemistry and performance in more detail.

2.2. 5′ Nuclease Chemistry (Using TaqMan® Probes)

TaqMan® probes are single-stranded DNA oligos that contain a reporter dye (6-FAM dye, Applied Biosystems, Foster City, CA) linked to the 5′ end of the probe and a nonfluorescent quencher (NFQ) at the 3′ end of the probe. When the probe is intact, the proximity of the reporter dye to the quencher results in suppression of the reporter fluorescence, primarily due to Förster-type energy transfer. The nonfluorescent quencher used in TaqMan® probes provides greater signal-to-noise ratios and thus increased sensitivity of detection relative to fluorescent dye–quenching adjuncts such as TAMRA (Applied Biosystems, Foster City, CA).

During PCR, the TaqMan® probe anneals specifically to a complementary sequence between the forward and reverse primer sites. Only probes that are hybridized to the complementary target are cleaved by the 5′ exonuclease activity of AmpliTaq Gold DNA polymerase (Applied Biosystems, Foster City, CA). Cleavage separates the reporter dye from the NFQ, resulting in a detectable increase in reporter dye fluorescence at each PCR cycle. This increase in fluorescent signal occurs only if the target sequence is complementary to the probe and is amplified during PCR (**Fig. 3**). The amount of fluorescence produced from the TaqMan® probe is measured at each amplification cycle, providing a real-time measurement of how many amplification targets are produced during each round of PCR. Identification of the PCR cycle when the exponential growth phase is first detectable (C_t) provides highly accurate quantitation of gene expression in the starting samples. The TaqMan® real-time PCR assay is sensitive (ability to detect 1 copy in 10 to 100 cells), reproducible (standard deviation <0.25), and has a very large dynamic range (7 to 8 logs) (**Fig. 4**).

2.3. TaqMan® Real-Time PCR Assay Design Strategies

One of the most critical elements in real-time PCR experimentation is designing highly specific assays for quantifying the transcript of interest. For TaqMan® real-time PCR assay, it typically involves design of a single TaqMan® probe and two unlabeled oligonucleotide PCR primers. The entire

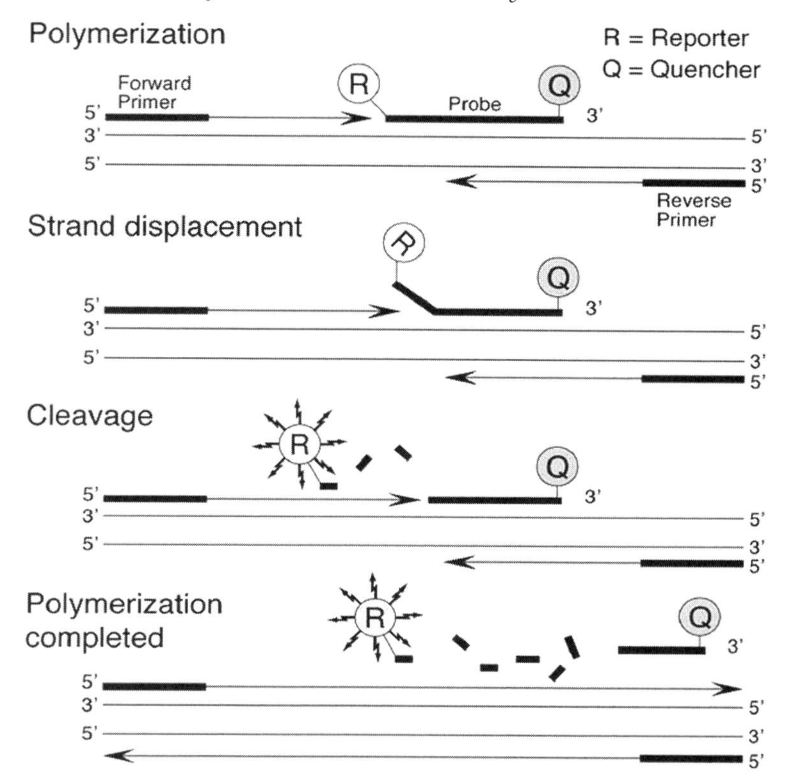

Fig. 3. 5′ nuclease reaction enables probe cleavage resulting in a fluorescent signal.

design process is automated using a highly sophisticated assay design pipeline that integrates comprehensive design rules, an extensive array of bioinformatics tools, and assay quality-control strategies (**Fig. 5**).

Based on information leveraged from both public and private genome databases, comprehensive sequence analysis is conducted to avoid (mask) SNPs (Single Nucleotide Polymorphisms), repeats, and areas of sequence discrepancy between the public and the private databases to identify suitable locations along mRNA transcripts for assay design. Problems associated with DNA contamination are minimized by designing PCR primers that bind in two distinct exons while placing the TaqMan® probe over the exon-exon boundary. To ensure the design of the most robust assays, selection criteria for primers and probes are based on both thermodynamic and chemistry parameters, including optimal melting temperature (T_m) requirements, GC-content, buffer/salt conditions, oligonucleotide concentrations, secondary structure, optimal amplicon size, primer-dimer minimization, and signal

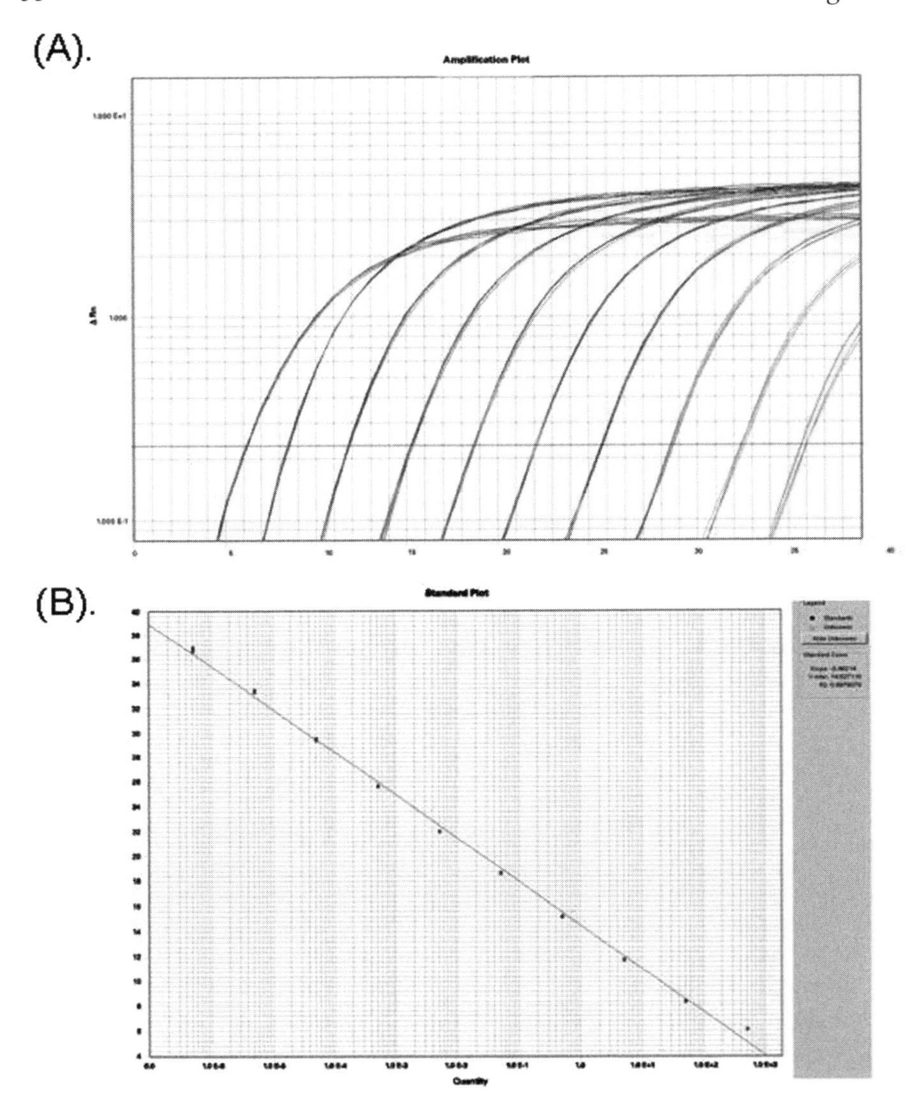

Fig. 4. Amplification of the Human Eukaryotic 18S rRNA target in 10-fold dilutions from 500 ng to 0.5 fg illustrating 9 logs of linear dynamic range. **(A)** Amplification plot showing the log of the change in fluorescence plotted versus cycle number. **(B)** Standard curve showing C_t values plotted versus the log of the quantity.

(A).

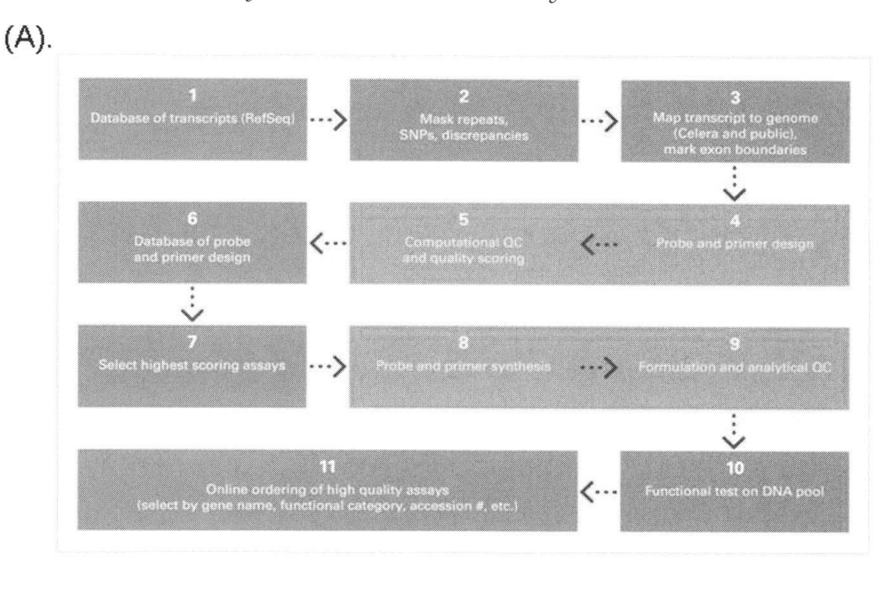

(B).

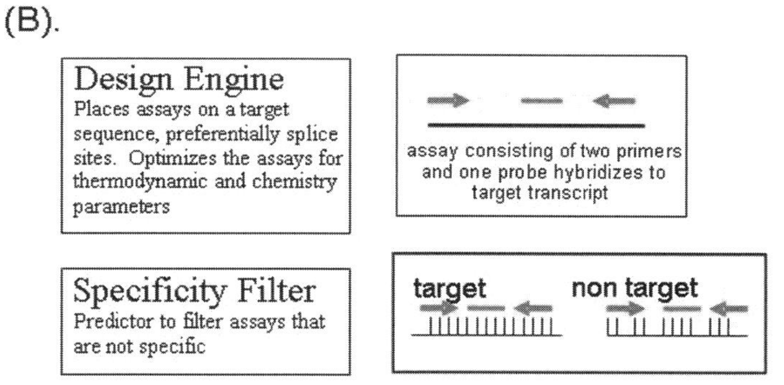

Fig. 5. **(A)** Schemes of automated pipeline for TaqMan® real-time PCR assay design. **(B)** Assay design and filtering strategy.

strength. All primers and probes are designed to work in a single set of standard PCR conditions, so all assays can be run with universal cycling and buffer concentrations. In addition, the design of TaqMan® probes is improved by incorporating both a minor groove binder (MGB) and the nonfluorescent quencher (NFQ) at the 3′ end of the oligonucleotide. There are several advantages of including the MGB moiety in the TaqMan® probe. First, efficient TaqMan® assays require the probe to have higher T_m than the PCR primers. The MGB moiety increases the T_m of the probe, enabling the use of shorter TaqMan® probes; second, shorter TaqMan® probes increase flexibility for designing assays in traditionally difficult sequence regions (i.e., AT-rich sequences); and finally, the relatively short MGB probes increase the probability of designing a probe over every exon-exon boundary of a multiexon gene. The T_m for each probe is calculated using nearest neighbor thermodynamics algorithms *(5)* that have been adapted to probes coupled to a MGB. Other parameters that are geared to optimize manufacturing quality control have been implemented in the design engine as well.

The design engine generates multiple assays for each target transcript. The second step of the design pipeline removes assays that are predicted to hybridize to transcripts other than the target. To that end, a program that both aligns assays to nontarget transcripts and scores the potential of generating a transcript specific signal was developed. The scoring system uses a position-specific alignment matrix and has been extensively validated experimentally (unpublished data). This extensively validated design and quality-control pipeline ensures that the assays that are designed are highly specific to their target and also allows the design of assays that discriminate between highly homologous gene families.

Currently, there is a comprehensive set of more than 680,000 predesigned TaqMan® probe and primer sets available for nearly all exon-exon junctions for transcripts from human, mouse, rat, *Arabidopsis*, *Drosophila*, and *Caenorhabditis elegans.* These predesigned assays are cost-effective and eliminate the labor barriers historically associated with designing and validating individual quantitative RT-PCR assays. The removal of these barriers has dramatically increased the ability of researchers to integrate quantitative real-time PCR technology much earlier and faster in their biomarker screening and validation workflow and greatly reduced the time to results.

2.4. Instrumentation

Real-time PCR instruments perform data acquisition and analysis by combining amplification, detection, and quantification in an integrated closed-tube system. This avoids problems of contamination and minimizes hands-on time. The entire process can be fully automated, which makes these instruments ideally suited for high-throughput screening applications. For example, the Applied Biosystems 7900HT real-time PCR (Foster City, CA) system can use a robotic plate loading and unloading device to provide a maximum throughput of up to 84 plates at a single time in unattended operation. This instrument uses user-interchangeable thermal cycler blocks that can accommodate 96- and 384-well plates, as well as the new 384-well TaqMan® Low Density Array format (discussed later). To detect fluorescence signal, samples are illuminated with an argon laser excitation source, and the resulting fluorescent emission is detected by a spectrograph attached to a charge-coupled device (CCD) camera. Emission wavelengths from 500 to 660 nm are monitored allowing the simultaneous detection of multiple fluorophores. Although this system is optimized for TaqMan® probe-based and SYBR Green I Dye assay chemistries, it also provides the features and capabilities to accommodate a wide range of other real-time chemistries.

2.5. *Protocols*

TaqMan® Gene Expression Assays (inventoried: part number 4331182; noninventoried: part number 4351372) consist of a 20X mix of unlabeled PCR primers and TaqMan® MGB probe (FAM dye labeled). These assays are designed for the detection and quantitation of specific genetic sequences in RNA samples converted to cDNA. Gene expression quantitation using TaqMan® Gene Expression Assays is performed as the second step in a two-step RT-PCR protocol on any ABI PRISM Sequence Detection System Instrument. (Applied Biosystems, Foster City, CA, USA) All TaqMan® Gene Expression Assays are optimized to work with either TaqMan® Universal PCR Master Mix, No AmpErase UNG (P/N 4324018), (Applied Biosystems, Foster City, CA, USA) TaqMan® Universal PCR Master Mix (P/N 4304437), or TaqMan® Gene Expression Master Mix Protocol (PN 4371135) and with complementary DNA (cDNA). These products use the universal thermal cycling parameters described in **Table 1**.

Table 1
Thermal Cycler Conditions

	Times and temperatures			
			Each of 40 cycles	
	Initial setup		Denature	Anneal/extend
Thermal cycler	Hold*	Hold	Cycle	
Sequence Detection Systems (7900HT, 7700, 7000)	UNG activation 2 min, 50°C	10 min, 95°C	15 s, 95°C	1 min, 60°C

*The 2-min, 50°C step is required for optimal AmpErase UNG activity when using TaqMan® Universal PCR Master Mix (P/N 4304437). This step is not needed when using the TaqMan® Universal PCR Master Mix, No AmpErase UNG (P/N 4324018).

To prepare the reaction components for a single 20-μL reaction (384-well plate) or a single 50-μL reaction (96-well plate), refer to **Table 1** for singleplex reactions.

2.5.1. *Prepare the PCR Reaction Mix* (Table 2)

1. Prior to use, mix the TaqMan® Gene Expression Master Mix thoroughly by swirling the bottle.
2. Calculate the volume of each component needed for all the wells in each assay, based on the number of reactions. Include extra volume to compensate for the volume loss that occurs during pipetting. (*Note*: Applied Biosystems recommends performing four replicates of each reaction.)
3. Cap the tube(s).
4. Vortex the tube(s) briefly to mix the solutions.
5. Centrifuge the tube(s) briefly to spin down the contents and eliminate any air bubbles from the solutions.

2.5.2. *Prepare the PCR Reaction Plate*

1. Transfer the appropriate volume of reaction mixture to each well of a plate:

 - MicroAmp (Applied Biosystems, Foster City, CA, USA) Fast Optical 48-Well Reaction Plate: 20 μL per well
 - MicroAmp Optical 96-Well Reaction Plate: 50 μL per well

Table 2
Singleplex PCR Reaction Mix using TaqMan® Universal PCR Master Mix*

Reaction component	Volume/well (20-μL volumereaction†)	Volume/well (50-μL volumereaction†)	Final concentration
TaqMan® Universal PCR Master Mix, No AmpEraseUNG (2X)‡ (ordered separately)	10	25	1X
20X TaqMan® Gene Expression Assay Mix	1	2.5	1X
cDNA diluted in RNase-free water	9	22.5	—
Total	20	50	

*No AmpErase UNG or TaqMan® Gene Expression Mater Mix can be used.
†If different reaction volumes are used, amounts should be adjusted accordingly.
‡Volumes should be the same if using TaqMan® Universal PCR Master Mix (2X) (P/N 4304437).

- MicroAmp Fast Optical 96-Well Reaction Plate: 20 μL per well
- MicroAmp Optical 384-Well Reaction Plate: 20 μL per well

2. Cover the plate with MicroAmp Optical Adhesive Film. For MicroAmp 96-Well Reaction Plates, MicroAmp Optical Caps may be used instead. (*Important*: Use a MicroAmp Optical Film Compression Pad with Optical Adhesive Film when using a MicroAmp 96-Well Reaction Plate on the ABI PRISM 7000 Sequence Detection System or the 7900HT Fast Real-Time PCR System.)
3. Centrifuge the plate briefly.

For further information on the plate set-up procedure and data analysis, refer to the user's manual for the appropriate Sequence Detection System Instruments (7900HT, 7700, 7500 and 7500 Fast, 7300, 7000).

2.5.3. Run the PCR Reaction Plate or Individual Tube

1. Set the thermal cycling conditions as described in **Table 2**.
2. In the plate document, select "Standard" mode and enter the correct sample volume (20 μL or 50 μL). (*Important*: Fast thermal cycling conditions are not for use with the TaqMan® Gene Expression Master Mix. For information, refer to the TaqMan® Gene Expression Master Mix Protocol.)

2.5.4. Analyze the Results

Data analysis varies depending on the instrument. Refer to the TaqMan® Gene Expression Master Mix Protocol (PN 4371135) and the instrument user guide for information on how to review and adjust threshold values and other details on analyzing your data.

Gene expression quantitation using TaqMan® Gene Expression Assays should be performed in separate wells (singleplex assay). We recommend that the endogenous control of choice be run in separate wells (singleplex) as this does not require any validation experiments. If performing multiplex experiments (**Note 1**), we recommend running multiplex and singleplex assays in parallel to confirm that the C_t values are not affected by multiplex PCR amplification.

For additional information regarding relative quantitation of gene expression experiments, refer to the ABI PRISM 7700 Sequence Detection System User Bulletin no. 2 (P/N 4303859).

2.5.5. Storage

Store between –15°C and –20°C; minimize freeze-thaw cycles. The 20X TaqMan® Gene Expression Assay may be diluted in TE buffer (Tris- EDTA buffer) (final concentration of TE should be 10 mM Tris-HCl/1 mM EDTA pH 8.0, use RNase-free water).

A more detailed protocol for TaqMan® Gene Expression Assays (P/N 4333458F) is available at the Applied Biosystems Web site (http://www.appliedbiosystems.com/).

3. APPLICATIONS

3.1. Biomarker Validation After Microarray Analysis

Advances in high-throughput biotechnology, in particular microarray technology, have provided a huge leap forward in the discovery phase of biomarker development. The microarrays' tremendous capability for global gene expression monitoring makes it an ideal starting point for discovering candidate clinical markers and potential therapeutic targets. However, an obvious corollary of increasing the rate at which potential biomarkers are discovered is an increase in the need to validate them in a cost- and time-effective manner. Biomarker validation generally includes two distinct phases: (1) technical validation to verify candidate biomarkers using an independent technology, and (2) biological validation to verify the predictive/classification power of the candidate biomarkers in a larger number of samples in both retrospective and prospective clinical studies. The performance capabilities and ease-of-use of real-time PCR chemistries and instrumentation has led to the widespread use of TaqMan® Gene Expression Assays as the preferred method for quantifying gene expression *(6)* as well as for independent validation of candidate biomarkers after microarray analysis *(7)*.

3.1.1. Technical Validation of Candidate Biomarkers

Although microarrays are an excellent tool for initial target discovery, the reliability of microarray results have been questioned because of the existence of different technologies and nonstandard methods of data analysis and interpretation. To critically examine these issues, large-scale TaqMan® Gene Expression Assay–based real-time PCR experimental data sets have been generated and used as the reference to evaluate the performance of various commercial microarray platforms *(8,9)*. These studies concluded that microarrays have acceptable sensitivity and accuracy in detecting differential expression, especially for genes with high and medium expression levels and for detecting greater than twofold changes. These studies also characterized some of the limitations of microarrays, in particular the ratio compression phenomena. As shown in **Fig. 6A** (see Color Plate 2, following p. 230) microarrays produce more compression for low or high expressed genes when compared with TaqMan® assays. Other microarrays limitations revealed by these studies include the significant decrease in

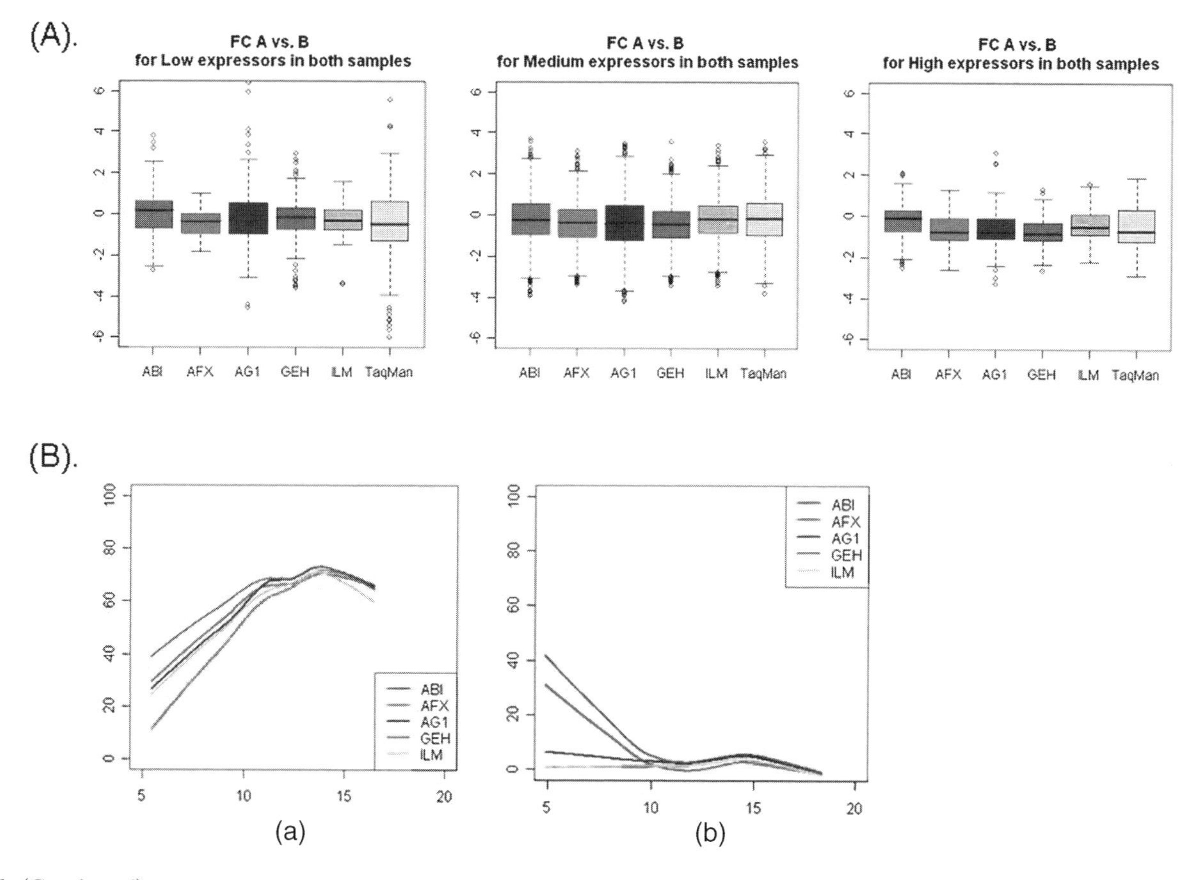

Fig. 6. (Continued).

overall accuracy of differential expression detection at low expression level (**Fig. 6B**) and the relatively poor sensitivity in detecting small fold changes (i.e., less than twofold) *(9)*. Although these limitations have been previously suspected by many, the large-scale TaqMan®-based real-time PCR "reference" data set provides a more quantitative view of these limitations for the first time and illustrates why the use of an orthogonal platform such as TaqMan®-based real-time PCR to validate data obtained by microarrays is essential.

TaqMan® real-time PCR assays have been widely applied as an independent validation of candidate biomarkers identified by microarray analysis. For instance, TaqMan® Gene Expression Assays have been used to validate potential prognostic biomarkers originally identified with three different microarray platforms *(7)*. In this study, 20 biopsy samples from early-stage breast carcinomas (T1/T2) were profiled using three different microarray platforms. Two main clinical-relevant subtypes, luminal A (associated with longer survival) and basal-like (associated with shorter

Fig. 6. Validation of microarray results using TaqMan® Gene Expression Assay data set as reference. **(A)** Fold change repression in microarray platforms: Genes were binned into low/medium/high according to TaqMan® assays C_t measurements (the cutoffs are set to 23:29:35). Only genes having expression level in the same bin in both sample A and B are included. Boxplots of the fold change for each microarray platform relative to TAQ (TaqMan® Gene Expression Arrays) is presented. **(B)** Assessment of true-positive rates and false-discovery rates using TaqMan® assays. **(a)** True-positive rate (TPR) and **(b)** false-discovery rate (FDR) assessment using TaqMan® assays. All common genes between TaqMan® assays and microarray platforms were used for the TPR analysis. TPR was defined as the percentage of differentially expressed genes in sample A compared with sample B detected by each microarray platform out of the ones detected by TaqMan® assays data as truth [TPR = TP/(TP + FN), FDR = FP/(TP + FP)], where TP is true positive, FN is false negative, and FP is false positive in microarray. Differential expression was detected by *t*-test, where FDR was controlled at the 5% level with no fold-change filters. For TaqMan® assays, genes were ordered according to the average signals of A and B, and for bins of 50 consecutive genes, we compared the significant difference calls between each microarray platform and TaqMan® assays. The FDR represents the percentage of differentially expressed genes detected only by microarray platforms out of all genes differentially expressed in microarray platforms. (Modified from Canales RD, Luo Y, Willey JC, et al. Evaluation of DNA microarray results with quantitative gene expression platforms. Nat Biotechnol 2006;24:1115–1122, with permission from *Nature Biotechnology*.) (*see* Color Plate 2, following p. 230)

survival), were identified in these carcinomas, substantiating the prognostic value of such expression-defined phenotypes in early breast cancer patients. Of 16,611 common genes among these three platforms, 319 genes were found to be differentially expressed between luminal A and basal-like tumors by all three technologies, which corresponds to an overall consistency of 30% (about 1000 to 1100 genes were identified to be differentially expressed by each platform separately). A minimal set of 54 genes that best discriminated the two subtypes was identified using the combined data sets generated from the three different array platforms. This set of 54 genes was further validated by TaqMan® Gene Expression Assays. Profile correlation analysis showed that the expression profiles across the 20 tumor samples determined by each array platform and by TaqMan® assays are highly correlated (median correlation coefficient $R > 0.9$). The rate of good correlation (Pearson correlation coefficient, $R > 0.8$) varied from 81% to 88% for the three array platforms (**Fig. 7A** and Color Plate 3, following p. 230). Hierarchical clustering analysis demonstrated not only excellent separation of the luminal A and basal-like subtypes but also that the same tumor samples analyzed by three microarray platforms and TaqMan® assays were clustered together by tumor sample rather than by method (**Fig. 7B**). Validated by multiple gene expression platforms, this set of 54 predictor genes defines a potential prognostic molecular set for breast cancer.

3.1.2. Biological Validation of Candidate Biomarkers

The second critical aspect of biomarker validation requires validation of the predictive value of the candidate biomarkers from the discovery phase on larger, independent populations of patient samples in a clinical research setting, which usually involves both retrospective study (patients with known outcomes) and prospective study (predict patient outcome). Because clinical studies are often done at multiple sites, a highly robust, easy-to-use, reproducible and standardizable technology is desirable. Difficulties with robustness, reproducibility, and normalization, combined with issues discussed previously, make microarrays a less practical method for this validation phase of biomarker development. TaqMan® Gene Expression Assays–based real-time PCR technology is highly reliable, easy to use and interpret, and therefore a much more practical tool for validating biomarkers in clinical research. For example, Lossos et al. ***(10)*** demonstrated the use of TaqMan® real-time PCR assays for developing gene expression tests that predict survival in diffuse large B-cell lymphoma (DLBCL) patients. In this retrospective clinical study, gene expression of preselected potential biomarkers was profiled in 66 DLBCL patients with known long-term follow-up histories. A panel of 36 genes was selected as testing targets

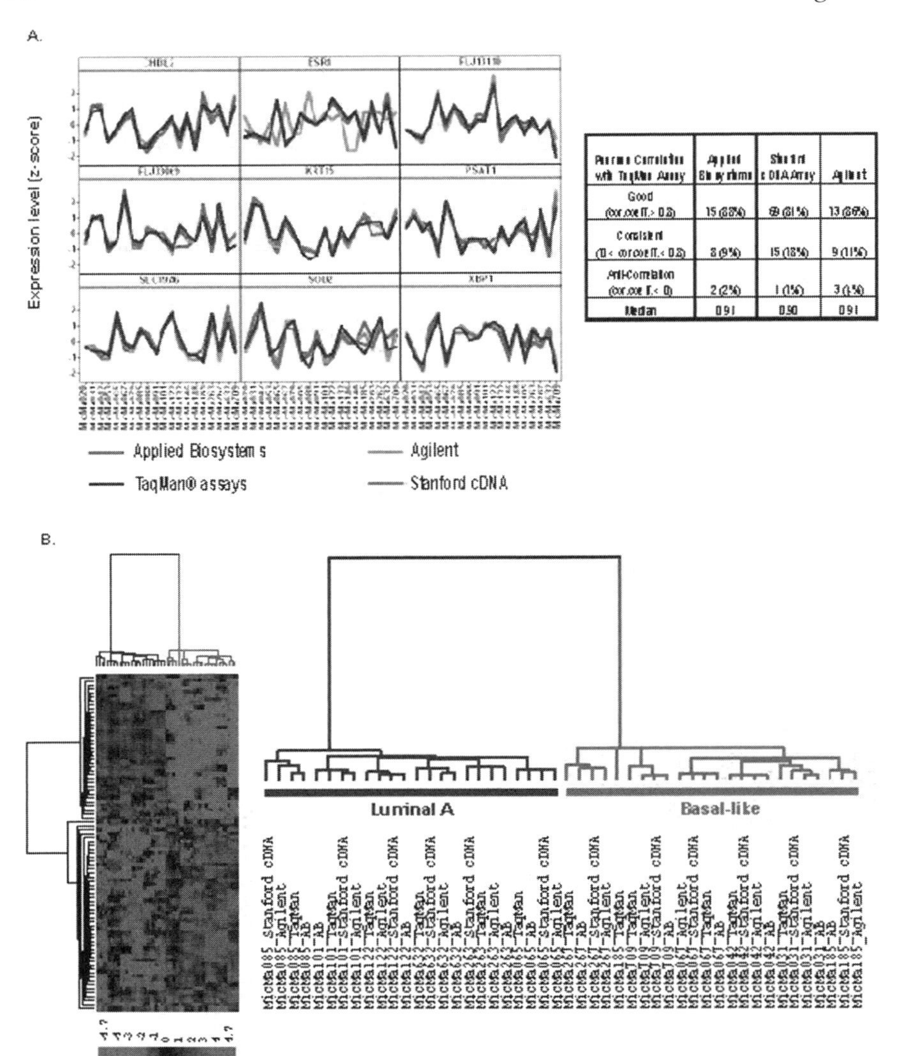

Fig. 7. Validation of potential prognostic markers by TaqMan® assay–based real-time PCR. Eighty-five marker genes including the minimal set of 54 genes identified to best distinguish the luminal A and the basal-like subtypes were validated by TaqMan® Gene Expression assays. **(A)** Profile correlation analysis showed the expression profile across the 20 tumor samples determined by the three microarray platforms and by TaqMan® Gene Expression assays are highly correlated with median correlation coefficient R > 0.9 and a rate of good correlation (R > 0.8) of 81% to 88%. **(B)** Hierarchical clustering analysis of expression data from the TaqMan® Gene Expression Assays and the three microarray platforms across the 85 marker genes. The same

from several sources of evidence, including reported signal prognostic genes identified for DLBCL and gene lists from three independent microarray studies. Using the TaqMan®-based gene expression validation data, the researchers performed a univariate analysis of genes that were correlated with overall survival, either positively or negatively. Six genes were demonstrated to be the strongest predictors: *LMO2*, *BCL6*, *FN1*, *CCND2*, *CCL3* (also known as *SCYA3*), and *BCL2*. A multivariate model based on the expression levels of the six genes was built and validated by applying it to previously published data from two independent sources. The researchers determined that the six-gene model was highly accurate and independent of the International Prognostic Index (IPI), a well-established predictor of patient outcome. They also further refined the survival predictions of patients within each IPI risk category (low, medium, and high), and by coupling the medium- and high-risk IPI scores with their high-risk expression profile, they were able to identify a group of patients whose survival was especially short (approximately 30% of all patients). Independently, the genes could not predict how long a patient survived after treatment; however, the six genes taken together were able to predict how long each of the 66 patients survived. This study demonstrated the TaqMan® real-time PCR assays to be a more suitable approach for biomarker validation and development in a clinical research setting.

To further streamline the work flow of TaqMan® real-time PCR and make it a more suitable and robust tool for biomarker development in clinical research setting, TaqMan® Low Density Arrays (TLDAs) have been developed to facilitate validation of biomarkers in clinical studies with large cohorts. TLDA is a 384-well micro fluidic card that enables researchers to perform up to 384 simultaneous real-time PCR reactions without the need to use liquid-handling robots or multichannel pipettors. It allows for 1 to 8 samples to be run in parallel against 12 to 384 TaqMan® Gene Expression Assay targets that

Fig. 7. (Continued) tumor sample analyzed by the three microarray platforms and the TaqMan® Gene Expression assays were clustered together except for one sample (MicMa185). For the Applied Biosystems microarrays, the mean expression values of the two microarray replicates were presented. The transformed z-scores were represented using a red-black-green color scale as shown in the key (green, below mean; black, equal to mean; red, above mean). (Modified from Sorlie T, Wang Y, Xiao C, et al. Distinct molecular mechanisms underlying clinically relevant subtypes of breast cancer: gene expression analyses across three different platforms. BMC Genomics 2006;7:127, with permission of *BMC Genomics*.) (*see* Color Plate 3, following p. 230)

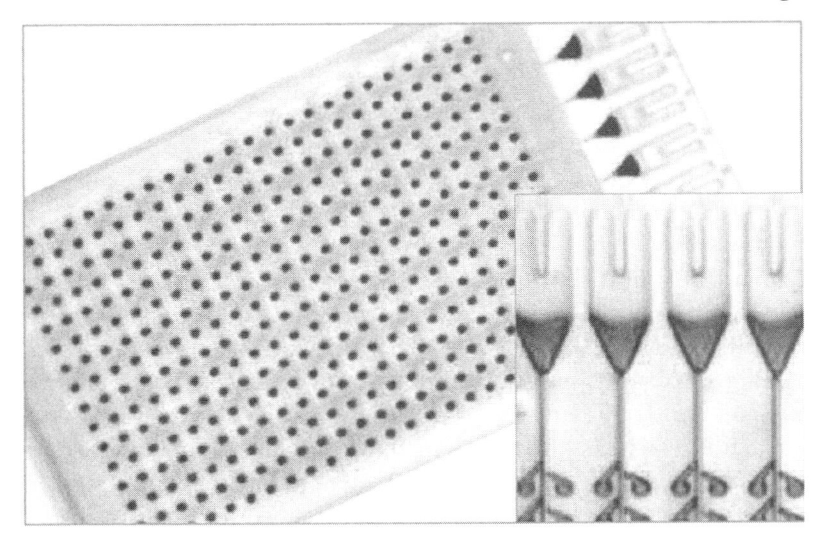

Fig. 8. TaqMan® Low Density Arrays (TLDAs).

are preloaded into each of the wells on the card (**Fig. 8**). The standardized format allows direct comparison of results between numerous samples from different labs, a critical feature required for clinical testing. The list of genes to be tested by TLDAs is completely customizable making them an ideal solution for biomarker validation in a simple, robust, and standardized format.

3.2. MicroRNA Expression Profiling

MicroRNAs (miRNAs) are an abundant class of endogenous, ~22-nucleotide, noncoding regulatory RNAs that have been shown to have critical functions in a wide variety of biological processes, including development, cell proliferation and death, and oncogenesis. The miRNA profiles have been shown to successfully classify developmental lineage and poorly differentiated tumors, highlighting the potential of miRNA profiling in cancer diagnosis ***(11)***. Furthermore, changes in miRNA gene expression between tumors and their normal counterparts could be exploited, not only for diagnostic and prognostic purposes in the clinic but also to investigate the specific roles of the various miRNAs in tumor development and to identify the genes and pathways that they target. Given that miRNAs regulate gene expression and enter readily into cells, they also have potential as therapeutics to treat cancers and other human diseases ***(12)***.

Because of their short size and the sequence similarity between family members, developing sensitive and specific assays for miRNA quantitation is rather challenging. miRNA arrays as well as bead-based approaches have been used for miRNA profiling; however, their detection sensitivity and specificity are typically low. A novel miRNA quantification method has been developed using stem-loop reverse transcription (RT) followed by TaqMan® PCR analysis (**Fig. 9**) *(13)*. Stem-loop RT primers are better than conventional ones in terms of RT efficiency and specificity. These TaqMan® miRNA assays are specific for mature miRNAs and discriminate among related miRNAs that differ by as little as one nucleotide. Like standard TaqMan® gene expression assays, TaqMan® miRNA assays exhibit a dynamic range of seven orders of magnitude. Quantification of five miRNAs in seven mouse tissues showed variation from less than 10

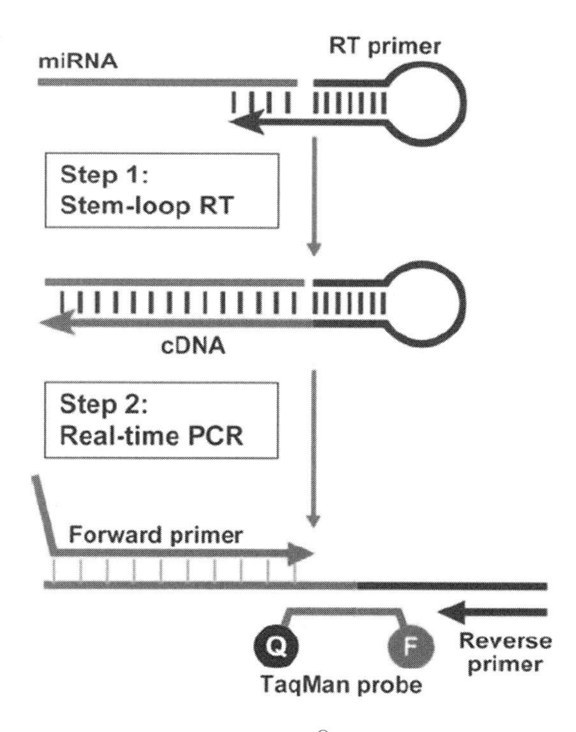

Fig. 9. Schematic description of TaqMan® miRNA assays. TaqMan®-based real-time quantification of miRNAs includes two steps: stem-loop RT and real-time PCR. Stem-loop RT primers bind to at the 30 portion of miRNA molecules and are reverse transcribed with reverse transcriptase. Then, the RT product is quantified using conventional TaqMan® PCR that includes miRNA-specific forward primer, reverse primer, and dye-labeled TaqMan® probes. The purpose of tailed forward primer at 50 is to increase its melting temperature (T_m).

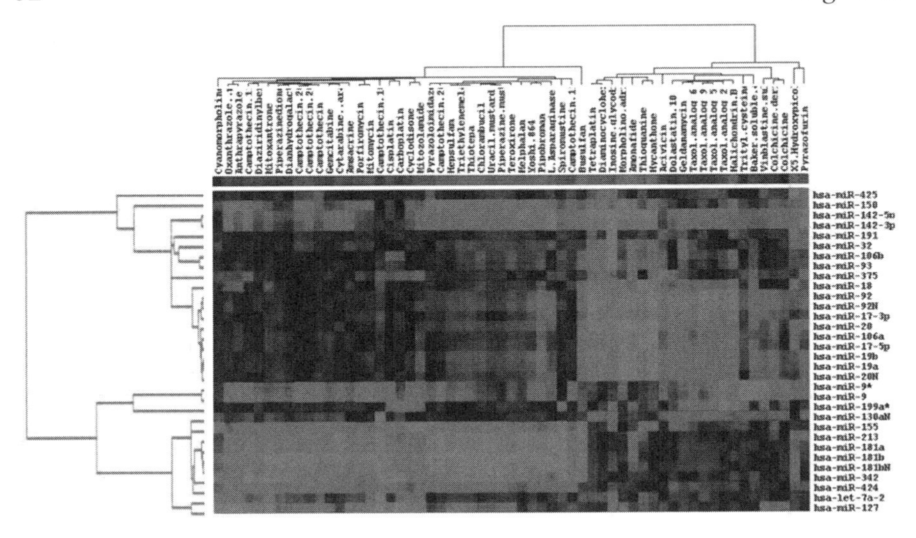

Fig. 10. Correlation between miRNA expression profile and drug potency. One hundred sixty-two TaqMan®-based miRNA assays were used to profile the level of miRNA expression in 57 of the 60 NCI cancer cell lines. Targets were selected based on the size of the nucleotide, abundance, and ubiquitous nature of the small RNAs. Growth inhibition data for the 57 human tumor cell lines from 119 standard anticancer drugs expressed as negative log of the molar concentration calculated in the NCI screen (–log10 [GI_{50}] values). Pearson correlation coefficients and their significance are calculated for each miRNA-drug pair. Hierarchical clustering based on correlation coefficients was performed using only miRNAs showing significant correlations ($p < 0.001$) with drugs potency (Barbacioru et al., unpublished data).

to more than 100,000 copies per cell. This method enables fast, accurate, and sensitive miRNA expression profiling and can identify and monitor potential biomarkers specific to tissues or diseases.

We have conducted a study to explore the possibility of using miRNA biomarkers to predict drug potency. A total of 236 TaqMan® miRNA assays were used to profile 57 of the NCI-60 cancer cell lines (unpublished data). Using these profiles, linear models were built to predict growth inhibition measurements (GI_{50}) for 119 standard anticancer drugs with which the 57 cancer cell lines were treated ***(14)***. We found that subsets of 7 to 10 miRNAs are sufficient for near optimal prediction (at least 85% accuracy) of drug potency against the NCI-60 cell lines (**Fig. 10**).

3.3. Gene Copy Number Detection

Gene copy number variation is now recognized as an important type of polymorphism in the human genome. Copy number polymorphisms have been associated with genetic diseases such as cancer, immunologic and neurologic disorders, as well as with variations in drug response. A new application of TaqMan® real-time PCR assays has been developed to quantify gene copy number. Quantitative measurement of gene copy number is achieved by duplexing TaqMan® real-time PCR assays, with one FAM dye-based assay designed to detect the genes of interest and one VIC (Applied Biosystems, Foster City, CA) dye-based assay against a reference gene with known two copies per diploid genome (i.e., RNaseP). For example, we have conducted a copy number analysis on five important drug metabolizing enzyme (DME) genes (*CYP2D6*, *CYP2E1*, *CYP2A6*, *GSTM1*, and *GSTT1*.) using TaqMan® real-time PCR assays (Li et al., unpublished data). Survey of 270 DNA samples from the International HapMap Project revealed large variation in gene copy number for these five genes within the HapMap sample panel; furthermore, significant differences in copy number frequency of these genes were found to be associated with different populations (**Fig. 11**). Combined with the TaqMan® genotyping assays, the newly developed TaqMan® copy number variation assays provide

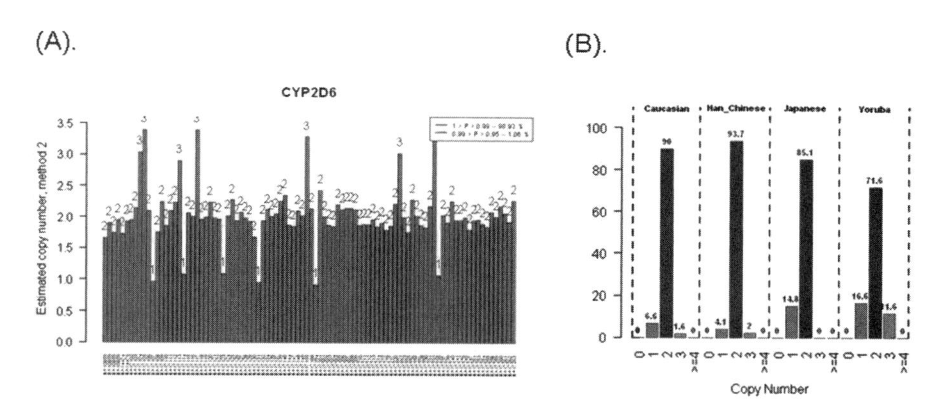

Fig. 11. Gene copy number quantitation using TaqMan® gene copy number assays. CYP2D6 is used as an example of copy number determination for the HapMap panel. (**A**) Copy number is determined for individual samples and copy numbers of one, two, and three are predicted. The majority of individuals have two copies. Around 10% and 5% of samples have one or three copies, respectively. For all the samples, the confidence level in predicted copy number is above 0.95, and furthermore, 99% samples have confidence between 0.99 and 1. (**B**) Distribution of copy number across the four populations: Caucasian, Han Chinese, Japanese, and Yoruba. There are significant differences in frequency of the CYP2D6 gene copy number across the four populations (chi-square test p value = 0.005) (Li et al., unpublished data).

powerful tools to fully characterize genotypes of the genes of interest and their roles in disease or drug response.

4. CONCLUSION

We have outlined the principles behind a specific quantitative real-time PCR technology; TaqMan® Gene Expression Assays, and the various applications of that technology from biomarker validation, including regular coding gene markers as well as the new class of noncoding miRNAs, to copy number detection, which is an emerging field with promising applications. This technology is particularly well suited to pharmacogenomic applications such as patient stratification and response to drug treatments and may well pave the way for the upcoming era of personalized medicine.

NOTES

1. Multiplex PCR is the use of more than one primer pair in the same tube. Refer to the ABI PRISM Sequence Detection System User Bulletin no. 5 (P/N 4306236) for information regarding multiplex reactions.)

REFERENCES

1. Morrison TB, Weis JJ, Wittwer CT. Quantification of low-copy transcripts by continuous SYBR Green I monitoring during amplification. Biotechniques 1998;24:954–958, 960, 962.
2. Bonnet G, Tyagi S, Libchaber A, Kramer FR. Thermodynamic basis of the enhanced specificity of structured DNA probes. Proc Natl Acad Sci U S A 1999; 96:6171–6176.
3. Tyagi S, Kramer FR. Molecular beacons: probes that fluoresce upon hybridization. Nat Biotechnol 1996;14:303–308.
4. Wittwer CT, Herrmann MG, Moss AA, Rasmussen RP. Continuous fluorescence monitoring of rapid cycle DNA amplification. Biotechniques 1997;22: 130–131, 134–138.
5. SantaLucia J, Jr. A unified view of polymer, dumbbell, and oligonucleotide DNA nearest-neighbor thermodynamics. Proc Natl Acad Sci U S A 1998; 95:1460–1465.
6. Ramaswamy S. Translating cancer genomics into clinical oncology. N Engl J Med 2004;350:1814–1816.
7. Sorlie T, Wang Y, Xiao C, et al. Distinct molecular mechanisms underlying clinically relevant subtypes of breast cancer: gene expression analyses across three different platforms. BMC Genomics 2006;7:127.
8. Canales RD, Luo Y, Willey JC, et al. Evaluation of DNA microarray results with quantitative gene expression platforms. Nat Biotechnol 2006;24:1115–1122.

9. Wang Y, Barbacioru C, Hyland F, et al. Large scale real-time PCR validation on gene expression measurements from two commercial long-oligonucleotide microarrays. BMC Genomics 2006;7:59.
10. Lossos IS, Czerwinski DK, Alizadeh AA, et al. Prediction of survival in diffuse large-B-cell lymphoma based on the expression of six genes. N Engl J Med 2004;350:1828–1837.
11. Lu J, Getz G, Miska EA, et al. MicroRNA expression profiles classify human cancers. Nature 2005;435:834–838.
12. Croce CM, Calin GA. miRNAs, cancer, and stem cell division. Cell 2005;122: 6–7.
13. Chen C, Ridzon DA, Broomer AJ, et al. Real-time quantification of microRNAs by stem-loop RT-PCR. Nucleic Acids Res 2005;33:e179.
14. Dai Z, Barbacioru C, Huang Y, Sadee W. Prediction of anticancer drug potency from expression of genes involved in growth factor signaling. Pharm Res 2006;23:336–349.

5

SAGE Analysis in Identifying Phenotype Single Nucleotide Polymorphisms

Sandro J. de Souza, I-Mei Siu, Janete M. Cerutti, and Gregory J. Riggins

Summary

Serial analysis of gene expression (SAGE) has been extensively used to evaluate gene expression in a variety of biological contexts. In this review, we provide a general overview of the SAGE methodology and its major applications. Furthermore, we provide a detailed, day-by-day based protocol for SAGE.

Key Words: bioinformatics; cancer; gene expression; SAGE; transcriptomics

1. INTRODUCTION

The full integration of genetic and phenotype data will only be achieved when a complete catalog of all genetic and epigenetic features of the human genome is available. A crucial step in this huge enterprise is the complete characterization of the human transcriptome. This includes (i) the identification of all human genes, (ii) the identification of all transcripts for these genes, including variants, and (iii) the characterization of their spatial-temporal pattern of expression.

Various strategies have been developed to explore the transcriptome. With the advent of molecular biology in the 1970s, Northern blot started to be used routinely to identify genes expressed in several conditions, both normal and pathologic. More recently, the development of reverse transcription–polymerase chain reaction (RT-PCR) represented a significant advance over the Northern blot, especially in the past few years when automated, robot-dependent protocols have appeared. In the early 1990s, the automated DNA

From: *Methods in Pharmacology and Toxicology: Biomarker Methods in Drug Discovery and Development*
Edited by: F. Wang © Humana Press, Totowa, NJ

sequencers allowed the development of a strategy to evaluate expressed sequences in a large scale. Expressed sequence tags (ESTs) were developed by Craig Venter's group with the aim of generating an expression fingerprint for a given tissue or cell ***(1,2)***. The strategy is a huge success with the resulting database, dbEST (maintained by the National Center for Biotechnology Information; NCBI), containing today more than 40 million sequences from a variety of organisms (dbEST press release, January 19, 2007).

However, in terms of throughput, none of these technologies are comparable with the ones more widely used today including serial analysis of gene expression (SAGE) ***(3)***, Massively Parallel Signature Sequencing (MPSS) ***(4)***, and microarray ***(5)***. Furthermore, these last methods are intrinsically quantitative, a feature normally missed by the EST data set in which most of the libraries are normalized.

Although these methods are primed primarily toward the transcriptome, they are also useful in the identification of single nucleotide polymorphisms (SNPs). Specific software using different statistical approaches has been developed with the aim to identify SNPs in the EST data set ***(6,7)***. Furthermore, SNPs that affect the tag-to-gene assignment in SAGE and MPSS experiments can be easily genotyped by scanning the respective libraries (this issue will be further discussed later).

In this review, we intend to describe the SAGE methodology and the bioinformatics approaches developed to explore the data derived from SAGE. Furthermore, we will discuss the use of SAGE data in the identification of biologically meaningful SNPs.

2. SAGE

SAGE uses automated DNA sequencing to count a large number of mRNA transcripts from basically any type of biological sample. The deeper coverage of the transcriptome achieved by SAGE, when compared with EST sequencing for example, comes from the fact that SAGE increases the number of genes to be counted by minimizing the portion of the transcript sequenced. This is achieved because SAGE works by cloning and sequencing a small (10 to 17 bp) tag of the cDNA at a defined position near the 3′ end of the transcript. This defined position corresponds mostly with the fragment immediately downstream of the 3′ most NlaIII site. Tags are ligated, cloned end-to-end, and sequenced allowing the serial analysis of multiple transcripts. The number of times a particular tag is observed in a given library is directly correlated with transcript abundance.

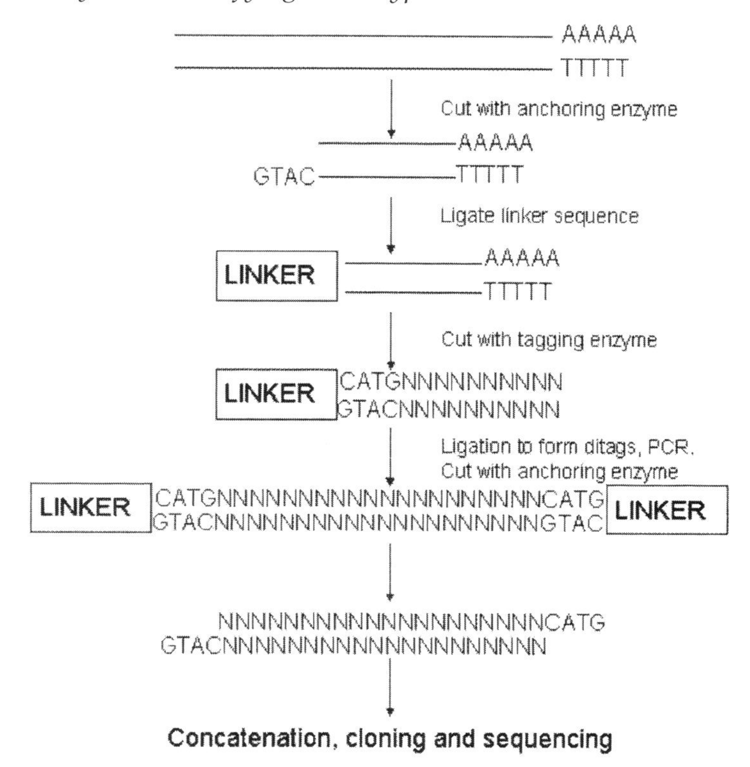

Fig. 1. Schematic view of the SAGE methodology.

A schematic procedure for library construction is shown in **Fig. 1**. To generate the tag data, polyA+ RNA is extracted and used as a template for cDNA synthesis using biotinylated polyT oligos. This allows the recovery of 3′ most cDNA fragments after digestion with an anchoring enzyme (mostly NlaIII) and affinity purification using streptavidin. These fragments are divided in half and ligated to two different linkers. Both linkers contain a site for the tagging enzyme (BsmF1) and the anchoring enzyme overhang. Each linker contains a specific prime site for PCR amplification. Fragments containing the linker plus a transcript sequence (tag) are obtained by digestion with BsmF1. The isolated linker tags are blunt ended, and the two different sets of fragments, each with a different linker, are ligated. Linker-ditags-linker constructs are amplified using primers specific to each linker, and the ditags are released by digestion with the anchoring enzyme. These ditags are purified, concatenated, cloned, and sequenced.

There are two major methodological versions of SAGE: the conventional one, described above, which generates 10-bp-long tags, and the Long SAGE *(8)*, a very similar strategy that generates 17-bp tags. The longer length of these latter tags makes them less ambiguous. Because of that, the genome sequence, when available for the corresponding species, can be used for mapping purposes (a detailed description of the genome mapping strategy is presented later).

Several different methods have appeared based on the same principles that govern the SAGE technology. **Table 1** lists the major methods with differences between them. As mentioned before, there is MPSS, a proprietary technology that goes deeper in the transcriptome coverage through the parallel sequencing on beads of transcripts' tags *(4)*. This approach can generate up to a million tags in a single experiment, which represents roughly a 5X coverage of a cell transcriptome. The length of the tag (16 to 18 bp) is longer than the conventional SAGE but shorter than Long SAGE. Methods that target the 5′ end of transcripts have emerged in the past few years. Cap analysis of gene expression (CAGE) is based on the construction of high-quality, full-length cDNA libraries. Instead of using the polyA tail as target, CAGE targets the 5′ end of cDNA using the cap-trapper technology. Tags are then generated, concatenated, and sequenced. CAGE has been extensively used for a variety of purposes. For more details, see Kodzius et al. *(9)*. Wei et al. *(10)* developed the 5′ and 3′ long serial analysis of gene expression (LS). The major aspect of both 5′ and 3′ LS is the extraction, concatenation, sequencing, and mapping of terminal transcripts' tags onto the genome sequence to define the transcription initiation site and the polyadenylation site.

There is a wide range of applications for SAGE. The most powerful and attractive feature of SAGE is its ability to measure the expression of thousands of genes in a quantitative way and without *a priori* knowledge of

Table 1
Comparison of Different Methods Using the Same Principle as SAGE

Method	Target	Anchoring enzyme	Tagging enzyme	Tag length(bp)
SAGE	3′	NlaIII	BsmF1	14
Long SAGE	3′	NlaIII	MmeI	21
MPSS	3′	DpnII	BbuI	16–20
CAGE	5′	Not used	GsuI/MmeI	16–20
Super SAGE	3′	NlaIII	EcoP1SI	26
LS	3′/5′	Not used	MmeI	20

the gene sequence. One is then able to perform expression profiling of normal and tumor tissues and identify messages that are differentially expressed in cancer (for a review, see Ref. ***(11)***). The use of Long SAGE has also allowed the identification of new transcribed loci in the human genome.

3. BIOINFORMATICS OF SAGE

Bioinformatics is fundamental for the processing and interpretation of any SAGE experiment. First, specific software is needed to process the sequence reads extracting the correct tags from the concatenated sequence and calculating their frequency in a given library. This complete catalog of tags and their frequencies is not fully informative due to the short length of the tags. Therefore, much effort has been devoted to the development of protocols that aim to assign tags to genes. Because tags are, in principle, located immediately downstream of the 3′ most NlaIII site, it is possible to deduce a "virtual" tag from the cDNA sequence of all known genes. A database of these virtual tags can then be searched using the "real" tags as queries. The first attempt to build a computational protocol for the annotation of SAGE tags came from NCBI with their SAGEMap ***(12)***. SAGEMap constructs a "reliable" mapping by selecting the best connection between a tag and a gene above a specific threshold. The score of this connection is given by an equation that takes into account the diversity of the nucleotide sequence types (linked to a given tag) and the number of sequences agreeing on the connection.

SAGE Genie ***(13)*** is an additional protocol for the mapping of tags to genes and vice versa. The basis of SAGE Genie is a differential scoring of databases according to the representation of the 3′ most SAGE tag. For instance, a tag matching a more reliable database, like the one represented by sequences from the Mammalian Gene Collection (MGC) initiative, gets a higher score than a tag matching the EST database. SAGE Genie takes into account as well the effect of several phenomena on the assignment. Alternative polyadenylation can generate an alternative tag through the deletion of the original 3′ NlaIII site (in the case of a shorter transcript) or the creation of a new 3′ most site (in the case of a longer transcript). An artifact commonly found in cDNA libraries is internal priming, in which an artifactual cDNA is synthesized through the annealing of a polyT primer to an internal stretch of As (a common feature in the UnTranslated Regions (UTRs) of mammalian genes). Both features were evaluated in the original publication of SAGE Genie. More recently, da Silva et al. ***(14)*** evaluated the effect of SNPs on the tag-to-gene assignment. This will be further discussed later. All the data from SAGE Genie is available through a portal on the World Wide Web (http://cgap.nci.nih.gov/SAGE/). In the same portal, several computational tools are available for the analysis of the data.

A serious shortcoming from both SAGE Genie and SAGEMap is the absence of a service that allows the user to submit a batch of sequences for processing. Very recently, Galante et al. ***(15)*** developed Automatic Correspondence of Tags and Genes (ACTG), a Web-based tool that allows a large-scale tag-to-gene mapping using several reference databases. ACTG can map data from SAGE, MPSS, and Sequencing By Synthesis (SBS)-based technologies.

Long SAGE and MPSS data are also suitable, due to their length, for genome mapping. This is important to place the tag into a genomic context. Nowadays, it is relatively common for a given species to have its genome sequenced without, however, many corresponding cDNAs. In this case, with a longer tag, one can map the tag to the genome, a process that helps the tag-to-gene assignment.

4. INFLUENCE OF SNPS ON THE INTERPRETATION OF SAGE AND MPSS EXPERIMENTS

In principle, alternative tags can be associated with SNPs present within the tag sequence or within the restriction site used for SAGE and MPSS library construction. Using a computational strategy based on the comapping of tags, SNPs, and genes onto the genome sequence, da Silva et al. ***(14)*** performed a genome-wide analysis on this association. The identification of SNP-associated alternative tags was categorized in three different types. The first category encompass all cases in which a SNP generated a new restriction enzyme site downstream to the original tag, producing in this way a new 3′ most tag. The second type corresponds with those cases in which the presence of a SNP disrupted the 3′ most restriction site associated with the original tag making the immediate upstream site to become the 3′ most site. Finally, a SNP could be located within the tag sequence itself producing a SNP-associated tag with one single mismatch when compared with the original tag. We found more than 2000 SNP-associated tags using publicly available SAGE and MPSS data. A fraction of these tags were validated using the Generation of Long cDNA fragments for Gene Identification-Massively Parallel Signature Sequencing (GLGI-MPSS) methodology ***(16)***. Furthermore, we were able to genotype the HB4a cell line for the presence of the SNP associated with the alternative tag.

5. A DETAILED PROTOCOL FOR MICROSAGE

Here we provide a day-by-day procedure for the construction of MicroSage libraries. The length of the procedure can be longer if stopping points (usually precipitation steps) are used.

5.1. Day 1

5.1.1. Total RNA Isolation

1. Start with at least 100,000 cells in order to obtain 1 μg total RNA, the absolute minimum amount of starting material. Optimally, 10 μg total RNA would be used for MicroSAGE.
2. Use RNA isolation kit/protocol of your choice.
3. Check quality of RNA by gel.

5.1.2. Kinase Linkers

Perform the following protocol in PCR-free conditions.

1. Dilute Linker 1B, 2B, 1A, and 2A to 350 ng/μL (linker sequences/information can be found in **Appendix D**).
2. Make the following mixes to phosphorylate Linkers 1B and 2B:

Contents	Tube 1	Tube 2
Linker 1B (350 ng/μL)	9 μL	–
Linker 2B (350 ng/μL)	–	9 μL
LoTE (recipe in **Appendix B**)	6 μL	6 μL
10X kinase buffer (NEB)	2 μL	2 μL
10 mM ATP	2 μL	2 μL
T4 polynucleotide kinase (10 U/μL)	1 μL	1 μL

3. Incubate at 37°C for 30 min.
4. Heat inactivate at 65°C for 10 min.
5. In one tube (labeled “Linker 1AB”), mix 9 μL Linker 1A (350 ng/μL) to the 20 μL phosphorylated Linker 1B prepared above (final concentration 200 ng/μL); in a second tube (labeled “Linker 2AB”), mix 9 μL Linker 2A (350 ng/μL) to the 20 μL phosphorylated Linker 2B prepared above (final concentration 200 ng/μL).
6. In order to anneal the linkers, perform the following incubations for both tubes:

   95°C, 2 min
   65°C, 10 min
   37°C, 10 min
   Room temperature, 20 min
   Store at –80°C

7. Phosphorylation reaction should be tested by self ligation as follows (both Linker 1AB and Linker 2AB should be tested):

Anneal mix	2 μL (from **step 6**, above)
5X ligase buffer	2 μL
T4 ligase (5U)	1 μL
ddH_20	5 μL
Total	10 μL

8. Incubate at 37°C for 30 min.
9. Analyze on a 1.5-mm, 12% PolyAcrilamide Gel Electrophoresis- Tris-AcEtate (PAGE-TAE) gel (for recipe see **Appendix C**). Analyze the size of the bands by comparing to 1-kb and 100-bp DNA markers. Run the gel until the purple bromophenol band gets near the bottom of the gel. The ligated dimers should run at 80 bp and the monomers at 40 bp.
10. Load all of the self-ligated linkers and 1 μL of the annealed mixes (as negative controls). Phosphorylated linkers should allow linker-linker dimers (80 to 100 bp) to form after ligation, whereas dephosphorylated linkers will prevent self-ligation. Only linker pairs that self-ligate >70% (with <30% of reaction unligated) should be used in subsequent steps.

5.2. Day 2

5.2.1. Isolation of polyA+ RNA and cDNA Synthesis

Recipes for the most frequently used buffers are given in **Appendix B**.

Thoroughly resuspend Dynabeads oligo (dT)25 (from mRNA direct kit; Dynal).
Transfer 100 to 120 μL to an RNase-free, 1.7-mL Eppendorf tube.
Place on magnetic Eppendorf tube holder (Dynal).
After ~30 s, remove the supernatant. Always place the pipette tip at the opposite side of the tube, push the pipette tip to the bottom, and pipette very slowly, so you won't disturb, and thus lose, the beads. (This is a washing step.)
Resuspend the beads in 500 μL lysis/binding buffer (from kit) to preserve the beads until ready to add the RNA. In all the washing steps, add solution to the tube while it is still on the magnetic stand in order to minimize "drying out" the beads, then close the cap and remove it from the magnet. Resuspend the beads by "flicking" the tube or by gentle pipetting. Leave the beads in this buffer until ready to add them to the RNA.
Mix 1 to 10 μg of total RNA with 1 mL lysis/binding buffer—the appropriate volume of RNA depends on the RNA concentration, so be sure that RNA volume does not exceed 100 μL.
Remove the supernatant from the Dynabeads.
Add the RNA/buffer mixture to these prewashed Dynabeads.
Incubate at room temperature for 5 min with constant agitation.
Place the tube on the magnet for 2 min or until solution is completely clear, then remove supernatant (the supernatant can be saved and used for DNA prep).
Wash 2 × 1 mL Buffer A/washing buffer containing Lithium Dodecil Sulfate (LIDS) + 20 μg/mL glycogen (add 1 μL 20 mg/mL stock glycogen for 1 mL of

buffer)—remember to wash by pipetting up and down, placing the tube in the magnet, and removing the super-natant once the solution is clear (removes non–polyA$^+$ RNA and genomic DNA).
Wash 1 × 1 mL Buffer B + 20 μg/mL glycogen. This buffer does not contain LiDS, so the beads will not conglomerate on the magnet as quickly and cleanly as the previous washes—be careful not to remove any beads!
Wash 4 × 100 μL 1X first strand buffer (dilute 5X first strand buffer from cDNA synthesis kit [Invitrogen] 1:5 with Di Ethyl Pyro Carbonate (DEPC) dH_20).
Remove the supernatant and resuspend beads in the following first strand synthesis mix (components from the cDNA synthesis kit) in the following order:

DEPC dH_20	54 μL
5X First strand buffer	18 μL
0.1 M DTT	9 μL
10 mM dNTP	4.5 μL
Total	85.5 μL

Place the tube at 42°C for 2 min.
Add 3 μL SuperScript RTII.
Incubate at 42°C for 1 h and mix the beads every 10 min by gently tapping the tube.
After incubation, place the tube on *ice* to terminate the reaction.
Immediately, with the tube still on ice, add the following components for the second strand synthesis (components from the cDNA synthesis kit), in the order shown below:

DEPC dH_20 (prechilled)	302 μL
10X second strand buffer	75 μL
10 mM dNTP	15 μL
E. coli DNA ligase*	3 μL
E. coli DNA pol I	12 μL
E. coli Rnase H	3μL
Total	500 μL

Incubate at 16°C for 2 h, mix beads every 10 min by flicking the tube gently.
During this incubation period, prepare the solutions for the rest of the protocol (recipes in **Appendix B**).
After incubation, place the tubes on ice and terminate the reaction by adding 40 μL 0.5 M EDTA, pH 8.0.
Place the beads on the magnet and remove the supernatant.

Wash the beads 1 × 0.5 mL 1X BW + 1% SDS, made as follows (if necessary, heat the buffer at 75°C for 2 min to dissolve any precipitated SDS or BSA):

2X BW buffer	500 µL
20% SDS	50 µL
ddH_20	450 µL
Total	1 mL

Resuspend the beads in 200 µL 1X BW + 1% SDS and heat at 75°C for 20 min.
Wash 4 × 200 µL 1X BW + 2X BSA, made as follows:

2X BW buffer	500 µL
100X BSA	20 µL
ddH_20	480 µL
Total	1 mL

Wash 2 × 200 µL 1X Buffer 4 (NEB Biolabs) + 2X BSA (transfer beads to a new tube after the first wash) made as follows:

Buffer 4	100 µL
100XBSA	20 µL
ddH_20	880 µL
Total	1 mL

We use 2X BSA in all the washing buffers because more BSA seems to reduce bead stickiness and improve the efficiency of the washes and the quality of the library. After the SDS washing/heating step, the beads are sticky until the first BSA wash, but then they are okay.

5.2.2. *Cleavage of cDNA with Anchoring Enzyme (NlaIII)*

Resuspend the beads in the following mix and incubate at 37°C for 1 h (mix every 15 min):

LoTE	171 µL
BSA (100X, NEB)	4 µL
10X Buffer 4 (NEB)	20 µL
NlaIII (NEB, no. 125S, in –80°C in lab)	5 µL

5.2.3. Ligating Linkers to cDNA

1. After incubation, remove the supernatant and wash the beads 2 × 200 µL 1X BW + 0.1% SDS + BSA. Prepare as follows:

2X BW	500 µL
20% SDS	5 µL
100X BSA	20 µL
ddH_20	475 µL
Total	1 mL

Place on a hot block because SDS and BSA precipitate. (The supernatant precipitates to a certain degree in the SDS buffer, but it is worth using, otherwise the beads stick too much leading to bead loss and inefficient washes.)

2. Wash 4 × 200 µL 1X BW + 2X BSA (freshly made):

2X BW buffer	500 µL
100X BSA	20 µL
ddH_20	480 µL
Total	1 mL

3. Wash 2 × 200 µL 1X T4 ligase buffer, which is prepared right before use as follows:

5X T4 ligase buffer (thawed just before use)	100 µL
Water	400 µL
Total	500 µL

4. Divide the beads into two tubes and place them on a magnet.
5. Remove the last wash and resuspend the beads as follows:

	Tube 1 (linker 1)	Tube2 (linker 2)
LoTE	6 µL	6 µL
5X T4 ligase buffer	2 µL	2 µL
Linker 1AB (diluted 1:10 in ddH_2O)	1 µL	0
Linker 2AB (diluted 1:10 in ddH_2O)	0	1 µL

6. Heat tubes at 50°C for 2 min then let sit at RT for 15 min.
7. Add 1 µL 5 U/µL T4 ligase (high concentration; Invitrogen) to each tube (10 µL reaction).
8. Incubate at 16°C for 2 h. Mix beads every 10 to 15 min by gently tapping the tube.

5.2.4. Release Tags Using Tagging Enzyme (BsmF1) of cDNA

1. After ligation, wash each sample 2 × 200 μL 1X BW + 0.1% + 2X BSA (freshly made as follows). (*Note*: pool tube 1 and tube 2 together after first wash in order to minimize loss in subsequent steps.)

2X BW	500 μL
20% SDS	5 μL
100X BSA	20 μL
ddH_20	475 μL
Total	1 mL

2. Wash 4 × 200 μL 1X BW containing 2X BSA (freshly made as follows):

2X BW buffer	500 μL
100X BSA	20 μL
ddH_20	480 μL
Total	1 mL

3. Wash 2 × 200 μL 1X Buffer 4 + 2X BSA (freshly made as follows) and transfer to new tubes after the first wash:

Buffer 4	100 μL
100X BSA	20 μL
ddH_20	880 μL
TOTAL	1 mL

4. Resuspend the beads in the following mix:

LoTE	170 μL
10X Buffer 4	20 μL
100X BSA	4 μL
BsmF1 (NEB, no. 572S)	2 μL
Total	196 μL

   BsmF1 produces a 5′ sticky end of 4 bases (10 bp + 4 bp into unknown sequence).

5. Incubate at 65°C for 1 h, mix intermittently.
6. Centrifuge at >20,000 × *g* for 2 min at 4°C.

7. Transfer supernatant to new tube (not siliconized!). *Keep the supernatant—this contains your actual tags!*
8. Wash beads once with 40 µL LoTE.
9. Pool the 200 µL supernatant and the 40 µL wash together (240 µL final volume).
10. Add 240 µL PC8 (recipe in **Appendix B**).
11. Vortex, centrifuge for 5 min at >11,000 × *g* at 4°C.
12. Transfer the upper aqueous phase to a new tube.
13. High-concentration EtOH precipitate by adding the following components, in order, to the aqueous phase: *Save sample in* –80°C freezer if stopping here, otherwise precipitate by placing in a dry ice/methanol bath for at least 15 min.

7.5 M ammonium acetate	133 µL
20 g/L glycogen	5 µL
100% EtOH	1000 µL

14. Centrifuge for 30 min at >20,000 × *g* at 4°C.
15. Remove the supernatant and wash pellet twice with 200 µL 75% EtOH.
16. Centrifuge again after the last wash, carefully remove the residual liquid with pipette tip, and let dry until all the residual liquid has evaporated.
17. Resuspend pellet in 10 µL LoTE.

5.2.5. Blunt Ending Released Tags

1. Now we have the tags with a protruding end that needs to be filled in to be able to blunt end ligate to form the ditags. Add, in order, the following mix to the resuspended pellet from the previous step:

ddH_20	25 µL
5X Second strand buffer	10 µL
25 mM dNTPs	1 µL
100X BSA	1 µL
Klenow (2 U/µL)	1 µL

2. Incubate at 37°C for 30 min.
3. Add 190 µL LoTE (240 µL final volume).
4. Add 240 µL PC8 (now have 480 µL total).
5. Vortex and then centrifuge for 5 min at >20,000 × *g at 4°C.*
6. Remove 200 µL of the upper aqueous phase containing the nucleic acids and aliquot into a tube labeled "ligase+" and aliquot the rest of the aqueous upper phase into a tube labeled "ligase–" (you can pool several "ligase–" samples

together, but make sure to equalize the volume of "ligase+" and "ligase–" samples by adding 160 μL LoTE to the "ligase–" sample).

7. High-concentration EtOH precipitate both the "ligase+" and "ligase–" tubes by adding, in order, to each tube the following:

7.5 M ammonium acetate	133 μL
Glycogen	5 μL
100% EtOH	900 μL

8. Centrifuge for 30 min at >11,000 × *g* at 4°C.
9. Remove the supernatant but keep it just in case it contains unprecipitated sample.
10. Wash 2 × 200 μL 75% EtOH.
11. Carefully remove residual liquid with pipette tip and air-dry for 5 min (do not overdry).
12. Resuspend the "ligase+" reaction in 5 μL LoTE.
13. Resuspend the "ligase–" reaction in 3 μL LoTE.

5.2.6. Ligating Tags to Form Ditags (Part 1)

1. Prepare 2X mixes as follows:

	2X "ligase+" mix	Double "ligase+" mix	2X "ligase–"mix
3 mM Tris pH 7.5	1.25 μL	2.5 μL	2.25 μL
5X ligase buffer	1.5 μL	3.0 μL	1.5 μL
T4 ligase (5 U/μL)	1 μL	2.0 μL	0 μL
Total		7.5 μL	3.75 μL

Prepare 3 mM Tris by adding 30 μL 100 mM Tris, pH 7.5, to 1 mL ddH_20.

2. Add 5 μL 2X "ligase+" mix to the "ligase+" tube. Add 3 μL 2X "ligase–" mix to the "ligase–" tube. Therefore, the total volume of "ligase+" is 10 μL, and the "ligase–" is 6 μL.
3. Incubate overnight at 16°C.

5.3. Day 3

5.3.1. Ligating Tags to Form Ditags (Part 2)

1. Add 10 μL LoTE to the "ligase+" tube (20 μL total volume).
2. Add 14 μL LoTE to the "ligase–" tube (20 μL total volume).
3. Aliquot the 20 μL into four aliquots of 5 μL each so that it isn't necessary to take the entire amount out of the freezer at once in order to reduce degradation.

5.3.2. PCR Amplification of Ditags

1. In order to identify the correct dilution of ditags to use for large-scale PCR amplification, a test PCR must be performed. Dilute 1 μL of each of the "ligase+" and "ligase–" reactions as follows:

1/50 "ligase+"	1/100 "ligase+"	1/200 "ligase+"
1 μL (+)cDNA → + 49 μL ddH_20	10 μL 1/50 "ligase+" → + 10 μL ddH_20	10 μL 1/100 "ligase+" + 10 μL ddH_20

Reaction mix only.

1/50 "ligase–"	1/100 "ligase–"	1/200 "ligase–"
1 μL (-)cDNA → + 49 μL ddH_20	10 μL 1/50 "ligase–"→ + 10 μL ddH_20	10 μL 1/100 "ligase–" + 10 μL ddH_2O

Therefore, we have 8 tubes total—use the small 8 well connected PCR tubes for the test PCR.

2. Perform PCR using the following mix:

	1 reaction	10 reactions
(TEST)		
10X PCR buffer (Appendix A)	5 μL	50 μL
25 mM dNTP (GibcoBRL)	3 μL	30 μL
DMSO (Sigma, no. D-2650)	3 μL	30 μL
ddH_20	35 μL	350 μL
Primer 1 (350 ng/μL)	1 μL	10 μL
Primer 2 (350 ng/μL)	1 μL	10 μL
Taq polymerase (5 U/μL)	1 μL	10 μL
Template (various dilutions)	1 μL	10 μL
	50 μL	500 μL

3. For the Hybaid MBS thermocycler, use the following optimized reaction scheme:

Cycles	Temperature/time
1 cycle	94°C, 1 min
26 cycles	94°C, 30 s; 55°C, 1 min; 70°C, 1 min
1 cycle	70°C, 5 min at end

Use 27 cycles for the "ligase+" reactions and 35 for the "ligase–" reactions. The appropriate cycle number is critical for isolating an adequate amount of ditag DNA for SAGE. Too few cycles will result in a low yield and may cause problems with subsequent steps. Too many cycles will give erratic results and can also result in low yields (see **Appendix E**). Therefore, we recommend trying various cycle numbers (e.g., 26, 28, 30) to determine the optimal number.

4. Remove 10 μL from each reaction and mix with 1 μL loading dye.
5. Electrophorese the samples on a 1-mm 12% polyacrylamide gel (recipe in **Appendix C**). Use 5 μL of a 10-bp ladder and 5 μL of a 100-bp ladder as markers. Electrophorese the gel at 160 V until the purple bromophenol blue band is at the bottom of the gel. Amplified ditags should be 102 bp in size. A background band of equal or lower intensity occurs around 80 bp. All other background bands should be of substantially lower intensity. The "ligase–" samples should not contain any amplified product of the size of the ditags even at 35 cycles.
6. After the PCR test to determine the appropriate dilution, perform large-scale PCR of 300 × 50 μL PCR reactions of the optimal dilution to pool the products. For instance, if the 1/50 dilution works the best and you need 300 reactions, then you will need 6 μL of the "ligase+" dilution + 294 μL LoTE = 1 μL per reaction × 300 reactions = 300 μL. We usually use a 300 reaction PCR premix that we aliquot into three 96-well plates with 50 μL PCR reaction/well.
7. Therefore, you will perform large-scale PCR as follows:

	1 Reaction	300 Reactions (large scale)
10X PCR buffer (**Appendix A**)	5 μL	1500 L
25 mM dNTP	3 μL	900 μL
DMSO	3 μL	900 μL
ddH_20	35 μL	10,500 μL
Primer 1 (350 ng/μL)	1 μL	300 μL
Primer 2 (350 ng/μL)	1 μL	300 μL
Regular Taq polymerase	1 μL	300 μL
Template (dilution)	1 μL	300 μL
	50 μL	15 mL (50 mL Falcon tube)

The PCR conditions for the large-scale amplification are as follows:

Cycles	Temperature/time
1 cycle	94°C, 1 min
26 cycles	94°C, 30 s; 55°C, 1 min; 70°C, 1 min
1 cycle	70°C, 5 min at end

(Use one less cycle for the large scale)

5.3.3. Isolation of Ditags

1. After PCR is complete, centrifuge the 96-well microplates in the swinging bucket centrifuge at 2500 rpm for 10 min.
2. If possible, use a multichannel pipette to pool all the samples in a reagent reservoir to collect all of the PCR products (otherwise use a regular pipette) and then transfer the pooled PCR reactions into a 50-mL Falcon tube.
3. Add an equal volume of PC8 and centrifuge in a swinging bucket rotor for 10 min at 2500 rpm.
4. Transfer the upper aqueous phase to a new 50-mL conical tube and EtOH precipitate as follows:

7.5 M ammonium acetate	5.1 mL
Glycogen	191 μL
Pooled PCR products	11.5 mL (use all of the PCR products)
100% EtOH	38.3 mL

5. Invert the tube several times. (At this point, you can keep the PCR products as a precipitate at –80°C and proceed to **step 7** the next day or incubate in a dry ice/methanol bath for 15 min and continue).
6. While PCR products are precipitating, make four 1.5-mm-thick 12% acrylamide TAE gels (recipe in **Appendix B**).
7. Divide the sample into two polypropylene/Teflon tubes and balance up to two decimals with 100% EtOH.
8. Centrifuge in a fixed-angle rotor at 11,952 × *g* at 4°C for 30 min.
9. Wash with 5 mL 75% EtOH and centrifuge for 5 min in the fixed-angle rotor at 11,952 × *g* at 4°C.
10. Air-dry the pellet.
11. Resuspend each pellet in 200 μL LoTE, pool the two tubes into one 400-μL sample, and place it on ice.
12. Add 40 μL 10X loading dye to the purified PCR products.
13. Apply a total of 110 μL to each of the four 12% acrylamide TAE gels.
14. Mix 5 μL of 10-bp DNA ladder with 5 μL 100-bp DNA ladder and load this mixture into a single well.
15. Electrophorese the gel for approximately 1.5 h at 160 V until the blue xylene band is near the bottom of the gel and the purple bromophenol blue band has run off the gel.
16. Stain the gel using SYBR Green I at a 1:10,000 dilution (add 10 μL SYBR Green I to 40 mL 1X TAE buffer) in a foil-wrapped container (two gels per box); let the gel soak in the stain for 15 min on the shaker.
17. Visualize the gel on a UV box using the yellow SYBR Green filter. Protect the DNA by putting an EtOH-cleaned glass plate on the UV box, and then put the gel on the glass plate.

18. Cut out only the amplified ditag (102 bp) band from the gel. Be sure to not take the markers and 80-bp background band.
19. Place one third of each band in an 0.5-mL microcentrifuge tube (12 tubes total) processed as described in **Appendix C, no. 3**.
20. Place the 0.5-mL microcentrifuge tubes in 2-mL round-bottom microcentrifuge tubes and centrifuge in microfuge at >20,000 × *g* for 5 min at RT (this serves to break up the cut-out-bands into small fragments at the bottom of the 2.0-mL microcentrifuge tubes). If most of the gel is not contained in the larger microfuge tube, then centrifuge for 5 more min.
21. Discard 0.5-mL tubes and add 250 μL LoTE and 50 μL 7.5 M ammonium acetate to each 2.0-mL tube.
22. Tubes can remain at this point at 4°C overnight (do not leave for more than one night or the acrylamide can break down too much, and do not store at –80°C).
23. Vortex each tube, place at 65°C? for 15 min, and centrifuge briefly to collect the condensation at the bottom of the tube.
24. Place 5 μL LoTE on the center of the membrane of each of 12 Spin-X tubes.
25. Transfer contents of each tube to a Spin-X tube by mixing the contents well and using a P1000 tip that has been trimmed to have an opening of approximately 2 mm to transfer to Spin-X tubes. Use a spatula to scoop out and transfer any residual gel pieces.
26. Centrifuge each Spin-X tube for 5 min at >20,000 × *g* at RT and throw away Spin-X filter.
27. EtOH precipitate the eluates in new 1.7-mL tubes as follows:

    300 μL sample
    133 μL 7.5M NH_4OAc
    5 μL glycogen
    1000 μL 100% EtOH

28. Centrifuge in a microcentrifuge at >20,000 × *g* for 30 min at 4°C.
29. Wash 2 × 200 μL 75% EtOH; for each wash, centrifuge for 5 min at 4°C.
30. Resuspend DNA in 10 μL LoTE/tube.
31. Pool samples into one tube (120 μL total).
32. Remove 1 μL for cDNA quantitation. The total amount of DNA at this stage should be 10 to 20 μg. If the concentration is not this high, redo the large-scale PCR prep/precipitation and then combine both large-scale 102-bp band preps. If the DNA concentration is too high, then the sample can be split in half and processed as two samples on day 4. If the sample is split in half, then the reaction should be brought to volume with LoTE.

5.4. Day 4

5.4.1. Purification of Ditags

1. Digest the 102-bp DNA with NlaIII by adding the following to the same sample tube:

PCR products in LoTE	120 μL
LoTE	196 μL
10X NEB Buffer 4	40 μL
100X BSA	4 μL
NlaIII (10 U/μL)	40 μL
Total	400 μL

Throw away the last of the NlaIII because it is necessary that the enzyme works perfectly, and if the enzyme is frozen and dethawed again, it will loose some activity.

2. Incubate 1 h at 37°C.
3. Extract with equal volume PC8 (400 μL).
4. Vortex, centrifuge for 5 min at >20,000 × *g* at 4°C. (Note: be very careful in the following steps because now we are working with small 24- to 26-bp DNA segments. Keep reactions on ice.)
5. Transfer the upper aqueous phase into two tubes (200 μL each) and then EtOH precipitate as follows:

   200 μL sample
   66 μL 7.5 M NH_4OAC
   5 μL glycogen
   825 μL 100% EtOH

   (*Note*: Can stop here to precipitate at –80°C overnight if desired. If doing so, the dry ice/methanol bath is not needed for the quick DNA precipitation. Continue the next day with centrifugation.)
6. Place in dry ice/methanol bath for 15 min.
7. Warm the tubes at RT for 2 min until solution has melted.
8. Centrifuge the tubes at 4°C for 30 min at >20,000 × *g*.
9. Remove the supernatant.
10. Wash 1 × 200 μL cold 75% EtOH. The EtOH is cold to protect the 26-bp ditags from denaturation. The melting point of 26-bp DNA is below room temperature.
11. Remove all EtOH by pipetting and briefly air-drying the pellet.
12. Resuspend the pellet in each tube in 10 μL of cold TE (not LoTE). This high-concentration Tris buffer is needed to protect the 26-bp DNA ditags.
13. Pool the resuspended DNA into one tube (20 μL total) and place on ice.
14. Add 2 μL 10X loading dye/sample buffer (22 μL total volume).
15. Load 5 μL of 10-bp and 100-bp ladders in separate wells on both sides of the four lanes of sample. Leave an open lane between the sample lanes and the marker lanes. Load 5.5 μL of this sample into four lanes of a 1.5-mm, 12% polyacrylamide TAE gel (10-well) and electrophorese at 100 V to 140 V. Note: do not run the voltage too high or your 26-bp product could melt. Do not run both loading dyes off the gel: the bromophenol blue front (runs about 45 bp) should be approximately 1.5 to 2 cm from the bottom of the gel.

16. Stain the gel using SYBR Green I stain, at a 1:10,000 dilution.
17. Cut out the 24- to 26-bp band from the four lanes (the band is between 25 bp and 30 bp). Remember to put an EtOH-cleaned glass plate on top on the UV box and then place the gel on the plate.
18. Place two excised bands in an 0.5-mL microcentrifuge tube (two tubes total) as made in **Appendix C, no. 4**. More 0.5-mL microcentrifuge tubes can be used if more lanes are loaded with 26-bp ditags or if the bands are large. Centrifuge in the microfuge at >20,000 × *g* at RT for 5 min or until the excised band is completely collected in the 2-mL tubes.
19. Discard the 0.5-mL tubes, and add 50 μL 7.5 M ammonium acetate and 250 μL LoTE (in this order) to the 2-mL tubes.
20. Vortex the tubes, and place at 37°C (*not* 65°C) for 20 min.
21. Use four Spin-X tubes to isolate the eluate. Spin for 5 min at >20,000 × *g* at RT.
22. EtOH precipitate in three tubes (200 μL each) as follows:

    200 μL sample
    66 μL 7.5M NH_4OAc
    5 μL glycogen
    825 μL 100% EtOH

23. You can stop here and incubate overnight at –80°C to increase the amount of DNA precipitated. If you do this, the next day start at the centrifugation step and skip the dry ice/methanol bath step.
24. Place in a dry ice/methanol bath for 15 min.
25. Centrifuge at 4°C at >20,000 × *g* for 30 min.
26. Wash 2 × 200 μL of cold 75% EtOH.
27. Air-dry to remove all residual EtOH.
28. Resuspend each DNA sample in 2.5 μL cold LoTE (be sure the total volume is 8 μL after resuspending the pellet).
29. Remove 1 μL of the purified ditags for quantitating while ligating.

5.5. Day 5

5.5.1. Ligation of Ditags to Form Concatamers

Length of ligation time depends on quantity and purity of ditags. Typically, several hundred nanograms of ditags are isolated and produce large concatamers when the ligation reaction is carried for 1 to 3 h at 16°C (lower quantities or less pure ditags will require longer ligations).

1. Mix the following:

Pooled purified ditags	7 μL
5X ligation buffer (BRL)	2 μL
T4 ligase (5 U/μL)	1 μL

Thaw the ligation buffer slowly because it contains ATP, which degrades easily. If the volume of pooled purified ditags is high, the reaction volume can be increased (e.g., 10 μL ditags + 3 μL 5X ligase buffer + 1 μL ligase + 1 μL LoTE).

2. Incubate for 1 h 10 min if you started with 10 μg of total RNA at 16°C. (Fewer ditags require a shorter incubation.)
3. Heat the sample 65°C for 5 min.
4. Place it on ice for 10 min.
5. Add 1.5 μL 10X sample buffer/loading dye (5 μL 10X dye and 5 μL water) to ligation reaction.
6. Electrophorese a 1-mm, 8% polyacrylamide TAE gel electrophoresis (with a 10-well comb) to separate the concatamers. Have this gel ready by the end of the ligation. In the first lane of the 8% polyacrylamide, load 10 μL of a 1-kb ladder (25 ng/μL) as a marker. Load entire concatenated sample into the fifth well (i.e., one lane). Load as follows:

   (1) empty
   (2) 100-bp ladder
   (3) 1-kb ladder
   (4) empty
   (5) entire sample
   (6) empty
   (7) 1-kb ladder

7. Samples are electrophoresed at 130 V until the purple bromophenol blue dye reaches the bottom of the gel. Keep an eye on the gel because the running speed depends on the amount of buffer added to the buffer chamber.
8. Stain the gel with SYBR Green I 1:10,000 dilution. (This gel is thin and prone to rip. Be careful when taking it apart—wet the gel with buffer to take it off the plastic plate.)
9. Visualize on UV box using the yellow SYBR Green filter. Concatamers will form a smear on gel with a range from about 100 bases to several kb. We usually isolate region 800–1500 bases and 1.5–3.0 kb.
10. Place each of these gel pieces into an 0.5-mL microcentrifuge tube with a needle-pierced bottom (two tubes total).
11. Place the tubes in a 2.0-mL round-bottom microcentrifuge tube and centrifuge at >20,000 × *g* for 5 min at RT.
12. Discard the 0.5-mL tubes and add 300 μL LoTE to the 2-mL tubes.
13. Vortex the tubes, and place at 65°C for 20 min.
14. Transfer the contents of each tube into one Spin-X microcentrifuge tubes (two Spin-X tubes total).
15. Centrifuge the Spin-X tubes in microcentrifuge for 5 min at >20,000 × *g* at 4°C.
16. EtOH precipitate both the high and low weight concatamers as follows:

300 μL eluate
133 μL 7.5 M NH_4OAc
3 μL glycogen
1000 μL 100% EtOH

17. Can stop here and precipitate at –80°C overnight, or precipitate in an EtOH/dry ice bath for 15 min and continue.

5.5.2. Cloning Concatamers

Prepare the pZero cloning vector and test on a gel. pZero is aliquoted in 2.5-μL aliquots and stored at –20°C. (If the pZero preparation is done in the morning, then the ligation of the concatamers into the pZero can be done in the afternoon for an overnight incubation.)

Digest as follows:

pZero (1 μg)	1 μL
NEB buffer 2	1 μL
SphI	1 μL
Sterile H_2O	7 μL
Total	10 μL

Incubate for 20 min at 37°C.

Remove 1 μL of the digestion to electrophorese on a 1% agarose TAE gel. Electrophorese 1 μL of the digestion on a 1% agarose TAE gel beside a 1-kb ladder, and pZero vacant. pZero should migrate in the gel at approximately 2.5 kbp.

Add 30 μL LoTE to the rest of the sample and heat it at 70°C for 10 min to inactivate the enzyme. It is now 25 ng/μL, and 1 μL can be used per ligation.

Precipitate the concatamers in a methanol/dry ice bath for 15 min.

Centrifuge the concatamers and pZero at >20,000 × *g* for 30 min at 4°C.

Wash the concatamers and pZero with 2 × 200 μL 75% EtOH, centrifuge, and remove EtOH.

Resuspend each tube of the purified concatamer DNA in 6 μL LoTE.

Resuspend the pZero in 30 μL LoTE (approx. 30 ng/μL).

Ligate concatamers into pZero as follows [the following volumes are in oligo (dT)25]:

Fraction	High fraction	Low fraction	Vector only	No ligase added
Vector	1	1	1	1
5X ligase buffer	2	2	2	2
T4 ligase (5 U/μL)	1	1	1	0
Dwater	0	0	6	7
Concatamers	6	6	0	0

Incubate overnight at 16°C in 1.7-mL Eppendorf tubes. (*Note*: Before setting up the ligation, make sure that you can continue the next day.)

5.6. Day 6

5.6.1. Electroporation of Ligation Products

1. Add 190 µL LoTE to the ligation mix to bring the sample volume to 200 µL.
2. Add 200 µL PC8.
3. Briefly vortex and centrifuge for 5 min at >20,000 × *g* at RT.
4. Transfer the upper aqueous upper layer to a new tube.
5. EtOH precipitate the aqueous phase in 1.7-mL Eppendorf tubes as follows:

   200 µL sample
   133 µL 7.5 M ammonium acetate
   5 µL glycogen
   777 µL 100% EtOH

6. Place EtOH precipitating DNA in the –80°C freezer until ready to electroporate.
7. Make the LB media and the LB/Zeocin plates (**Appendix C**).
8. Centrifuge at >20,000 × *g* at 4°C for 30 min.
9. Wash 4 × 200 µL 70% EtOH by centrifuging for 5 min at >20,000 × *g* at 4°C.
10. After the last wash, remove EtOH and air-dry pellets.
11. Resuspend each pellet in 10 µL LoTE—these are your ligation mixtures—and place on ice.
12. For electroporation, place the required number of 0.1-cm electroporation cuvettes and 1.7-mL microcentrifuge tubes on ice (four for one SAGE library).
13. Thaw one vial (100 µL) of ElectroMAX DH10B cells on ice. Before use, mix by pipetting the cells up and down several times and pipet 25 µL of the competent cells into each 1.7-mL microcentrifuge tube.
14. Pipet 1 µL of the ligation mixture into the competent cells of the appropriate 1.7-mL microcentrifuge tube. Keep the rest of ligation mixtures at –80^{0}C in case a second electroporation is required.
15. Pipette the ligation mixture-competent cells into a chilled disposable cuvette. Keep the cuvettes on ice.
16. Set the *E. coli* electroporator pulser apparatus to 1.8 kV when using the cold 0.1-mm cuvettes (Do not touch the metal portion of the cuvette that touches the power supply.)
17. Pulse once. Hold the current for 4 s. Check that the electroporation worked for the required amount of time by pressing the left two buttons at the same time. The display should read 5.0 if the electroporation took place for 4 s. After electroporation, do not place the sample back on ice.
18. Remove the cuvette from the chamber and *immediately* add 1 mL LB medium and quickly but gently resuspend the cells.
19. Transfer the cells to a 15-mL conical tube and incubate while shaking at 100 rpm at 37°C for 30 min to 1 h. Only close the Falcon tubes to the first closure point so the culture can "breathe" while shaking.

20. Plate the transformation mixture onto fresh LB-Zeocin plates (100 μg/mL Zeocin). Various dilutions of both the high and low concatamer fractions are pipetted onto the LB-Zeocin plates as follows:

    100 μL undiluted electroporated competent cell culture (two plates)
    10 μL electroporated competent cell culture diluted with 100 μL LB media (two plates)
    100 μL of the two negative controls (vector only, no ligase) (two plates)

21. The cells are spread by using glass beads. Use standard sterile technique, including flaming all bottles when opening and closing.
22. Incubate overnight at 37^{0}C. Make sure to incubate the plates upside down. Keep rest of transformation mixtures at 4^{0}C. Insert-containing plates should have hundreds to thousands of colonies while control plates should have 0 to <20 colonies.

5.7. Day 7

5.7.1. Colony PCR

1. Check insert sizes of the clones by colony PCR. Check at least 20 colonies from each ligation dilution plate to calculate the average insert size. Make the appropriate amount of PCR master mix in the PCR free conditions as follows:

	Per reaction
10X PCR buffer (see **Appendix A**)	2.5 μL
DMSO	1.25 μL
25 mM dNTPs	0.5 μL
M13F (350 ng/μL)	0.5 μL
M13R (350 ng/μL)	0.5 μL
ddH$_2$0	19.25 μL
Taq (5 U/μL)	0.5 μL
Total	25 μL

2. Aliquot the PCR reactions a 96-well plate (25 μL/well) in PCR free conditions.
3. For each colony, use a sterile pipet tip or a toothpick to gently touch one colony, and then dip the tip into the PCR mix.
4. Perform PCR as follows (optimized for HYBAID OMNE machine):

1 cycle	95°C, 2 min
30 cycles	95°C, 30 s; 56°C, 1 min; 70°C, 1 min
1 cycle	70°C, 5 min

5. Load a 1-kb and 100-b ladder to determine the insert size. Electrophorese 5 μL of each reaction on a 1.5% agarose TAE gel at 100 V until the purple bromophenol blue band is near the bottom of the gel. A library is good if >90% of the colonies have inserts of at least 700 bp after subtracting the 226 bp (represents no insert). When estimating the average insert size, include the blank lanes.

6. CONCLUSION

The identification of alleles associated with specific traits will enter a revolutionary period with the new sequencing technologies. Genome and cDNA sequencing will become even more common. The acquisition of both a personal genome and phenome will represent the effective pillar for personalized medicine. We envisage that the strategy developed by us ***(16)*** will be particularly informative as it allows the association of expression and genotype data. Furthermore, it can quantify the expression of the different alleles in a heterozygous subject by simply counting the number of the respective tags.

REFERENCES

1. Adams MD, Kelley JM, Gocayne JD, Dubnick M, Polymeropoulos MH, Xiao H, et al. Complementary DNA sequencing: expressed sequence tags and human genome project. Science 1991;252:1651–1656.
2. Adams MD, Dubnick M, Kerlavage AR, Moreno R, Kelley JM, Utterback TR, et al. Sequence identification of 2,375 human brain genes. Nature 1992;355:632–634.
3. Velculescu VE, Zhang L, Vogelstein B, Kinzler KW. Serial analysis of gene expression. Science 1995;270: 484–487.
4. Brenner S, Johnson M, Bridgham J, Golda G, Lloyd DH, Johnson D, et al. Gene expression analysis by massively parallel signature sequencing (MPSS) on microbead arrays. Nat Biotechnol 2000;18:630–634.
5. Fodor SP, Read JL, Pirrung MC, Stryer L, Lu AT, Solas D. Light-directed, spatially addressable parallel chemical synthesis. Science 1991;251,:767–773.
6. Nickerson DA, Tobe VO, Taylor SL. PolyPhred: automating the detection and genotyping of single nucleotide substitutions using fluorescence-based resequencing. Nucleic Acids Res 1997;25:2741–2751.
7. Marth GT, Korf I, Yandell MD, Yeh RT, Gu Z, Zakeri H, et al. A general approach to single-nucleotide polymorphism discovery. (1999) Nat. Genet 23:452–456.
8. Saha S, Sparks AB, Rago C, Akmaev V, Wang CJ, Vogelstein B, et al. Using the transcriptome to annotate the genome. Nat Biotech 2002;20:508–512.
9. Kodzius R, Kojima M, Nishiyori H, Nakamura M, Fukuda S, Tagami M, et al. CAGE: cap analysis of gene expression. Nat Methods 2006;3:211–222.

10. Wei C–L, Ng P, Chiu KP, Wong CH, Ang CC, Lipovich L, et al. 5′ Long serial analysis of gene expression (LongSAGE) and 3' LongSAGE for transcriptome characterization and genome annotation. Proc Natl Acad Sci U S A 2004;101:11701–11706.
11. Cerutti JM, Riggins GJ, de Souza SJ. What can digital transcript profiling reveal about human cancers? Braz J Med Biol Res 2003;36:975–985.
12. Lal A, Lash AE, Altschul SF, Velculescu V, Zhang L, McLendon RE, et al. A public database for gene expression in human cancers. Cancer Res 1999;59:5403–5407.
13. Boon K, Osorio E, Greenhut SF, Schaefer CF, Shoemaker J, Polyak K, et al. An anatomy of normal and malignant gene expression. Proc Natl Acad Sci U S A 2002;99:11287–11292.
14. Silva APM, de Souza JES, Galante PAF, Riggins GJ, de Souza SJ, Camargo AA. The impact of SNPs on the interpretation of SAGE and MPSS experimental data. Nucleic Acids Res 2004;32:6104–6110.
15. Galante PA, Trimarchi J, Cepko CL, de Souza SJ, Ohno-Machado L, Kuo WP. Automatic correspondence of tags and genes (ACTG): a tool for the analysis of SAGE, MPSS and SBS data. Bioinformatics 2007;23:903–905.
16. Silva AP, Chen J, Carraro DM, Wang SM, Camargo AA. Generation of longer 3′ cDNA fragments from massively parallel signature sequencing tags. Nucleic Acids Res 2004;32:e94.

APPENDIX A: SPECIAL REAGENTS TO ORDER

RNA Reagents

Superscript Choice System cDNA Synthesis Kit, Invitrogen no. 18090-019
oligo (dT)25mRNA Direct kit, Dynal no. 610.11

Magnetic Beads

Dynabeads M-280 streptavidin slurry, Dynal no. 112.05
Magnet, Dynal no. 120.04

Enzymes

T4 Polynucleotide Kinase, 10 U/μL, NEB no. 201S
BsmF1, NEB no. 572S
NlaIII, NEB no. 125S (ship on dry ice and store at –80°C; degrades easily)
Sph1, NEB no. 182S
Klenow (2 U/μL), Pharmacia/USB no. 27-0929-01
T4 Ligase High Concentration (5 U/μL), Invitrogen no. 15224-041
Taq, Invitrogen no.10966-034

Miscellaneous

Glycogen, Boehringer Mannheim no. 901-393
pZERO-1 plasmid, Invitrogen no. K2500-01 (unstable so aliquot into separate tubes)

10 mM dNTP mix, Invitrogen no. 18427-013
DMSO, Sigma no. D2650
7.5 M NH_4OAc, Sigma no. A2706
Spin-X columns, Corning Costar no. 8160
Primers and linkers (**Appendix D**)
ElectroMAX DH10B, Invitrogen no. 18290-015
19:1 acrylamide:bis, Bio-Rad no. 161-0144
37.5:1 acrylamide:bis, Bio-Rad no. 161-0148
50X Tris Acetate Buffer, Quality Biological no. 330-008-161
Select agar, Invitrogen no. 30391-023
Zeocin, Invitrogen no. R250-01

APPENDIX B: SOLUTIONS

2X BW Buffer (Make as a 50-mL Stock)

10 mM Tris-HCl (pH 7.5)
1 mM EDTA
2.0 M NaCl
Bring to volume with water
Store at room temperature

LoTE (Make as 50-mL Stock)

3 mM Tris-HCl (pH 7.5)
0.2 mM EDTA (pH 7.5)
Bring to volume with ddH_20
Store at room temperature

PC8 (Make in the Hood, Cover with Aluminum Foil)

Make a 1:1 mixture of Tris-HCl saturated phenol and chloroform in 50 mL Falcon tubes. Make four at a time (one for 4°C and three to freeze).

10X PCR Buffer

166 mM $(NH_4)_2SO_4$
670 mM Tris pH 8.8
67 mM $MgCl_2$
100 mM β-mercaptoethanol
Aliquot into 0.5-mL aliquots and store at –20°C

Low-Salt LB Media/Zeocin LB Agar Plates

1. For 1 L, dissolve 10 g tryptone, 5 g yeast extract, and 5 g NaCl in 950 mL ddH_20.
2. Adjust the pH of the solution to 7.5 with 5 M NaOH, add 15 g agar, and bring the volume to 1 L.
3. Autoclave for 20 min on the liquid cycle.

4. Let the agar cool to ~55°C in a 60°C water bath.
5. Thaw the 100X Zeocin stock solution and add to a final dilution of 1:100.
6. If using a cell line carrying a lacI^q gene, add IPTG to a final concentration of 1 mM.
7. Pour into 10-cm Petri plates. Cover the plates with foil and let the plates harden.
8. Once agar has solidified, move them to a clean hood (lights off) to dry 30 min.
9. Zeocin is stable for less than a week, covered and stored at 4°C.

Handling Zeocin

- High salt, high pH, or low pH inactivates Zeocin. Reduce the salt in bacterial medium and adjust the pH to 7.5 to keep the drug active (see Low-Salt LB Media/Zeocin LB Agar Plates, above).
- Store Zeocin at –20°C protected form light and thaw on ice before use.
- Zeocin is light sensitive. Store the drug and plates or medium containing the drug in the dark.
- Wear gloves, a laboratory coat, and safety glasses when handling Zeocin-containing solutions.
- Do not ingest or inhale solutions containing the drug.
- Be sure to bandage any cuts on your fingers to avoid exposure to the drug.

APPENDIX C: GEL ELECTROPHORESIS AND MISCELLANEOUS METHODS

1. 12% PAGE (for isolating PCR products and ditags)

40% polyacrylamide (19:1 acrylamide:bis)	10.5 mL
ddH_20	23.5 mL
50X Tris acetate buffer (Quality Biological no. 330-008-161)	700 µL
10% APS	350 µL
TEMED	30 µL

Mix above and add to vertical gel apparatus. We currently use XCell (Invitrogen/Novex). Add comb and let gel sit at least 30 min to polymerize. Electrophorese the gel at 160V.

The first three components can be mixed together ahead of time and kept wrapped in aluminum foil at 4°C. Add 10% APS and TEMED (in 4°C) right before gel is ready to be poured.

Electrophorese the gel in 1X TAE.

To a 10-µL sample add 1 µL 10X loading buffer dye containing xylene cyanol and bromophenol blue.

For one gel make as follows:

- 10 mL (component mixture of 1–3) add 100 μL 10% APS and 10 μL TEMED. Shake back and forth to mix (do not vortex).
- 500 ng of DNA marker is sufficient for visualization on a gel. Therefore, 5 μL is needed of the diluted mix.
- 1-mm and 1.5-mm gels require at least 10 mL gel solution.
- After gel is set, clean wells with *ddH_2O*, remove white sticker off the bottom so current can travel through, mark slots so wells are easy to visualize during loading, fill inner and outer chambers with TAE buffer, and wash wells with 1X TAE and a syringe to clean the wells.
- Be careful not to get any bubbles in the gel, especially when using combs with one large well.

2. 8% PAGE (for separating concatamers)

40% polyacrylamide (37.5:1 acrylamide:bis)	10 mL
ddH_20	39 mL
50X Tris acetate buffer	1 mL
10% APS	500 μL
TEMED	50 μL

- Make a 50-mL stock of the first three components because the acrylamide goes bad.

3. Mincing gel bands
   (1) Pierce the bottoms of 0.5-mL Eppendorf tubes with a 21-gauge needle two times.
   (2) Place 0.5-mL tube in a 2-mL round-bottom centrifuge tube.

APPENDIX D: LINKER AND PRIMER SEQUENCES

Linker 1A (Obtain Gel-Purified)

5′ TTT GGA TTT GCT GGT GCA GTA CAA CTA GGC TTA ATA GGG ACA TG 3′

Linker 1B (Obtain Gel-Purified)

5′ TCC CTA TTA AGC CTA GTT GTA CTG CAC CAG CAA ATC C[amino mod. C7] 3′

Linker 2A (Obtain Gel-Purified)

5′ TTT CTG CTC GAA TTC AAG CTT CTA ACG ATG TAC GGG GAC ATG 3′

Linker 2B (Obtain Gel-Purified)

5′ TCC CCG TAC ATC GTT AGA AGC TTG AAT TCG AGC AG[amino mod. C7] 3′

Primer 1

5′ GGA TTT GCT GGT GCA GTA CA 3′

Primer 2

5′ CTG CTC GAA TTC AAG CTT CT 3′

M13 Forward

5′ GTA AAA CGA CGG CCA GT 3′

M13 Reverse

5′ GGA AAC AGC TAT GAC CAT G 3′

Note: High-quality linkers are crucial to several steps in the SAGE method. Linkers 1A, 1B, 2A, and 2B should be obtained gel-purified from the oligo synthesis company.

APPENDIX E: FREQUENTLY ASKED QUESTIONS AND ANSWERS

Q:I've tried the SAGE method and I do not obtain 102-bp ditag-containing PCR products. Why not?

A:The SAGE method is composed of a series of straightforward enzymatic reactions, each of which can fail for a number of reasons. Listed below are the most common causes of failure of the method:

1. Poor starting material. Insufficient, poor quality, or degraded RNA. Check RNA on denaturing gel before cDNA synthesis. Use no less than 2.5 μg of high-quality polyA RNA.
2. Poor-quality or insufficient cDNA synthesis. Check cDNA quality by gel electrophoresis (cDNA synthesis should result in several micrograms of cDNA and should range in size from several hundred base pairs to more than 10 kb).
3. Poor-quality linkers. Obtain linkers and biotin-oligo dT gel purified. Check linker kinasing by self-ligation and oligo dT biotinylation by biotin gel shift experiment (see **Appendix B** for details).
4. Poor-quality reagents. Obtain fresh reagents from recommended sources. Note that the half-life of NlaIII is less than 3 months at –20°C.
5. Not following the detailed protocol. The specific steps in the protocol have been optimized, so try to keep to them as much as possible (e.g., only use the PCR buffer and cycle temperatures in the protocol as other conditions may not work).

6. Concentration of dNTPs is too high in the 102-bp PCR amplification. Increasing the concentration of dNTPs in the PCR reaction causes a dose-dependent increase in PCR product until a threshold is reached. A slight increase beyond this threshold results in a rapid and complete loss of PCR product. Because the maximum dNTP concentration varies with each lot purchased, we strongly recommend titrating the dNTPs before doing the large scale. To simply check if a PCR product exists, we recommend starting with a low but safe dNTP concentration (e.g., 0.5 to 1.0 mM).
7. Too many PCR cycles. When doing the large-scale PCR amplification, be sure to try a range of cycle numbers. Too many cycles can result in loss of the 102-bp PCR product.

Q:I can see a 102-bp product, but when I try to digest with NlaIII it only partially digests or it doesn't digest at all.

A:This is a common problem that many people have noticed. There are several reasons why this may occur.

1. NlaIII is inactive. When ordering NlaIII, we ask NEB to send us the enzyme on dry ice. If not specified, the enzyme, which is normally sent on an ice pack, will often arrive at room temperature. We also store NlaIII in aliquots at –80°C, which helps prolong its activity. Make sure the enzyme is fresh—its half life is only a few months.
2. Not enough enzyme in the digestion reaction. On occasion, with certain less-active batches of NlaIII and high yields of PCR product, it may help to add more enzyme by scaling up the volume of the digestion reaction. NEB is currently redefining what a unit of NlaIII should be.
3. Perhaps the most important reason for lack of digestion is that the CATG site of the PCR product is missing and therefore cannot be cut. This occurs because of exonuclease contamination during early steps of the SAGE protocol. The exonuclease activity appears to come from the DNApol1 used in the second strand reaction of the cDNA synthesis. Subsequently, the exonuclease activity of this enzyme can chew back the CATG ends of both the linkers and NlaIII cut cDNAs resulting in partial or complete blunt end ligation of the linkers. Although the original SAGE protocol uses phenol extraction to remove the enzyme, a small amount may still remain. We now recommend doing two sequential phenol extractions for large-scale SAGE users. For microSAGE users, the problem has been worse as phenol extractions are not used at this step. In this case, inactivation of the enzyme with SDS seems to be the most effective way to remove residual exonuclease activity.

Q:Although I can obtain 24- to 26-bp ditags, I cannot generate large concatamers. How can I improve this?

A:There are essentially two parameters that affect the size of the concatamers. One is the quantity of the ditags present in the ligation reaction. We routinely perform 100 to 200 50-μL PCR reactions using the PCR protocol specified in the SAGE protocol. This generates 500 ng to more than 1 µg of ditags.

The second parameter is the purity of the ditags. After the NlaIII digest, the resulting products contain large amounts of linker material (around 40 to 42 bp) or other DNA products as smears in the background. Although the ditag is gel purified, these other DNA fragments may contaminate that portion of the gel and serve to "poison" the ligation reaction. The linkers, for example, if present in the ligation reaction, could ligate the end of a concatamer but prevent its extension and cloning. Removal of linkers with an extra purification step using biotinylated primer 1 and primer 2 and streptavidin beads overcomes this problem. Isolation of several hundred nanograms of ditags using this procedure is sufficient to generate large concatamers.

The ideal length of ligation reactions will vary depending on the quantity and quality of the ditags. For several hundred nanograms of ditags, ligation times of 30 min to 3 h are optimal.

Q:How do I find out about current SAGE applications and recent publications?

A:Access the SAGE home page on the Internet: http://www.sagenet.org/.

6

Single Nucleotide Polymorphisms and DNA Methylation Analysis Using Pyrosequencing Methods

Jinsheng Yu and Sharon Marsh

Summary

The data generated from the Human Genome Project has led to an explosion of technology for low-, medium-, and high-throughput genotyping methods. Pyrosequencing is a genotyping assay based on sequencing by synthesis. Short runs of sequence around each polymorphism are generated, allowing for internal controls for each sample. Pyrosequencing can also be used to identify triallelic, indel, and short-repeat polymorphisms, as well as determining allele percentages for DNA methylation or pooled sample assessment. Assays details for pyrosequencing of clinically relevant polymorphisms and DNA methylations are described in this chapter.

Key Words: DNA methylation; Pyrosequencing; single nucleotide polymorphism

1. INTRODUCTION

The explosion of genetic information available after the completion of the Human Genome Project has proved a gold mine for pharmacogenetic research to identify biomarkers. Early estimates predicted more than 1.42 million single nucleotide polymorphisms (SNPs) are present in the human genome *(1)*. The 126 build of the public SNP repository, dbSNP (completed by May 2006, see http://www.ncbi.nlm.nih.gov/SNP/index.html), contained 27,846,394 submissions for the human genome, of which 5,646,244 were validated SNPs (20%). Polymorphisms in coding and control regions of genes can cause significant interindividual variation in the resulting protein function and activity, leading to important differences in disease susceptibility and drug metabolism. This expansion in evaluable SNPs has led to a number of detection methods *(2,3)*. Pyrosequencing (Biotage AB, Uppsala, Sweden) is sequencing by synthesis, a simple-to-use technique

From: *Methods in Pharmacology and Toxicology: Biomarker Methods in Drug Discovery and Development*
Edited by: F. Wang © Humana Press, Totowa, NJ

for accurate and quantitative analysis of DNA sequences, and it produces specific sequence data in the form of peaks on a pyrogram where the light signal collected is transformed and presented based on the DNA base composition of interest. It does not require the presence of a restriction enzyme site, and polymerase chain reaction (PCR) product and internal primer sites can vary in size and position. In addition, it can also be used to identify triallelic, indel, and short-repeat polymorphisms, as well as to determine allele percentages for DNA methylation ***(4)*** or pooled sample assessment ***(5)***. The DNA methylation assay by Pyrosequencing is to detect a special "C to T" polymorphism as after treatment with sodium bisulfate, unmethylated cytosine residues are converted to thymine, whereas methylated cytosine residues are retained as cytosine. Being an intrinsic feature of the Pyrosequencing method, the availability of sequence directly adjacent to the polymorphisms allows internal quality control checks to be made for each sample. Pyrosequencing is performed on a 96-well platform, and in an average day, more than 3000 individual genotypes can be measured. This method has been used to genotype a number of clinically relevant polymorphisms ***(6–8)*** and DNA methylation status ***(9,10)***. Included in these methods are assay details for commonly analyzed clinically relevant polymorphisms in the cytochrome P450 and adenosine triphosphate (ATP)-binding cassette transporter genes CYP3A4 and ABCB1, and also for DNA methylation levels in the DNA mismatch repair and fluorouracil catalytic enzyme genes MLH1 and DPYD.

2. MATERIALS

2.1. DNA Template

DNA from any source can be used in Pyrosequencing assays (**Note 1**). Commonly used kits for DNA extraction, including Gentra (Minneapolis, Minnesota, U.S.A.) and Qiagen (Hilden, Germany), do not inhibit the assay. For use in methylation assays, the original genomic DNA must be converted through the widely used bisulfate conversion method (**Section 3.1**). After treatment with sodium bisulfate, unmethylated cytosine residues are converted to thymine; whereas methylated cytosine residues are retained as cytosine.

2.2. Bisulfate Conversion of Genomic DNA and SssI Methylase Treatment

1. 4 M sodium bisulfate (Sigma, St. Louis, Missouri, U.S.A.) at pH 5 (freshly prepared).
2. 10 mM hydroquinone (Sigma) (freshly prepared).

3. 2 M NaOH.
4. Mineral oil from Sigma.
5. Heating block for DNA incubation.
6. DNA clean-up kit from Qiagen.
7. *Sss*I methylase from New England Biolab.

2.3. Polymerase Chain Reaction

1. Primer design software. (Any free or custom-purchased software is appropriate. Pyrosequencing also sells custom primer design software.)
2. 1 to 5 ng DNA template (**Note 2**).
3. PCR mastermix, for example: 30 mM Tris-HCl, 100 mM potassium chloride, pH 8.05, 400 μM dNTP, and 5 mM magnesium chloride (**Note 3**).
4. Hot start Taq polymerase (**Note 4**).
5. DNase- and RNase-free 18.2 mΩ water.
6. DNA oligonucleotides (primers).
7. Unskirted 96-well PCR trays.
8. 96-well sealing film or silicon mat for covering 96-well plates in a thermocycler.
9. Thermocycler with 96-well capacity and heated lid.

2.4. Agarose Gel Electrophoresis

1. Agarose.
2. 50X TAE buffer: for 1 L, add 242 g Tris base, 57.1 mL glacial acetic acid, and 18.6 g ethylenediamine tetraacetic acid to 18.2 mΩ water. Store at room temperature. Dilute to 1X with water prior to use.
3. Microwave.
4. Ethidium bromide: 4 μL of 10 mg/mL ethidium bromide/100 mL agarose; add *after* heating.
5. 6X loading dye: for 100 mL: 30 mL glycerol, 70 mL water plus a pinch of bromophenol blue and a pinch of xylene cyanol FF (amount can be varied depending on the desired color). Store at room temperature.
6. Gel apparatus: casting tray, gel tank, lid, and power supply.
7. UV gel documentation system with thermal printer.

2.5. Processing PCR for Pyrosequencing

1. Centrifuge with rotor/buckets to handle 96-well plates.
2. 2X binding buffer: for 1 L, add 1.21 g Tris, 117 g NaCl, 0.292 g ethylenediamine tetraacetic acid to water, pH 7.6, with 1 M HCl. Sterile filter then add 1 mL Tween-20.
3. Bead mix: 240 μL streptavidin coated Sepharose beads, 4560 μL 2X binding buffer, and 3600 μL 18.2 mΩ water per 96-well plate (the magnetic bead processing protocol for a PSQ96 or PSQ96MA is described elsewhere) ***(11)***. Excess Sepharose/binding buffer mix can be stored in a glass bottle at 4°C.
4. 96-well plate shaker (e.g., Eppendorf thermomixer; Fisher Scientific, Hampton, NH).

5. Vacuum prep tool and troughs (Biotage, Uppsala, Sweden).
6. 70% ethanol in 18.2 mΩ water (**Note 5**).
7. 0.2 M NaOH in 18.2 mΩ water.
8. Washing buffer: for 1 L, add 1.21 g Tris to water, pH 7.6 with 4 M acetic acid. Sterile filter.
9. Annealing buffer: for 1 L, add 2.42 g Tris, 0.43 g magnesium acetate tetrahydrate to water, pH 7.6 with 4 M acetic acid. Sterile filter.
10. Pyrosequencing primer mix: 12 μL of 0.3 μM Pyrosequencing primer in annealing buffer per well dispensed into a 96-well Pyrosequencing plate (Biotage).
11. Heating block capable of at least 80°C
12. Pyrosequencing plate adaptor set (base and iron) (Biotage).
13. Adhesive sealing film for 96-well plates.

2.6. *Pyrosequencing*

1. PSQhs96 or PSQhs96A Pyrosequencer with Pyrosequencing 96A version 1.1 or 96MA software or higher. A detailed protocol for the PSQ96 or PSQ96MA has been described previously ***(11)***.
2. PSQ cartridge, capillary dispensing tips or nucleotide dispensing tips, and reagent dispensing tips for hsPSQ96 and hsPSQ96A (Biotage).
3. Pyrosequencing HS reagent kit (Biotage).
4. DNase and RNase free 18.2 mΩ water.
5. Microcentrifuge.

3. METHODS

Pyrosequencing is based on sequencing by synthesis. The assay takes advantage of the natural release of pyrophosphate whenever a nucleotide is incorporated onto an open 3′ DNA strand. The released pyrophosphate is used in a sulfurylase reaction releasing ATP. The released ATP can be used by luciferase in the conversion of luciferin to oxyluciferin. The reaction results in the emission of light, which is collected by a CCD camera and recorded in the form of peaks, known as pyrograms (**Fig. 1**). When a nucleotide is not incorporated into the reaction, no pyrophosphate is released, and the unused nucleotide is removed from the system by degradation through apyrase. This four-enzyme process is performed in a closed system in a single well.

3.1. *Bisulfate Conversion of Original Genomic DNA*

1. Before the bisulfate conversion, a normal genomic DNA or low-methylation sperm DNA can be treated with the *Sss*I methylase to generate a methylation positive control DNA sample (i.e., high-methylation DNA): 1 μg DNA is incubated with 1 unit of *Sss*I methylase and 160 mM *S*-adenosylmethionine at 37°C overnight.

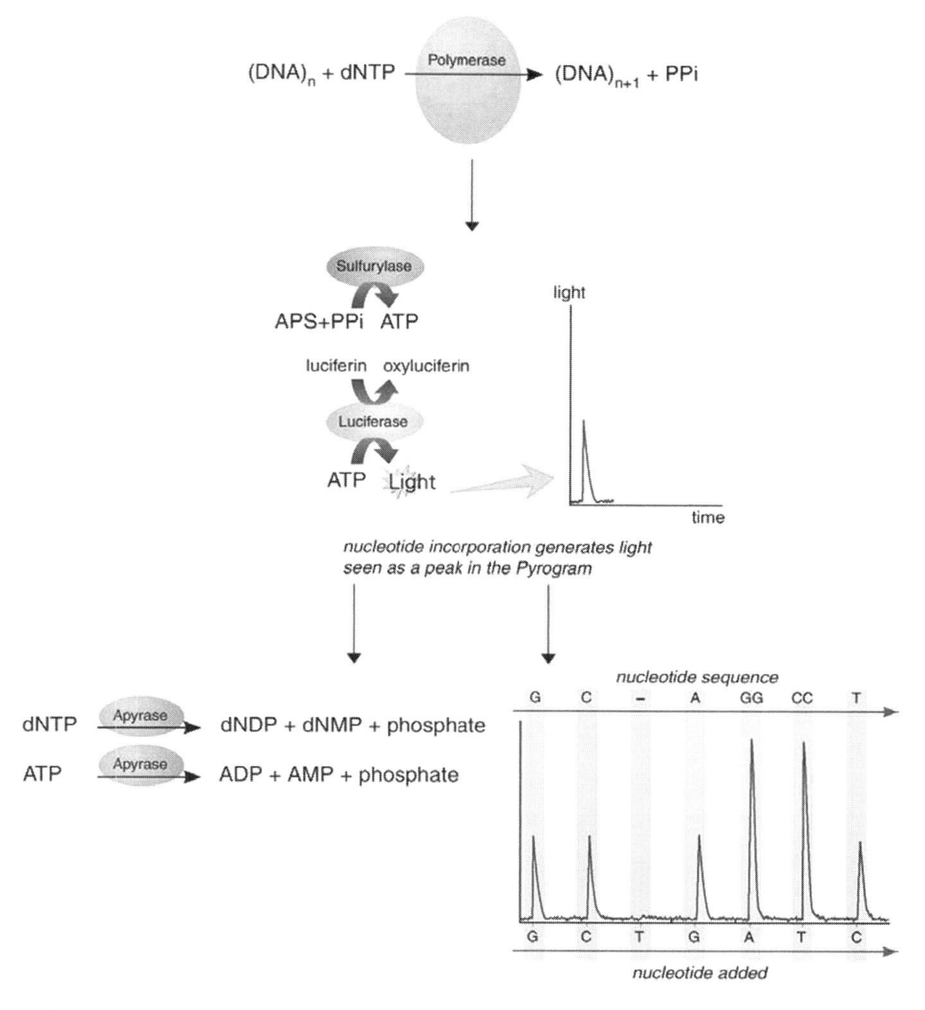

Fig. 1. The Pyrosequencing reaction. A modified ATP is used for the nucleotide dispensations to prevent its direct use by luciferase in the reaction. (Reprinted from *Methods in Molecular Biology*, vol. 311, Marsh S. et al.: *Pyrosequencing of Clinically Relevant Polymorphisms*, p. 100, 2005, with permission from the Humana Press Inc.).

2. Treatment of genomic DNA with sodium bisulfate (NaHSO): Add 5 μL 2 M NaOH (0.2 M final) into 45 μL DNA solution (0.5 to 1 μg) and incubate tube at 37°C for 15 min; and then add 550 μL methylation solution composed of 520 μL 4 M NaHSO (pH 5.0, final 3.5 M) and 30 μL 10 mM hydroquinone (final 0.5 mM), and invert tube to mix well. Slowly add 300 μL mineral oil to avoid evaporating and incubate at 55°C for 14 to 16 hours under dark light.

3. Clean-up the DNA with Qiagen kit: Place one GFX column (Amersham-Pharmacia Biotech, Uppsala, Sweden) in a collection tube and add 500 μL capture buffer to the GFX column. Remove the mineral oil from DNA sample tube and transfer the posttreatment DNA sample to the GFX column (dividing 600 μL of each sample into three columns). Mix thoroughly by pipetting the sample up and down 4 to 6 times and then centrifuge at 14,000 rpm for 30 s. Discard the flow-through by emptying the collection tube. Place the GFX column back inside the collection tube and add 500 μL wash buffer to the column. Centrifuge at 14,000 rpm for 30 to 60 s, discard the collection tube, and transfer the GFX column to a fresh 1.5-mL tube. Apply 50 μL elution buffer (1X Tris-EDTA buffer, pH 8.0) (or DNase- and RNase-free water) directly to the top of the glass fiber matrix in the GFX column and then incubate the sample at room temperature for 1 min. After centrifugation at 14,000 rpm for 1 to 2 min, the purified DNA is recovered.
4. Final modification of DNA with NaOH: Add 9 μL 2 M NaOH into 47 to 50 μL eluted DNA at room temperature for 5 min, and then add 400 μL 100% ethanol and 15 μL 3 M sodium acetate (pH 5.5), invert tube to mix well for rich precipitation of DNA. Keep the DNA sample on ice for 30 min, and then centrifuge at 14,000 rpm for 20 min at 4°C. After centrifugation, pour off solution, and resuspend DNA pellet in 50 μL water.
5. Store the purified bisulfate-converted DNA in –20°C for the methylation PCR assay.

3.2. PCR Primer Design

1. Any primer design software, freely available or custom purchased, may be used to design PCR primers for Pyrosequencing. The polymorphism may be in any position of the PCR amplicon except the PCR primer sequence itself, from only 1 base away from the 3′ end of the PCR primer sequence to the center of PCR amplicon. SNPs, indels, repeats, and so forth, do not require specific PCR primer design modifications. For the DNA methylation assay, the primer design strategy is just like designing a C/T SNP, but it can include as many CpG dinucleotide loci as the assay capability allows. Currently, the Pyrosequencing run can be carried out for up to 70-bp sequence of interest covering ~20 CpG loci.
2. Primers should be between 15 and 30 bases long, with an optimum size of 20 bases, ideally with a GC:AT ratio around 50% (although not essential, as you are at the mercy of the location of the polymorphism).
3. Amplicon sizes less than 300 bp are optimal, amplicon sizes 300 to 500 bp give acceptable results, amplicon sizes greater than 500 bp will require optimization during the processing of the PCR product for Pyrosequencing. Amplicon sizes of 100 to 200 bp are suitable for most template sources, including fragmented DNA.
4. Care should be taken to avoid any possible template loops from primers or the single-stranded amplicon doubling back on themselves, as these can lead to background problems during the Pyrosequencing assay (**Note 6**).

5. Optimum primer melting temperature (T) is 60°C; however, again, the position of the polymorphism determines the ability to design optimum primers and 50°C to 69°C will work. The individual primers should ideally have T values within 2°C of each other to allow effective optimization of the PCR.
6. Primer specificity should be checked by screening the primers across available human genome sequence using the NCBI Blast program (http://www.ncbi.nlm.nih.gov/blast/). Extra care should be taken when designing assays for gene family members (e.g., cytochromes) or genes with known pseudogenes (e.g., DHFR), as cross-hybridization of primers can lead to high background, reduced signal, and/or false-positive results.
7. One primer needs to be biotinylated at the 5′ end. Which primer to biotinylate is dependent on the Pyrosequencing primer orientation.
8. Examples of primers for MLH1 and DPYD methylations are shown in **Table 1**.

3.3. Pyrosequencing Primer Design

1. The entire PCR amplicon sequence, including forward and reverse primer sequences, is required to generate the optimum Pyrosequencing primer. It is recommended that the users purchase the custom Pyrosequencing primer design software.
2. Unless multiplexing is required (**Note 7**), the software should be defaulted to find both forward and reverse primers to improve the likelihood of obtaining the optimum primer sequence. The software will list all possible forward and reverse primers by score. A score of 100 is "perfect," scores between 90 and 100 are considered "high." Often, "medium" scores yield usable primers, as certain scoring parameters are more critical than others (**Note 8**). Template loops likely cause background will be highlighted by the software (**Note 9**).
3. The orientation of the Pyrosequencing primer will determine the PCR primer to be biotinylated. Forward Pyrosequencing primers require a biotinylated reverse PCR primer; and reverse Pyrosequencing primers require a biotinylated forward PCR primer.

3.4. PCR Optimization

1. Primer optimization of magnesium concentration and temperature should be carried out in advance for new assays. Ideally, a gradient PCR with different magnesium concentrations should be performed, if a thermocycler with a gradient block is available. If a pre-made PCR mix is used, only temperature optimization need be performed (**Note 10**). An example gradient setup based on a 96-well PCR block with gradient function follows:

   Mastermix (**Note 11**)

   130 μL AmpliTaq Gold PCR mastermix (Applied Biosystems, Foster City, CA)
   Forward primer (10 pM final concentration)

Table 1
Primer Details for MLH1 and DPYD Methylations

Gene	PCR stage	Orientation	PCR primers (5′–3′)*	Sequence to analyze
MLH1	First round	Forward	tttTtTaaTtTtgtgggttgTtggg	
		Reverse	AAaAAccacaaAaAcaAAAccaa	
	Second round	Forward	**5′-Biotin**/TtgTTcgTtaTTtagaaggatatg	
		Reverse	tctActcctattAActAAatatttc	
	Pyrosequencing	Reverse	TcgTtaTTtagaaggatatg	5′-CGGACAGCGATCTCTAACGCG CAAGCG-3′
DPYD	First round	Forward	tttagTagtttagagattaaaggTTagt	
		Reverse	AAAccatAAcaAtAcctacaAtc	
	Second round	Forward	ggTtgaaTtgggaagg	
		Reverse	**5′-Biotin**/aAtctAccaAtAacaaaccctc	
	Pyrosequencing	Forward	ggTtgaaTtgggaagg	5′-CGGGCTGCGGCTGCCGTGCC TGGCGCGGGAGCCG-3′

*Uppercase letters denote a transition of C to T or G to A.

Reverse primer (10 pM final concentration)
13 μL DNA
Up to 260 μL with 18.2 mΩ water

Add 20 μL mastermix to row of a 96-well plate or 12 0.2-mL tubes and place the gradient block (ensure samples cover a continuous row).

PCR Program [based on an MJ Research (Reno, NV) gradient block]

93°C for 20 min (or appropriate temperature/time to activate Taq)
30 cycles of:

94°C for 30 s
55°C to 72°C for 30 s
72°C for 30 s
Then:

72°C for 5 min
Store at 4°C.

2. The gradient PCR should be visualized using a 1% or 2% agarose gel. The optimal temperature should give the brightest single band at the appropriate amplicon size. Care should be taken to avoid temperatures where a smeared or multiband product can be seen, as these can increase Pyrosequencing background or reduce specificity if coamplifying a different DNA region. Where several temperatures of equal band intensity are available, the highest temperature should be picked to ensure specificity.

3.5. PCR for Pyrosequencing

1. Care should be taken to avoid contamination. The bench area should be swabbed with 70% ethanol or 5% bleach solution before each PCR set-up and barrier tips should be used for all pipetting steps (if a robot workstation with fixed tips is used, tips should be cleaned in 5% bleach solution between every DNA dispensation and between every 96-well plate mastermix dispensation).
2. 1 μL of (1 to 5 ng) DNA (depending on source; **Note 2**) should be dispensed into an unskirted 96-well PCR tray (**Note 12**). At least one well should not contain DNA to act as a negative control (**Note 13**).
3. A 20-μL PCR is ideal for Pyrosequencing; however, if the PCR product is especially strong or wide-peak pyrograms occur, a 10-μL reaction will work well. For a 20-μL reaction based on ABI AmpliTaq Gold PCR mastermix (Applied Biosystems):

10 μL ABI AmpliTaq Gold PCR mix
Forward PCR primer (10 pM final concentration)
Reverse PCR primer (10 pM final concentration)

Up to 19 μL with 18.2 mΩ water
1 μL template

4. The PCR plate should be well sealed using a silicon mat or adhesive film. The following PCR program should be run (**Note 14**):

   93°C for 10 min (or relevant temperature/time for Taq activation)
   55 cycles of:

   95°C for 30 s
   X°C for 30 s (based on gradient-derived annealing temp)
   72°C for 30 s
   Then:

   72°C for 5 min
   Store at 4°C.

5. It is possible to directly use the PCR product for Pyrosequencing; however, it is advisable to check the product and the negative control on a 1% to 2% gel to ensure the reaction has been performed successfully and no contamination is present. Contamination is identifiable at the Pyrosequencing stage; however, it is cheaper and faster to run an agarose gel than to process and run a contaminated/failed Pyrosequencing plate. 96-Well plates should be briefly centrifuged and the lid removed with care to prevent sample aerosol and inadvertent cross-contamination. Typically, 5 μL of the negative control and 5 μL of 5 to 6 wells should give an idea of the success of the PCR. The Pyrosequencing will not be affected by the reduction in volume in these wells. Because of the unusually large number of PCR cycles, some smearing may be visible on a gel, even if the optimum annealing temperature has been used. At this stage, the smearing typically does not affect the Pyrosequencing reaction if the PCR primers are specific and the negative control does not contain product.
6. The PCR product can be stored at 4°C until needed. PCR trays should be briefly centrifuged as condensation may occur on the lid, which is a possible source of post-PCR contamination.
7. In the DNA methylation assay, the PCR products may be obtained by a two-round nested PCR because of the insensitivity of PCR primers to the DNA methylation status of starting DNAs *(12)*. Specifically in the examples of MLH1 and DPYD methylation assays, the first-round external PCR reaction was carried out for 40 cycles using AmpliTaq Gold PCR master mix (Applied Biosystems), 5 pmol of each primer, and 10 ng of the bisulfate-converted genomic DNA in a 20-μL reaction. The external reaction was run at 94°C, 55°C, and 72°C for 1 min each, and the second-round nested reaction was run for 55 cycles at 94°C, 60°C, and 72°C for 1 min each for *MLH1* and 30 s each for *DPYD*. The nested reaction was carried out with 5′-biotinylated reverse or forward primers in a 20-μL reaction containing 2 μL of 1:5 diluted first-round PCR products and 4 pmol of each primer.

3.6. PCR Processing for Pyrosequencing

This protocol assumes the use of a streptavidin/Sepharose bead setup for Pyrosequencing on an hsPSQ96 or an hsPSQ96A system. The magnetic bead processing method for the PSQ96 or PSQ96MA is described elsewhere *(11)*.

1. A 96-well Pyrosequencing plate containing Pyrosequencing primer mix should be set up as described in **Section 2.4** (see **Note 15**).
2. The small volume readily evaporates, if the setup time is longer than 10 to 15 min; cover the plate with adhesive film. Primer plates can be aliquoted in advance and stored at 4°C. It is advisable to allow them to reach ambient temperature and briefly centrifuge them before use after storage.
3. Add 70 μL of Sepharose bead mix as described in **Section 2.4** to each well of the PCR product. Replace silicon lid/adhesive film securely.
4. Shake the 96-well plate for 5 min at room temperature. If using the Eppendorf thermomixer, 1400 rpm is the optimum speed. This allows the streptavidin-coated Sepharose beads to anneal to the biotin tag on the PCR primer. Use the plate immediately; if the plate is allowed to sit, the beads will settle to the bottom of the wells and will not be accessible to the vacuum tool. If settling has occurred, briefly return the plate to the shaker to disperse the beads.
5. Align reagent troughs, PCR product/bead mix tray and Pyrosequencing primer tray as shown in **Fig. 2** (**Note 16**).
6. With the vacuum switched OFF, shake the vacuum tool tips into clean 18.2 mΩ water. Discard water, refill trough, and switch the vacuum on. Place filter tips into trough until all water has been removed (~30 s).

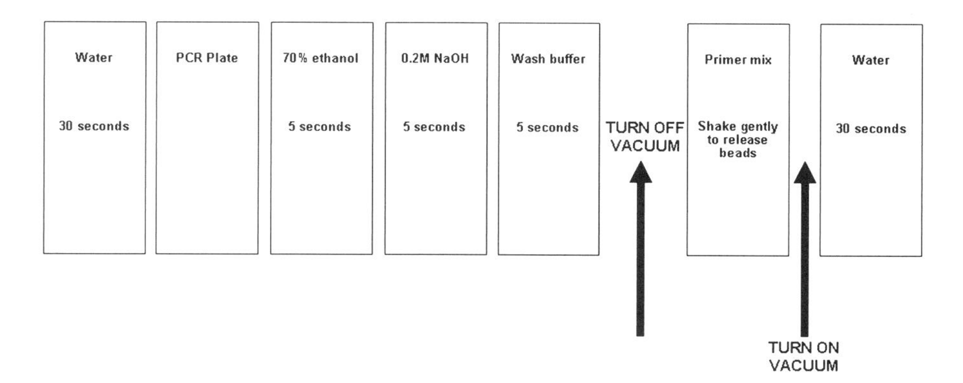

Fig. 2. Reagent layout for processing PCR plates for Pyrosequencing. The same orientation is important for the PCR and Pyrosequencing plates. (Reprinted from *Methods in Molecular Biology*, vol. 311, Marsh S. et al.: *Pyrosequencing of Clinically Relevant Polymorphisms*, p. 106, 2005, with permission from the Humana Press Inc.).

7. Place filter tips into the wells containing the PCR/bead mix. Ensure all liquid has been removed from the tray by slightly rocking the vacuum tool can prevent surface tension from causing liquid to remain in the wells. The beads attached to the biotin primer will prevent the PCR product from going through the filters.
8. With the vacuum still on, place the filer tips in the 70% ethanol. Wait a few seconds until a good flow of liquid is seen through the tubing and allow the tips to suck up ethanol for 5 s. Repeat with 0.2 M NaOH and washing buffer. The NaOH denatures the DNA; therefore, only single-stranded PCR product remains adhered to the filter tips.
9. Switch the vacuum off or remove the vacuum hose from the vacuum tool and place the filter tips into the Pyrosequencing plate containing the Pyrosequencing primer/annealing buffer mix. Residual vacuum will cause the primer mix to be sucked up through the tips so ensure it is fully off. Gently rock the tips in the wells to disperse the PCR product.
10. Place the Pyrosequencing plate onto a heating block at 80°C for 2 min. Ensure the plate sits on the Pyrosequencing plate adaptor with the corresponding lid (or "iron") placed over the plate to prevent evaporation. After 2 min, remove from heating block and place on a bench surface to cool. Once the plate is cool to the touch, cover with an adhesive seal (unless it will be run within 10 to 15 min) to prevent evaporation. If evaporation has occurred, adding 12 μL of annealing buffer will rescue the plate. Covering the plate while it is too hot will cause condensation on the lid, which can lead to cross-contamination of the wells.
11. Processed plates can be stored at 4°C until needed.

3.7. Pyrosequencing

3.7.1. Entering Assay Details

1. Open the Pyrosequencing software. A user name and password is required. This is usually set up with instrument installation. Individual or group-wide passwords can be used.
2. If the assay is not already entered into the software, on the left of the screen click "simplex entry" (**Note 17**). In the menu tree to the right of the simplex entry icon, scroll to the top, right click over "simplex entry," and select "new entry."
3. The required fields are a unique name for the assay (usually gene/SNP name) and a sequence to analyze (**Table 1**). Usually, 5 to 6 bases after the SNP position provides enough information for the assay. SNPs should be denoted as, for example, T/C (triallelic or tetra-allelic SNPs can also be entered, e.g., G/A/T or G/A/T/C) and indels as, for example, [GATC]. Short repeats should be entered as a series of indels, for instance, [TA][TA][TA]. Clicking "dispensation order" will automatically generate the least amount of nucleotide dispensations required for optimum genotype information. The dispensation order can be

edited manually by typing in the dispensation order field, which is useful for troubleshooting problem assays.

4. Select "show histograms" and the predicted pyrogram pattern will be displayed on the right. The default screens show both homozygous patterns and the heterozygous pattern. It is possible to scroll through histograms on the lower panel, useful if multiplex of multiple indels are to be analyzed, and so forth. Selecting individual or all predicted histograms on the box below the dispensation order and clicking "export" opens the histograms in a browser window where they can be printed or saved.
5. Click "save." At this stage, the parameters can no longer be altered, a duplicate setup with a unique name will need to be created for any alterations to the assay.

3.7.2. Entering an SNP Run

1. Select the "SNP run" icon on the far left of the screen.
2. On the menu tree, right-click over "SNP run" and select "new SNP run" (**Note 18**).
3. The essential parameters on the setup tab are a unique run name (e.g., gene/SNP/sample set/date) and the active well map. The default plate map is for a full 96-well plate. Individual wells can be selected (hold down control for nonadjacent wells), clicking the "activate wells" button will gray-out unused wells. In addition, instrument parameters must be selected from the drop-down menu. Usually, "instrument parameters" is a default file; however, care should be taken to ensure the appropriate parameters are selected for nucleotide or capillary dispensing tips, as they are not interchangeable. Parameter set-up instructions can be found with the dispensing tip packaging.
4. The essential parameters on the setup tab are to select the SNP assay by clicking on the drop-down menu under "simplex" and selecting the assay name entered in **Section 3.7.1** and to fill the plate map by clicking and dragging over the active (white) wells (**Note 19**).
5. Once the run has been set up, click "save." This can be edited post-save, and changes can be re-saved.
6. If multiple plates of the same assay are to be run, on the menu tree right click over the SNP run you have just entered and select "duplicate SNP run." The only parameter necessary is a unique run name.

3.7.3. Individual Plate Run for PSQhs96 and PSQhs96A

1. On the SNP run setup page described under **Section 3.7.2**, click the "view" tab and select "run." This will list the appropriate volumes of nucleotides, enzyme, and reagent needed for the individual run.
2. Set up the cartridge holder as shown in **Fig. 3**. It is essential that all nucleotide/capillary and reagent tips are clean before use. To check for blockages in the nucleotide and reagent tips, fill with 18.2 mΩ water and

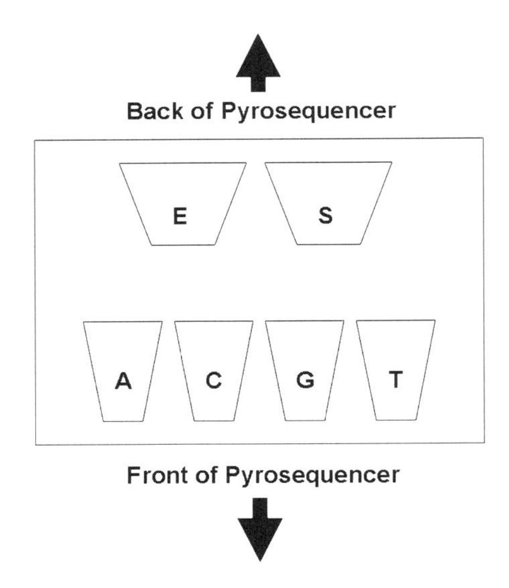

Fig. 3. Reagent and nucleotide cartridge orientation. E, enzyme; S, substrate; A, C, G, and T, nucleotides. A modification of dATP is used to prevent the nucleotide from being a direct source for the oxyluciferase. (Reprinted from *Methods in Molecular Biology*, vol. 311, Marsh S. et al.: *Pyrosequencing of Clinically Relevant Polymorphisms*, p. 108, 2005, with permission from the Humana Press Inc.).

apply pressure over the top of the tip. Water should squirt from the bottom of the tip. If this does not occur, try filling/emptying the tip several times with water and re-try forcing liquid through. If the tip remains blocked, discard. For nucleotide dispensing tips, do *not* force water through them. The hydrophobic disks may dislodge and prevent the tip from functioning. Rather, ensure the tip has been rinsed several times in water and has been stored in a clean, lint-free environment (**Note 20**).

3. Nucleotides, enzyme, and substrate are sold as a reagent kit. Each vial is clearly labeled. Nucleotides come as a solution, enzyme and reagent are lyophilized and should be resuspended with 18.2 mΩ water before use, and the volumes vary per kit and are clearly marked on the labels. The enzyme and substrate both dissolve rapidly and no mixing or shaking is required. Indeed, this should be avoided as air bubbles in the liquid could cause tip blockages or inconsistent dispensation. Unused resuspended enzyme and substrate can be stored at –20°C for future use.
4. If using the nucleotide dispensing tips, the nucleotides should be microfuged for 10 min and care should be taken to not aliquot from the bottom of the vial in case any precipitate is present, which could cause tip blockage. For all dispensing tips, it is recommended that non-barrier pipette tips are used as fibers can cause tip blockage.

5. If the capillary dispensing tips are used, the nucleotides should be diluted 1:1 with Tris-EDTA buffer (pH 8.0), and mixed well before use.
6. The nucleotide and reagent dispensing tips should be filled according to the volumes suggested by the software. Nucleotide dispensing tips should be filled by *doubling* the amount suggested by the software. Care should be taken not to pipette air bubbles and to gently angle the liquid down the sides of the tips. Capillary and reagent dispensing tips can allow minute air bubbles without affecting their performance. With nucleotide dispensing tips, it is extremely important to check all of the tips for air bubbles. These can usually be removed by gently tapping the sides of the tips until the air bubbles surface, or, if necessary, dislodging them with a clean pipette tip.
7. A test plate should be run after each cartridge refill. This is extremely important when using the nucleotide dispensing tips; and three or four test plates should be run in succession to ensure no blockages are present. The substrate reagent-dispensing tip is also prone to blockage if the substrate is allowed to sit in the tip at room temperature for any length of time. To run a test plate: place the cartridge in the pyrosequencer and the test plate in the 96-well plate platform. On the far left of the software screen, select the "instrument" tab, and then select "instrument" and "manage." Click "test." A warning will appear asking you to check that you have placed the test plate (**Note 21**) into the instrument. Click "ok." The test takes approximately 30 s. Remove the plate. In the center there should be six wells with liquid: four nucleotides, a reagent, and a substrate. If there are fewer than six wells with liquid, a blockage has occurred.
8. Remove the adhesive film carefully from the Pyrosequencing plate and place it in the pyrosequencer. Close all levers and click "run" on the plate run setup. The pyrosequencer will now automatically dispense enzyme, substrate, and nucleotides in the predetermined dispensation order. The progress of each individual well can be monitored at any time by selecting the relevant well on the 96-well plate map on the screen.
9. To automatically analyze the data once the run has completed, select "analyze all."

3.7.4. Batch Runs Using the PSQhs96A

1. SNP runs should be set up as described in **Section 3.7.2.**, saved, and closed.
2. Select the "Batch run" icon on the far left of the Pyrosequencing software, on the menu tree right click over "batch runs" and select "new batch run." One to 10 plates can be run in each batch. A unique name for each batch must be provided, and the instrument parameters must be selected for each batch. If barcoded plates are not used, uncheck the "barcode" field.
3. On the far left of the software, click on the "SNP runs" icon. From the menu tree, click and drag your SNP runs into the 1 to 10 slots on the batch window.
4. On the top menu bar, select "batch" and "setup information." This will open a browser window (may take a few seconds) with the total amount of nucleotides

(which should be *doubled* for the capillary dispensing tips), enzyme, and reagents needed for the entire batch.

5. The cartridge should be set up as described in **Section 3.7.3.** The dispensing tips should be cleaned between every batch, and a test plate should be run before every batch.
6. Remove the adhesive film from the Pyrosequencing plates and stack them (check that the plates can be lifted free without sticking to the lower plates, occasional warping may occur, causing plates to stick together, which jams the robotic arm). Place plates in the robot stacker unit. The correct plate orientation is shown on the top of the stacker unit. Ensure the plates lie flat on the base of the stacker unit and are between the grooves. Plate 1 on the batch setup should be on the top, plate 10 (or the last plate in the batch setup) should be on the bottom.
7. Ensure the stacker unit is firmly pushed into place. The nucleotides will not dispense if the unit is only partially home.
8. Click the "play" icon. Plates will automatically load and be discarded throughout the batch.
9. Plates will automatically be analyzed by the software when run in batch mode. They can be accessed from the batch setup window or from the individual SNP run files.

3.7.5. *Analysis of Pyrosequencing Results*

1. Once the Pyrosequencing run has been analyzed by the software, the 96-well plate map will be color-coded according to the result. Blue indicates a well in which the pyrogram matches one of the predicted histograms and a genotype can be accurately called. Orange indicates a possible match with a predicted histogram; however, human intervention is required to validate the call. Red indicates a failed well, where no match with a predicted histogram can be found.
2. The well(s) where no DNA was added in the PCR should automatically be scored failed (**Note 22**). Nonspecific peaks in the negative control(s) may be evident. These are likely to be caused by looping of the internal primer and can aid troubleshooting assays by identifying whether the internal primer is the culprit for background peaks.
3. Samples checked (orange) for human intervention can be edited by clicking on the specific well and opening up the predicted histograms from the "histogram" tab on the right. If a genotype consensus is reached, the sample call can be manually edited by right-clicking over the genotype above the pyrogram. Genotypes can be selected, and pass/check/fail can be altered. The well on the plate map will show a dark circle, indicating that manual editing has taken place.
4. The data can be exported as a report, as a tab delimited file, or as an XML file. Custom export options are also available. The export function can be accessed by selecting "report" and then saved as the appropriate file type. Selected wells

MLH1 Sequence to analyze: C/TGTTTGC/TGC/TGTTAGAGATC/TGTTGTTC/TG

Theoretical outcomes for 50% methylation: C/T C/T C/T C/T C/T

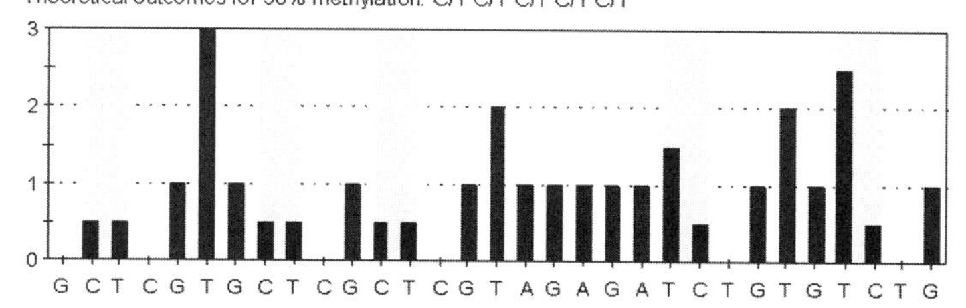

Experimental pyrogram for a methylated DNA sample:

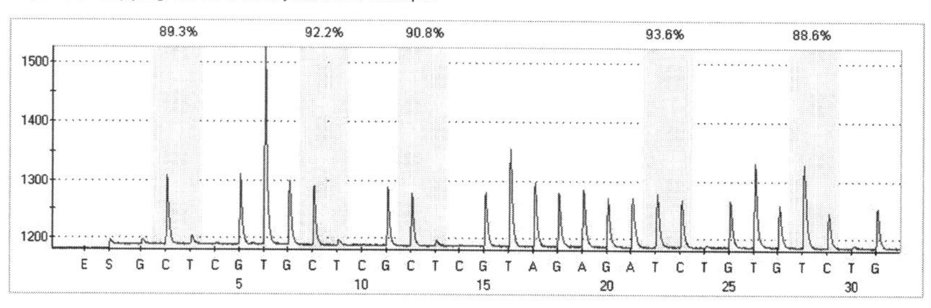

DPYD Sequence to analyze: TC/TGGGTTGC/TGGTTGTC/TGTGTTTGGC/TGC/TGGGAGTC/TGTAGGATC/TG

Theoretical histogram for 50% methylation C/T C/T C/T C/T C/T C/T:

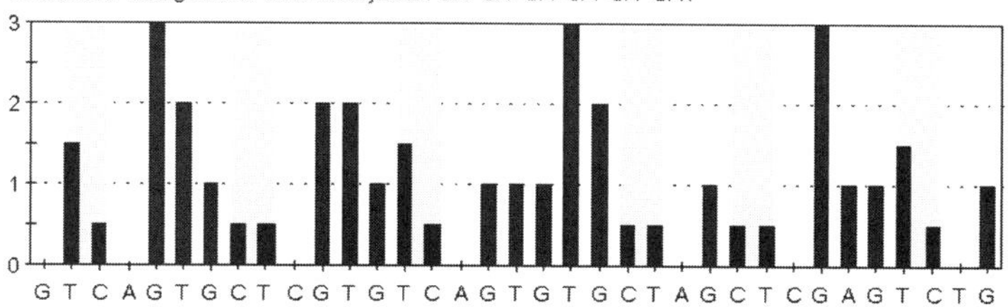

Experimental pyrogram for a methylated DNA sample:

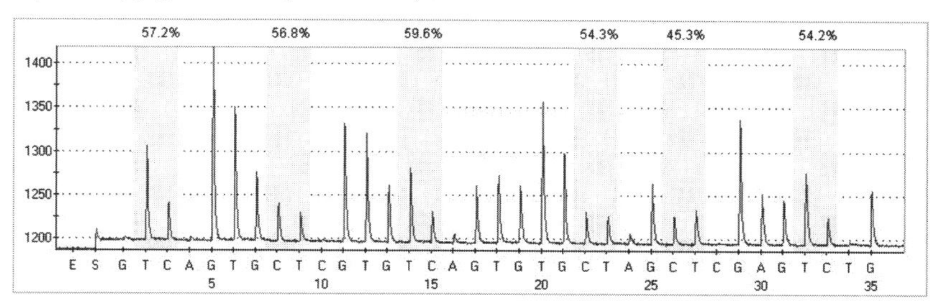

Fig. 4. Predicted histograms and typical pyrograms for MLH1 and DPYD promoter methylation assays.

or the entire plate can be saved/exported. Pyrograms (all or selected) can also be saved or printed, up to six per page (**Note 23**).

5. In the DNA methylation assay, the Pyrosequencing run should be analyzed by the Allele Quantification function within the software. The allele C and T peak percentage data can be exported along with the quality control marks (pass/check/fail) for each of the C/T loci (**Fig. 4** and **Note 24**). An average percentage of the C allele (i.e., C/T ratio) from three experiments can be used for the DNA methylation status at each CpG locus. For the overall DNA methylation status of a gene, the mean C/T ratio of all CpG loci tested for that gene can be calculated. When the mean C/T ratio is less than 0.05, it is defined as unmethylated (UM). Otherwise, 0.05 to 0.20 is denoted as low methylation (LM), 0.21 to 0.50 as medium methylation (MM), and greater than 0.50 as high methylation (HM). For example, in our study, overall there were 19 of 48 (39.6%) and 20 of 49 (40.8%) tumors that were defined as DNA-methylated (C/T ratio $\geqq 0.05$) in DPYD and MLH1, respectively. Among these methylated tumors, MM and HM were observed in 2 and 4 (12.5%) for DPYD and 5 and 9 (28.6%) for MLH1, and the methylation levels across multiple loci for these two genes was similar ($p > 0.1$) (i.e., so-called dichotomous trait) (**Table 2**).

Table 2
DNA Methylation Result of 52 Colorectal Cancer Patients

	CpG locus	No. of methylated DNA samples	Mean methylation level of methylated DNA samples	UM	LM	MM	HM
	1	13	0.37				
	2	18	0.29				
DPYD	3	22	0.35	29	13	2	4
(n = 48*)	4	24	0.24				
	5	13	0.29				
	6	13	0.35				
	1	15	0.56				
	2	23	0.44				
MLH1 (n = 49*)	3	15	0.60	29	6	5	9
	4	24	0.36				
	5	20	0.54				

UM, unmethylated; LM, low methylation; MM, medium methylation; HM, high methylation.

*Four and 3 samples failed in the DPYD and MLH1 assays, respectively.

4. NOTES

1. Pyrosequencing has been successfully performed on DNA from cell lines, blood, serum, plasma, paraffin-embedded tissue, frozen tissue, and whole-genome–amplified product. In addition, complementary DNA from various sources has also been successfully pyrosequenced.
2. The actual starting concentration of DNA depends on the quality of the template. DNA extracted from blood is highly accessible for PCR and consequently 0.5 to 1 ng can produce reliable, reproducible product. DNA from plasma, serum, frozen tissue, and whole-genome–amplified methods tend to be fragmented and more template may be necessary for optimum PCR. A test in advance of serial dilutions of the template DNA should be performed with the PCR primers to find the appropriate concentration that gives a clean, high-yield PCR product. In addition, for the quantitative methylation assay, the starting amount of the DNA should be larger than that for regular SNP assays. At least 10 ng of bisulfate-converted DNA should be used in the methylation PCR.
3. Pre-made mixes of buffer, magnesium, dNTPs, and Taq polymerase are recommended as they provide consistent results and minimize pipetting errors.
4. Non–hot-start Taq is also suitable; however, primer dimers are less of a problem with hot-start Taq and this is recommended.
5. All solutions should be made using 18.2 mΩ water. Solutions other than the NaOH and 70% ethanol should be sterile filtered prior to the addition of Tween-20. 10X washing buffer, annealing buffer, and NaOH can be made and stored at room temperature for dilution to the working concentrations. All solutions can be stored at room temperature.
6. Problem template loops will also be flagged in the Pyrosequencing primer design software.
7. This protocol is based on simplex assays; however, multiplexing with up to three internal primers can be performed, either from the same PCR product or from different PCR products. The primer design software can only determine one internal primer at a time; often the first-choice primers for each will not be useful in a multiplex assay where the combined sequence to analyze is best designed to generate unique SNP dispensations. In addition, the orientation of the primers is vital for multiplex assays as only one PCR primer can be biotinylated.
8. Critical scoring parameters on the Pyrosequencing primer report:

   Mispriming: If the internal primer can anneal to multiple positions within the amplicon, the 3′ ends of the annealed region can incorporate nucleotides leading to incorrect genotype calls or unacceptable background.

   Duplex Formation: If the internal primer can dimerize with itself, as for the mispriming, unacceptable background or reduced signal intensity owing to suboptimum primer annealing may result.

   Hairpin Loop: If the primer forms secondary structures, the amount of primer available for the reaction is diminished, and reduced signal can result.

Template Loop: Loops of more than approximately four to five GC-rich regions will be flagged by the software, and these primers should be avoided. Loops less than four bases should also be avoided if possible to reduce the likelihood of background.

Noncritical scoring parameters on the Pyrosequencing primer report:

Repeated Base at SNP Sequence: This reduces the score if there are a string of identical bases around the SNP. This is not something that can be controlled or optimized for as the SNP position is not movable. Typically, the pyrograms can accommodate as many as three bases in a row with no problems. Four to six bases may be difficult to read manually as the scale will be affected. More than six repeated bases are not recommended because distinguishing the peak heights becomes very difficult.

Primer Length: Primer lengths longer than 15 bases give a reduced score based on the expense of longer primers. The length of the primer is not critical to the reaction.

9. If an appropriate Pyrosequencing primer cannot be found as the critical scoring parameters are flagged, it is possible to "trick" the software to improve the search. As the software will only look five bases to either side of the SNP for a suitable primer, entering a fake SNP five bases before or after will extend the region searched. This may help to overcome mispriming and dimer problems. To eliminate template loops, adjusting the 5′ end of the PCR primer that would cause the loop will help (e.g., shifting the primer two to three bases to the left or right or trying a PCR primer in a slightly different region). As only one primer is likely to cause the loop problem, if a primer in the opposite orientation is available (even if not the highest score), this is often the easiest solution.
10. Pre-made PCR mixes are usually a fixed magnesium chloride concentration. If primer conditions are not optimized through temperature alone, extra magnesium chloride may be added to the PCR mix. In addition, problem assays may be improved by the addition of 5% to 10% dimethyl sulfoxide or 1 M Betaine. This will not affect the Pyrosequencing.
11. The mix is for 13 samples, allowing one extra sample for pipetting discrepancies.
12. If a larger volume of DNA is necessary, adjustments can be made to the PCR mastermix (reducing the water volume), or DNA may be dispensed into the plate and allowed to dry down overnight at room temperature. The DNA is reconstituted once the PCR mastermix is added.
13. For multiple primer sets/plate, at least one negative control/primer set should be included.
14. Fifty-five cycles are run to ensure all primers and nucleotides are exhausted and not available to cause background during the Pyrosequencing. If wide peaks occur in the program, reducing the number of cycles to 40 may help to prevent these.
15. Multiple assays can be run in a 96-well plate; indeed, each well could contain a different internal primer. The wells corresponding with the negative controls

from the PCR setup should contain internal primer, as this is a valuable troubleshooting method for program background issues.

16. A workstation platform is available from Pyrosequencing, which holds the reagent troughs and plates in specified positions. Any method to hold the reagent troughs stationary is appropriate, for example, rigid plastic tip box lids.
17. If a multiplex assay is to be set up, select the "multiplex entry" icon, right click over "multiplex entry" on the menu tree, and select "new entry." Type in the three separate dispensation orders for each internal primer. The computer-generated dispensation order will give a combined dispensation for the three SNPs. The field requirements here are the same as for the simplex entry except two or three sequences to analyze may be entered.
18. The menu tree for SNP runs can be organized into folders so multiple users can easily access their files. If this has been done, right click over the relevant folder and select "new run."
19. Each well can contain a different simplex/multiplex entry if desired, simply select the entry and click in the appropriate well until all active wells are filled.
20. Pyrosequencing provides specific storage boxes for the tips with the instrument, and more are available from the company if required.
21. To save on plate costs, attach adhesive film to the top of the test plate. The dispensation will occur on the film rather than in the wells, and this can be wiped off and the plate can be reused.
22. If multiple primer sets are used/plate, the negative controls for each primer set should be checked for contamination.
23. The report structure is available in forms readily transferable to most database/spreadsheet systems.
24. Pyrosequencing can generate quantitative data for DNA methylation level. However, the concept of the DNA methylation level can be confused in the literature, because the DNA methylation level may be interpreting either how many molecules with a gene sequence of interest in a DNA sample or how many CpG loci in a gene sequence for a certain DNA sample have been methylated. In fact, most DNA samples from tissue specimens have mixed genomes from a variety of cell types including cells not of interest to the study (e.g., stromal cells), and different cells or tissues can be methylated differentially at a specific CpG locus. The DNA mixture from different types of cells can mask the true level of DNA methylation at a specific CpG site for a specific type of cells. This is particularly important for quantitative DNA methylation studies; and laser capture microdissection and fluorescence-activated cell sorting can provide highly pure cells of specific types.

ACKNOWLEDGMENTS

The assistance of Derek Van Booven is greatly appreciated. This work is supported by the NIH Pharmacogenetics Research Network (U01 GM63340); http://pharmacogenetics.wustl.edu.

REFERENCES

1. Sachidanandam R, Weissman D, Schmidt SC, et al. A map of human genome sequence variation containing 1.42 million single nucleotide polymorphisms. Nature 2001;409:928–933.
2. Kwok PY. Methods for genotyping single nucleotide polymorphisms. Annu Rev Genomics Hum Genet 2001;2:235–258.
3. Fornage M, Doris PA. Single-nucleotide polymorphism genotyping for disease association studies. Methods Mol Med 2005;108:159–172.
4. Colella S, Shen L, Baggerly KA, Issa JP, Krahe R. Sensitive and quantitative universal Pyrosequencing methylation analysis of CpG sites. Biotechniques 2003;35:146–150.
5. Neve B, Froguel P, Corset L, Vaillant E, Vatin V, Boutin P. Rapid SNP allele frequency determination in genomic DNA pools by pyrosequencing. Biotechniques 2002;32:1138–1142.
6. Skarke C, Kirchhof A, Geisslinger G, Lotsch J. Rapid genotyping for relevant CYP1A2 alleles by pyrosequencing. Eur J Clin Pharmacol 2005;61:887–892.
7. Garsa AA, McLeod HL, Marsh S. CYP3A4 and CYP3A5 genotyping by Pyrosequencing. BMC Med Genet 2005;6:19–23.
8. Ahluwalia R, Freimuth R, McLeod HL, Marsh S. Use of pyrosequencing to detect clinically relevant polymorphisms in dihydropyrimidine dehydrogenase. Clin Chem 2003;49:1661–1664.
9. Tost J, Dunker J, Gut IG. Analysis and quantification of multiple methylation variable positions in CpG islands by Pyrosequencing. Biotechniques 2003;35:152–156.
10. Brakensiek K, Wingen LU, Langer F, Kreipe H, Lehmann U. Quantitative high-resolution CpG island mapping with pyrosequencing™ reveals disease-specific methylation patterns of the CDKN2B gene in myelodysplastic syndrome and myeloid leukemia. Clin Chem 2007;53:17–23.
11. Rose CM, Marsh S, Ameyaw MM, McLeod HL. Pharmacogenetic analysis of clinically relevant genetic polymorphisms. Methods Mol Med 2003;85: 225–237.
12. Horowitz N, Pinto K, Mutch DG, et al. Microsatellite instability, MLH1 promoter methylation, and loss of mismatch repair in endometrial cancer and concomitant atypical hyperplasia. Gynecol Oncol 2002;86:62–68.

7

The Application of Laser Capture Microdissection for the Analysis of Cell-Type-Specific Gene Expression in a Complex Tissue: *The Primate Endometrium*

William C. Okulicz

Summary

There are a number of approaches that have been used in the past to describe and analyze the hormonal influences and mechanisms that govern primate endometrial responses as well as the regulation of cellular responses in many complex heterogeneous tissues. These have included gross anatomy, morphology/histology, hormonal manipulation, steroid binding assays, and immunohistochemical and *in situ* hybridization analyses. Transmission and scanning electron microscopy as well as ultrasound have also been used to study uterine (endometrial) structure. Recently, a number of new and powerful cellular and molecular techniques have been added to our arsenal of approaches to study tissue and cell-type gene and protein expression. These include quantitative (real-time) PCR analysis of gene expression, differential display, gene microarray analyses, proteomic analysis, and laser capture microdissection (LCM). LCM is a new technology that has substantially expanded our ability to examine cell-type–specific and region-specific responses in complex cellular tissues. Importantly, this technique allows the retrieval of specific cells from specific morphologic units within the tissue of interest.

The primate endometrium is an example of a complex heterogeneous tissue that requires proper cellular maturation/differentiation to achieve a hospitable environment for embryo implantation. The genes and gene networks that are involved in this process are likely to be regulated in a temporal, spatial, and cell-type–specific context within this tissue. As an example, we present some of our recent work on the application of LCM in our studies on the hormonal regulation of gene expression in this complex tissue.

Key Words: cell-type–specific expression; differential display; endometrium; gene expression; gene microarrays; implantation; laser capture microdissection (LCM); progesterone; rhesus monkey

From: *Methods in Pharmacology and Toxicology: Biomarker Methods in Drug Discovery and Development*
Edited by: F. Wang © Humana Press, Totowa, NJ

1. INTRODUCTION

The changing pattern of estradiol (E) and progesterone (P) secretion during the primate menstrual cycle governs the hormonal regulation of endometrial growth, differentiation, shedding, and reconstruction that is an essential component of continued reproductive competence *(1,2)*. The focus of our laboratory has been on P-dependent regulation of endometrial response in the rhesus monkey, an appropriate model for human endometrial function. P is expected to regulate a wide variety of endometrial genes that include growth factors and their receptors, extracellular matrix proteins, and enzymes involved in cellular metabolism *(3–13)*. These features of P action most likely involve a cascade of signal transduction pathways that control the global maturation of the endometrium. One important mechanism that regulates the factors in such pathways is the activation or repression of their respective gene products at the transcriptional/translational level; a fate ultimately determined by specific promoter-binding transcription factors and associated coactivators/repressors including interference RNA (iRNA).

1.1. Structure of the Primate Endometrium

The rhesus endometrium has been characterized by Bartelmez *(3)*, using histologic criteria, as composed of four horizontal zones (**Fig. 1**): the transient functionalis is composed of zone I, the luminal epithelia and densely packed stroma, and zone II, the upper third segment of the glands; the germinal basalis is composed of zone III, the middle third of the glands, and zone IV, the deepest portion of the glands adjacent to the myometrium. A similar zonation of the endometrium is also apparent in the human *(4)*.

One of the most striking characteristics of the primate endometrium is its remarkable regenerative capacity *(14)*. Hartman *(15)* showed that in the rhesus monkey, the endometrium could regenerate completely after an endometrectomy that would leave only a few endometrial cells on the surface of the myometrium. This regenerative capacity of the endometrium is perhaps not surprising because of the central role it plays in menstruating primate reproduction. This feature of the primate endometrium has allowed us to use these very valuable animals several times depending on their health and an appropriate postsurgical recovery period.

1.2. Cell Type and Spatiotemporal Regulation of the Endometrium

In addition to the morphologic zonation described above, the endometrium's complexity is further defined by the number of different cell types that it harbors within these zones. These cell types include luminal

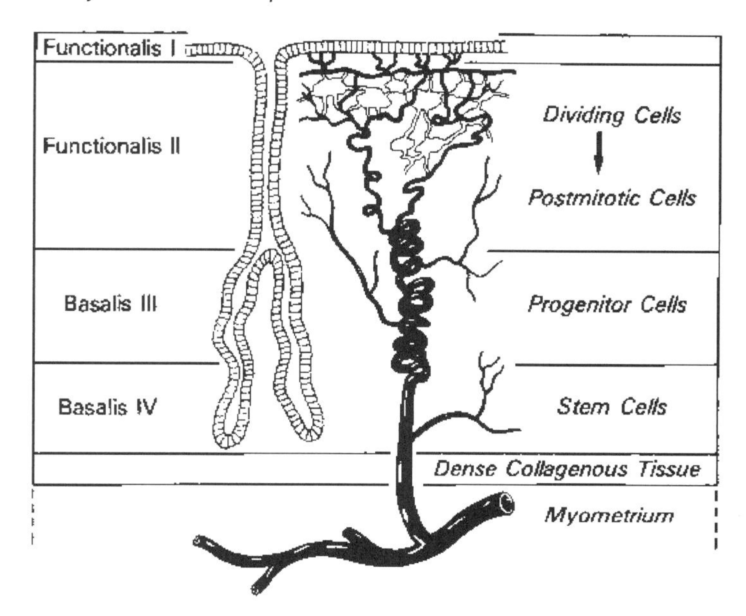

Fig. 1. Zonation of the primate endometrium *(24,68)*.

and glandular epithelia, stromal fibroblasts, vascular smooth muscle cells, endothelial cells, and cells of the lymphocytic system.

It has become increasingly clear that different cells or cell types within a target tissue can respond dissimilarly to the same hormonal milieu. In our laboratory *(16,17)* and others *(19–21)*, it has been shown by immunohistochemical techniques that P inhibition of nuclear estrogen receptor (ER) is most pronounced in zones I, II, and III while strong positive staining of glandular epithelia in zone IV is retained. Stromal cells in zones I, II, and III are also more rapidly affected by P demonstrating a temporal regulation of this inhibition. A recent study in the human has also shown that stromal cells are more sensitive to P downregulation of ER *(22)*. It has also been shown that there are zonal-dependent differences in proliferation during both E- or P-dominance during both natural menstrual cycles and artificial menstrual cycles in the rhesus monkey *(23–26)*. These studies and others *(27–29)* strongly support the concept that the primate endometrium contains distinctive microenvironments that can respond differentially to the same hormonal stimulation, and different cell types within these regions of the endometrium are also differentially responsive. This important concept is essential to the understanding of responses to stimuli in complex tissues such as the endometrium.

2. ANIMAL MODEL

The development and use of artificial menstrual cycles in the rhesus monkey was first described by Hodgen ***(30)***. These studies showed that simulation of the menstrual cycle by the timed insertion and removal of silastic implants of estradiol (E) or progesterone (P) was sufficient to allow the endometrium to support implantation and eventual delivery (IVF and surrogate transfer). Luteal-phase defects in women are purported to be the most common endocrinopathy in infertility and recurrent abortion wherein low secretory P levels are not sufficiently elevated to achieve appropriate endometrial maturation ***(2,31–34)***. Both short and inadequate luteal phases similar to those found in women have been described in the rhesus monkey ***(35)***. These latter studies also provide support for the usefulness of the rhesus monkey as a model for luteal phase defects in women ***(32)*** wherein low secretory P levels lead to retarded endometrial maturation. We have created our inadequate secretory phase P levels in accord with the levels determined to be inadequate based on these previous studies. Our previously published studies ***(16,25,36–38)*** describe in detail the protocols for creation of adequate and inadequate cycles. These studies showed that the hormone levels produced by these protocols are coincident with those observed in the natural menstrual cycle ***(36)*** as well as inadequate secretory phases ***(32)***. The profiles of serum E and P observed using the above protocols are shown in **Fig. 2** for adequate (A) and inadequate (B) secretory phases.

3. LASER CAPTURE MICRODISSECTION

Traditionally, cell-type–specific and region-specific regulation of genes and gene products has been analyzed using *in situ* hybridization and immunohistochemistry, respectively. Although these techniques remain important tools in our experimental arsenal, a new technology, laser capture microdissection (LCM), initiated by the National Institutes of Health (NIH), has substantially expanded our ability to examine cell-type–specific and region-specific responses ***(39,40)***. Microdissection using this technique allows the retrieval of specific cells from specific morphologic units within the tissue of interest. Material harvested using this approach can subsequently be used for a variety of genetic or proteomic analyses (e.g., Refs. ***(39,41–44)***).

Since the introduction of LCM technology, laboratories in a wide variety of scientific areas have capitalized on its power to harvest specific cell types within their tissue of interest. The heterogeneity of most tissues (e.g., cell types and regional areas) makes this technology particularly amenable to many studies. Perhaps not surprisingly, some of the first applications of

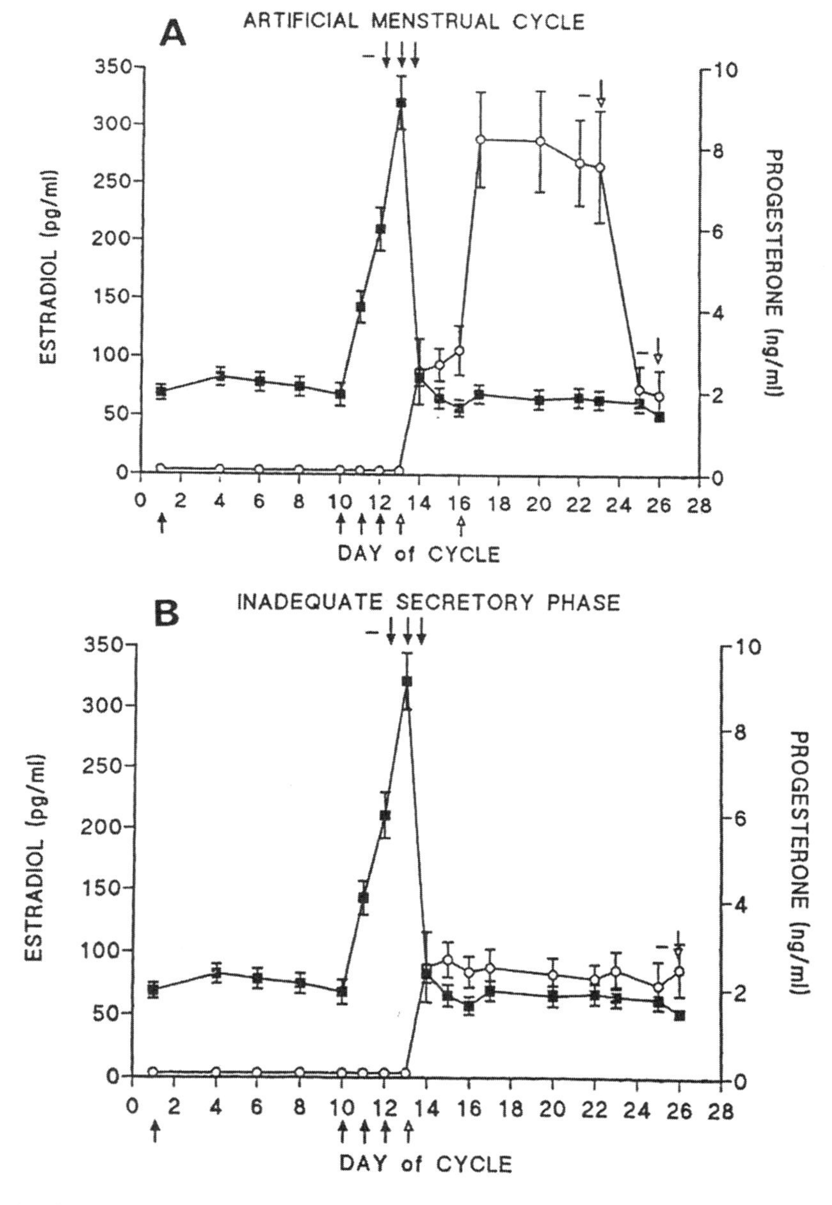

Fig. 2. Serum E and P levels during simulated **(A)** adequate or **(B)** inadequate menstrual cycles in the rhesus monkey. Closed squares and closed arrowheads represent the mean ± SEM (n = 6 to 8) of serum E levels and implant insertion and removal, respectively. Open circles and open arrowheads represent the mean ± SEM (n = 6 to 8) of serum P levels and implant insertion and removal, respectively ***(38,68,95)***.

LCM were in the area of molecular diagnostics ***(41,45,46)*** particularly in cancers because the cancer cells could constitute less than 5% if the total tissue and render analyses of whole tissue confusing or uninformative. Some of the initial studies focused on cancerous lesions ***(47)*** such as the prostate ***(45,48,49)***, breast ***(50)***, colon ***(51)***, lung ***(52)***, and follicular lymphoma ***(53)*** as examples. Other applications have included studies on spermatogenesis and the development of the fetal testis ***(54,55)*** as well as in neuroscience research (e.g., Refs. ***(56–58)***). Although studies on mammalian systems have dominated the literature, there have also been investigations on selective retrieval of plant cell types/tissues using this technology ***(59,60)***.

Our laboratory has used LCM to study the primate endometrium, a complex heterogeneous structure (see above) whose components are difficult at best or impossible to study in isolation. For example, separation of endometrial stroma from epithelia has been shown to dramatically alter epithelial response to hormonal signals ***(61)***. Microdissection can overcome limitations of traditional means of analysis and allow the application of powerful molecular methods of analysis on specific cell types within specific morphologic units in the tissue of interest ***(42)***. Our studies described herein will be used as an example of the power and application of this technology. We have used this approach in combination with PCR analysis, differential display and microarray analysis to study differential gene expression in different cell types within different regions of the primate endometrium ***(27–29,62)***.

3.1. Method of LCM Analysis of Endometrial Tissue

Details of our LCM procedures can be found in our previous publications (see above). We provide below the steps we have used to obtain appropriate cellular targets from primate endometrial tissue. Additional information on procedures and Arcturus LCM equipment can be found at www.moleculardevices.com.

Procedure

1. Endometrial tissue is obtained at laparotomy by endometrectomy from rhesus monkeys.
2. The tissue is oriented in a an ice-cooled, small aluminum foil cup and frozen immediately in Tissue Tek OCT (Fisher Scientific, Pittsburgh, PA) embedding compound and stored desiccated at –80°C prior to further processing.
   (*Note*: It is important to orient the tissue according to the plane of interest subsequent to cryostat sectioning. For example, we orient the tissue so the myometrium will be part of our sections in order to clearly establish different zones/regions within the endometrium. Dessication of the samples can limit ice

formation upon frequent opening and closing of the freezer and the resulting condensation that can affect tissue integrity.)

3. After mounting the tissue block, we obtain cryostat (–25°C) sections (6 μm) to determine proper tissue orientation microscopically using Toluidine Blue O (0.5% aqueous) staining (5 to 10 s, water wash). Once proper orientation or reorientation of the tissue has been established, tissue sections for LCM [cryostat (–25°C), 6 μm] are placed on untreated plain glass slides and immediately fixed in 70% ethanol (10 min).
   (*Note*: Slides are precooled to cryostat temperature prior to sectioning and very briefly mounted (finger warmth) within the cryostat. This procedure minimizes the condensation that can occur when sections are mounted outside of the cryostat. In addition, the fixation in 70% ethanol is also done within the cryostat to also limit water condensation.)
4. Slides are air-dried for 20 min to promote adhesion of the section.
   (*Note*: We have also used a desiccator to promote adhesion when humidity is elevated.)
5. Slides are stained with hematoxylin and eosin using the following sequential solutions (10 to 30 s each): Mayers hematoxylin, distilled water, bluing reagent, 70% ethanol, 95% ethanol, and Eosin Y.
6. Slides are then dehydrated with two 10-s washes using 95% ethanol, and two 10-s washes using 100% ethanol. Finally, the slides are placed in xylene twice for at least 5 min and then dried in a vacuum desiccator for 15 to 20 min prior to LCM.
   (*Note*: A fresh bottle of 100% ethanol and fresh xylene should be used routinely.)
7. Luminal or glandular epithelia or stromal cells from appropriate regions of the endometrium are identified morphologically and harvested using a PixCell II LCM System (Arcturus Engineering, Mountain View, CA; now distributed by Molecular Devices, Inc., Sunnyvale, CA).
   (*Note*: The 10× objective is used for tissue orientation and the 20× and 40× for specific cell-type harvesting depending on the application. Photographic documentation prior to and after microdissection of the cells and region of interest should be done routinely.)
8. The 15- or 30-μm-diameter beams are used depending on the application. Amplitude and pulse duration ranged from 35 to 50 mW and 3 to 5 ms, respectively.
9. Cells are collected on TF-100 caps (Arcturus Engineering) containing the transfer film from a minimum of two to three sections for each sample. Tissue samples from at least three different animals are analyzed individually or pooled for subsequent analyses where appropriate depending on the application.
10. Excess cellular/tissue debris, not part of the laser etched area, is removed by a brief treatment with a CaptureSure Pad (Arcturus Engineering, Mountain View, CA).
   (*Note*: The removal of this material can be visualized microscopically and documented photographically if required.)

11. The TF-100 caps are placed on ice-cooled Microfuge (Arcturus Engineering, Mountain View, CA) tubes of the appropriate size that contained the RNAqueous lysis/binding buffer (100 μL) (RNAqueous-4PCR Kit; Ambion, Austin, TX).
12. The tubes are inverted several times to remove the captured tissue form the transfer cap, vortexed, briefly centrifuged, and processed for RNA using the RNAqueous protocol (see below).

Four different cDNA populations were prepared from endometrial tissue harvested by LCM from adequate secretory cycles (days 21 to 23). These populations were glandular epithelia and stroma from the functionalis (FG and FS, respectively) and glandular epithelia or stroma from the basalis (BG and BS, respectively) ***(25–27)***. One of the first objectives of our studies was to assess the quality and potential usefulness of endometrial tissue harvested in this manner for subsequent gene expression studies. There are numerous steps in the preparation of suitable genetic material from laser microdissected tissue any one of which could compromise the quality of a sample ***(39)***. Because of the time and effort that is required for an analysis of gene expression using this approach, it is useful to have some guide to the relative quality of a sample.

3.2. Synthesis and Amplification of cDNA Populations

Tissue limitations for our laboratory and others may not allow traditional means of analysis of RNA integrity (e.g., relative size of the mRNA population by agarose gel electrophoresis). In an effort to overcome this drawback we have used an adapter-specific primer amplification approach to allow visualization of a cDNA smear [see below and ***(27,28,63)*** for further details]. This approach coupled with the detection of an appropriate housekeeping gene(s) can serve as a useful guide to estimate sample quality for those investigators faced with limited tissue/cells. The above approach also provides considerable material (cDNA) from a single round of amplification (approximately 75-fold) that will subsequently allow a number of comparative studies on gene expression to be performed ***(27)***.

RNA extraction was performed using the RNAqueous-4PCR Kit (Ambion). The LCM transfer caps were placed in Brinkmann microcentrifuge tubes with RNAqueous lysis/binding solution, which contained guanidinium thiocyanate. After vortexing and centrifuging the tubes, the caps were removed and RNA was isolated using the RNAqueous protocol. The DNAse I treatment that is part of this protocol has been shown previously to effectively remove genomic DNA. The Superscript Choice System (Life Technologies, Rockville, MD), was used for first strand cDNA synthesis using a mixture of both oligo(dT) and random hexamer primers according to the manufacturer's protocol. Second strand cDNA synthesis and adaptor ligation with EcoRI (*Not* I) adapters were performed using the same kit. cDNA populations were purified in Qiaquick spin columns and amplified by PCR in 100 μL containing 0.5-μm LINK-CUA primer, 0.25 mM dNTPs, 1.5 mM $MgCl_2$, 1x buffer, and 2 units Taq polymerase in a thermal cycler (94°C, 1 min; 50°C, 1 min; 72°C, 2 min) for 30 cycles. The LINK-CUA primer (5′-CUACUACUACUAAATTCGCGGCCGCGTCGAC-3′) is complementary to the EcoRI adaptor. After amplification of the cDNA populations, one-fiftieth (2 μL or 40 ng) was used as template in PCR reactions using primers specific for human 18S ribosomal RNA (Ambion) and G3PDH (Clontech, Palo Alto, CA). The results obtained with 18S ribosomal RNA were used to normalize the cDNA populations. Details of the above approach can be found in our previous publications ***(25,26)***.

In addition to cDNA smears, the relative size range of our cDNA populations was also estimated using the above primers for both 18S ribosomal RNA (product, 324 bp) and G3PDH (product, 983 bp). Our data suggested that the relative size range of a cDNA smear can influence the subsequent detection of these housekeeping genes (see below) ***(27,28)***.

3.3. Results with cDNA Amplification Strategy

With the use of primers for a housekeeping gene, it would be expected that its presence at the correct fragment size would provide evidence for the suitability of the genetic material for further analysis. Whereas this may in part be true, the absence of an appropriate PCR product could mean either that the preparation is poor (e.g., degraded) or that there is insufficient material for detection (false negative). Our data for 18S ribosomal RNA expression prior to and after amplification show that a false negative can be detected. For example, a detectable band of the correct size is only apparent following amplification for either 30 or 40 cycles (FS+ vs FS–) (**Fig. 3A,B**) ***(26)***.

The use of this amplification strategy cannot only allow detection of a false-negative result but also can provide some additional information with regard to the quality of the genetic material through cDNA smears. For example, the cDNA populations prior to or after amplification for glands from the functionalis did not show an 18S ribosomal RNA band (324 bp) despite amplification (FG+, **Fig. 3B**). These data would suggest that the quality of this material rather than the quantity is most likely at fault. The very low molecular weight smear of this cDNA population correlates with this result (**Fig. 3C**). cDNA smears after amplification of these laser microdissected samples can also be useful in the design of an appropriate expected fragment size for a given housekeeping gene or other gene of interest.

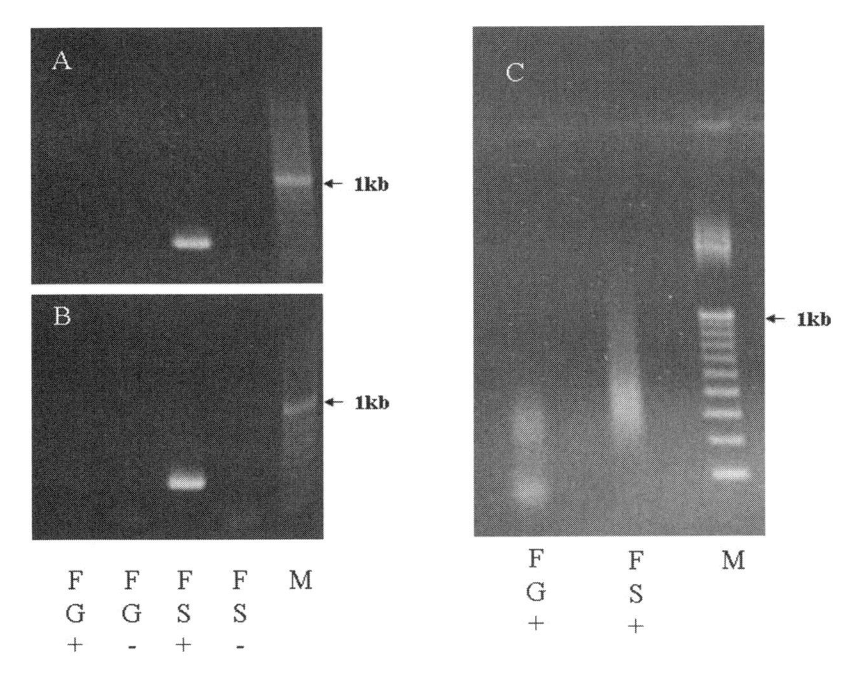

Fig. 3. Detection of 18S ribosomal RNA expression before and after amplification of cDNA. **(A)** Results after 30 cycles of PCR, **(B)** results after 40 cycles of PCR, **(C)** corresponding cDNA smears after amplification of adaptor-ligated cDNA *(25)*. "+" signs indicate cDNA that has been amplified after adaptor ligation, and "–" signs indicate double-stranded cDNA before amplification. FG, cDNA from functionalis glandular epithelia; FS, cDNA from functionalis stromal cells. A 100-bp ladder is shown with 1 kb band indicated by an arrow *(26)*.

4. DIFFERENTIAL DISPLAY

Differential display is an approach for the identification of mRNAs that show differences in expression level between two or more experimental groups or that are unique to a cell type, tissue or developmental stage *(64,65)*. The method allows random cDNAs to be amplified by a pair of short arbitrary PCR primers (10-mer) *(66,67)*. The sequence of the primers dictates which panel of cDNA fragments of the total will be amplified: only those gene fragments containing sequences complementary to the primers will be amplified. The use of different primer sets will result in different patterns of gene fragments that are amplified, and we have used this method to broaden the scope of genes to be analyzed. Because the fragments are small (approximately 400 bp), they can be cloned and quickly sequenced and subsequently compared by homology to GenBank database entries.

4.1. Method of Differential Display, Cloning, and Sequencing

DDRT-PCR was performed using the RNAimage kit (GenHunter, Nashville, TN). Two nanograms of cDNA is amplified in 20 μL reactions containing 1X buffer, 2.0 μM dNTPs, 20 Ci/mmol alpha-[^{33}P]ATP, 0.2 μM $HT_{11}A$ primer (5'-AAGCTTTTTTTTTTTA-3'), 0.2 μM H-AP1 primer (5'-AAGCTTGATTGCC-'"), and 0.05 units Taq polymerase (Qiagen, Valencia, CA). Reactions were carried out in a PTC-200 thermal cycler (MJ Research, Waltham, MA) at 94°C/1 min, 40°C/2 min, and 72°C/1 min for 40 cycles and analyzed by denaturing polyacrylamide gel electrophoresis, omitting the fixing stage. The autoradiogram and gel were aligned by needle punctures, and individual bands were carefully excised from the gel with a razor blade. Gel slices attached to filter paper were eluted by boiling in 100 μL water for 10 min, spun to remove debris, and the supernatant precipitated with glycogen. DNA fragments thus isolated were reamplified as described above except with 250 μM dNTPs in the absence of radiolabel. Products were directly cloned into the plasmid vector pCR 2.1-TOPO^R (Invitrogen, Carlsbad, CA) and sequenced (UMass Medical School Nucleic Acid Facility). Homology searches were performed against GenBank entries using BLAST programs (NCBI). Alternatively, the following arbitrary primers were used to provide additional patterns of gene expression:

Arbitrary Primers

ARB-1: CTGATCCATG
ARB-2: CTTGATTGCC
ARB-4: GTTGCGATCC
ARB-5: GACCGCTTGT
ARB-7: CTTTGGTCAG
ARB-8: CAAGCGAGGT
ARB-9: AACGCGCAAC

4.2. Differential Display Analysis During the Window of Receptivity

The different patterns (bands) of expression that are observed after electrophoresis and autoradiography of the radioactive fragments ***(67)*** must be subsequently confirmed by PCR analysis (false positives are common) and, if appropriate, sequenced. We had previously hypothesized different

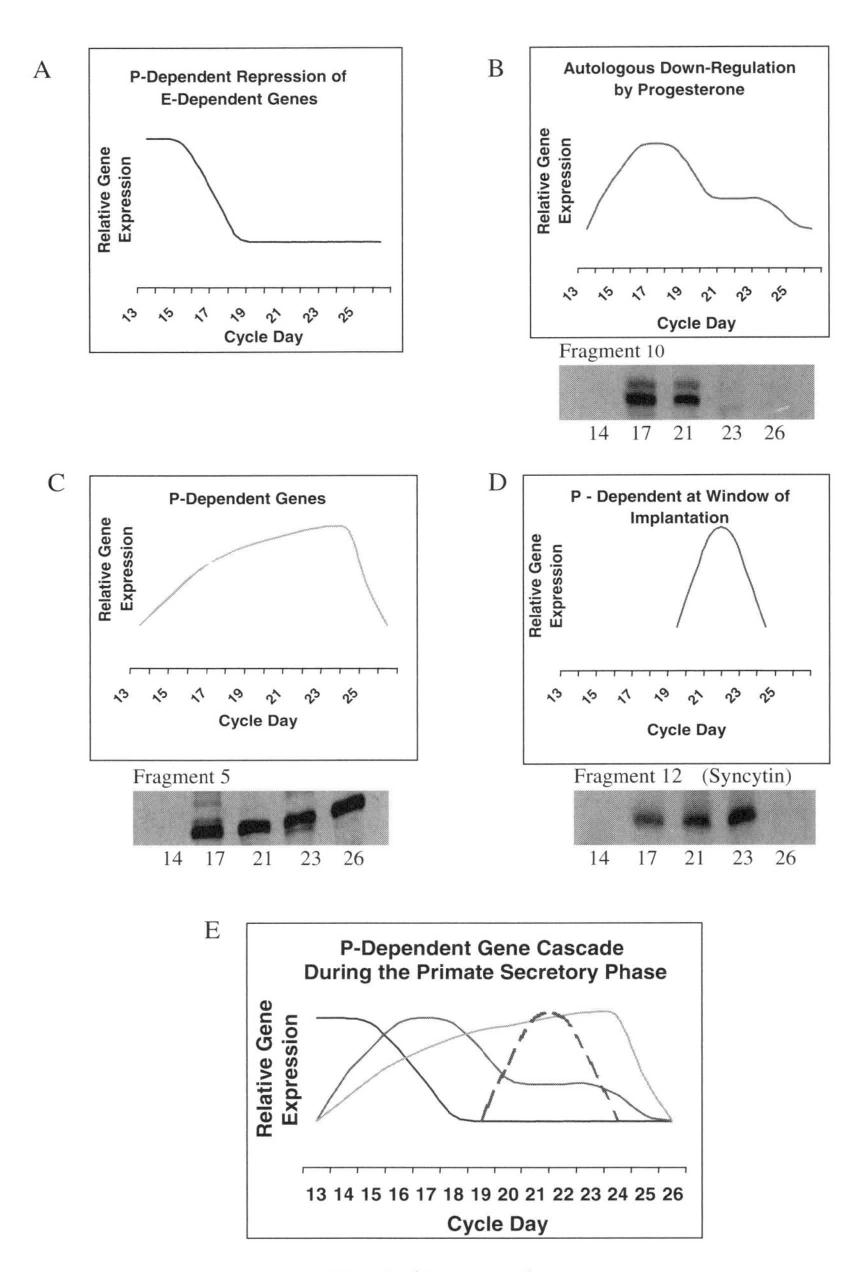

Fig. 4. (Continued)

patterns of gene expression during the secretory phase in the rhesus endometrium based on our work and other studies in the literature ***(68)***. The results of our studies (using whole endometria) are shown in **Fig. 4**, and the hypothetical patterns of expression are described in the legend (see also the composite panel E) ***(69)***. Of particular interest are the fragments that displayed a restricted expression during the expected window of receptivity (panel D). Two fragments exhibited high homology to previously characterized human genes, syncytin (the envelope gene) and secretory leukocyte protease inhibitor (SLPI). A third fragment, similar to BAT2 (KIAA1096), also displayed a restricted pattern of expression during the window of receptivity (see below) ***(69)***.

Syncytin is a highly fusogenic membrane glycoprotein that appears to be expressed specifically in the placenta. The protein induces the formation of giant syncytia and can mediate fusion of cytotrophoblasts into the syncytiotrophoblast layer, which is essential for pregnancy maintenance ***(70)***. Endogenous viral syncytin appears to have been sequestered to serve an important physiologic role during pregnancy and placental morphogenesis ***(71)***. Our results are the first to describe the expression of this gene in the primate endometrium. Its potential role in the non–pregnant uterus, however, remains to be elucidated. We speculate that syncytin could be a P-induced endometrial decidualization factor (it induces syncytia, or multinucleated cells). Interestingly, the expression of HERV-K (closely related to HERV-W) is stimulated in cultured human tumor cells by sequential E and P treatment, most likely mediated by the presence of a P receptor binding site situated in its long terminal repeat (LTR) ***(72)***. Because differentiation of P-induced endometrial stromal cells into decidual cells is essential for embryo implantation and placentation, expression of syncytin during the window of receptivity of the primate menstrual cycle is likely to play an important role in preparation of the endometrium for successful implantation.

SLPI is a neutrophil elastase inhibitor that also has antibacterial and anti-inflammatory properties. In addition to its antiprotease activity, secretory

Fig. 4. A schematic representation together with examples of relative gene expression patterns in the rhesus endometrium during the secretory phase ***(69)***. **(A)** Progesterone (P) repression of estradiol (E)-dependent genes. **(B)** Autologous downregulation of P-dependent genes (e.g., fragment 10). **(C)** P-induction of genes during the secretory phase (e.g., fragment 5). **(D)** P-induction of gene expression during the window of receptivity (e.g., fragment 12, syncytin). **(E)** A composite cascade of gene expression patterns **(A)** to **(D)**. The associated panel **(B–D)** for each fragment shows the relevant temporal portion of the gel after PCR analyses.

leukocyte protease inhibitor (SLPI) has been shown to regulate intracellular enzyme synthesis, epithelial cell growth by activation and repression of distinct growth-regulatory genes, and mediate normal wound healing *(73,74)*. King et al. *(75)* have shown by immunohistochemistry that SLPI is expressed in glandular epithelium of human endometrium from the mid to late secretory phase suggesting a direct or indirect regulation by progesterone. These studies are in agreement with our results that show increased expression of this gene from days 17 to 23 of adequate secretory phases in the rhesus monkey. Leukocytes infiltrate the endometrium prior to menstruation and may in part be responsible for the rise in secretory SLPI during the secretory phase. SLPI has been found to have antibacterial effects, and one region of the molecule has 37% homology with defensins, a family of antibacterial proteins *(75,76)*. SLPI also inhibits the NF kappa-B signal transduction pathway involved in inflammatory response *(77)*. The most likely role of SLPI is as a natural antibiotic and anti-inflammatory molecule. Infection ascending through the cervix could pose a threat to the implanting and developing conceptus *(75)*.

The third fragment initially represented a predicted human protein, KIAA1096, based on the open reading frame *(78)*, which was subsequently shown to be BAT2. BAT2 shares similarities with some transcriptional regulatory proteins containing zinc finger motifs and proline- or glutamine-rich regions *(79)*. As noted above, BAT2 also contains motifs typical of the integrin receptor family. Integrins (receptors) have been implicated in successful implantation in the human *(11)*. BAT2 has also been mapped within the class III region of the major histocompatibility complex (MHC), which encodes genes involved in immune function or MHC-associated disease susceptibility *(79)*.

4.3. Serum P Level and Gene Expression

An example of the relationship between the temporal change in serum P level (adequate cycle) and restricted gene expression is shown in **Fig. 5** for SLPI during the window of receptivity in an adequate secretory phase. Note that the pattern of rising and falling serum P is paralleled by the rise and fall of SLPI expression. Similar profiles for the genes (gene fragments) described above also displayed such a pattern (e.g., BAT2, syncytin, WFDC2). We have also studied the expression of these genes during inadequate secretory phases (cannot support implantation, see above) and have shown that SLPI, BAT2, and WFCD2 genes are strikingly underrepresented compared with their expression in adequate secretory phases (see below).

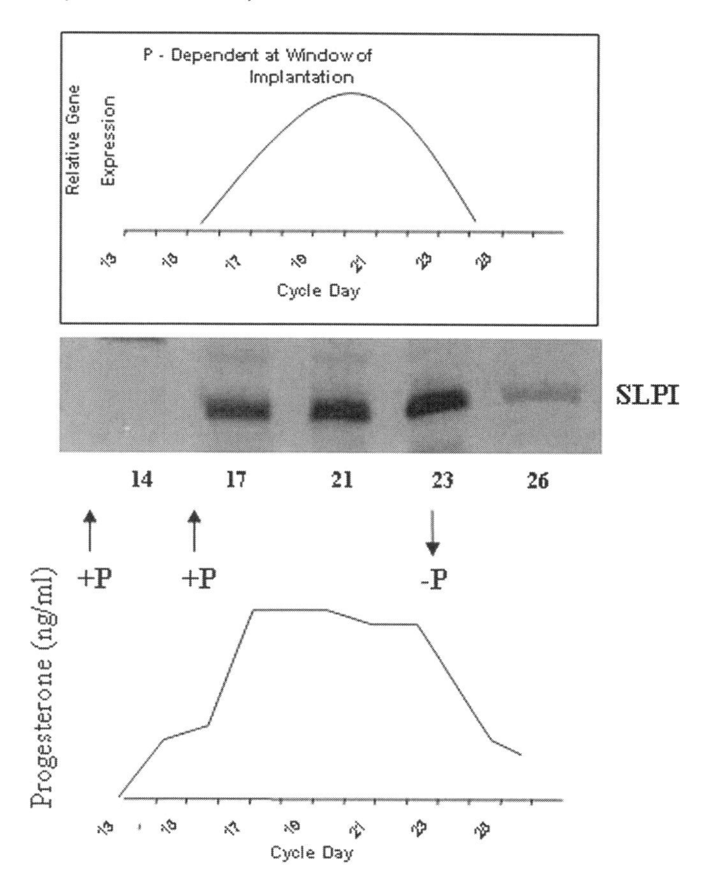

Fig. 5. SLPI expression and hormonal levels of progesterone (P) during the expected window of receptivity in the rhesus monkey endometrium ***(69)***. Upper panel shows the hypothetical expression pattern and the associated temporal pattern of SLPI expression as determined by PCR analyses. The lower panel shows the associated serum progesterone levels during this timeframe.

A restricted expression of genes during the window of endometrial receptivity/implantation provides support for the potential importance of these genes in maturation of the endometrium. All levels of regulation described in our working model (see above) are, however, considered important because of their potential to be linked. That is, preceding gene expression patterns may control and direct subsequent gene expression that will allow proper endometrial maturation necessary for embryo implantation. It is anticipated that refinements or additions to this working model will result from future studies in our laboratory and others.

5. GENE MICROARRAY ANALYSIS

We, as well as other investigators *(10,80–84)*, have used the powerful but often perplexing technique of gene expression (microarray) profiling to identify differentially expressed genes in the human and nonhuman primate endometrium. A complete discussion of these data is beyond the scope of this chapter, and our approach and methodology can be found in our previous publication *(83)*. Our studies focused on the identification of differentially expressed endometrial genes during a normal secretory phase (progesterone-dominant) versus the proliferative phase (estrogen-dominant) in the rhesus monkey *(83)*. Specifically, our data confirmed the elevated expression of SLPI (see above) and identified another member of the SLPI family of secretory proteins, whey acidic protein four-disulfide core domain 2 (WFDC2). Both WFDC2 and SLPI belong to a family of 14 WAP

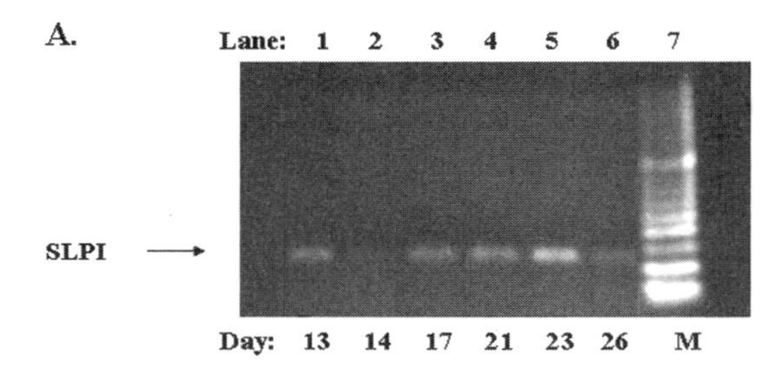

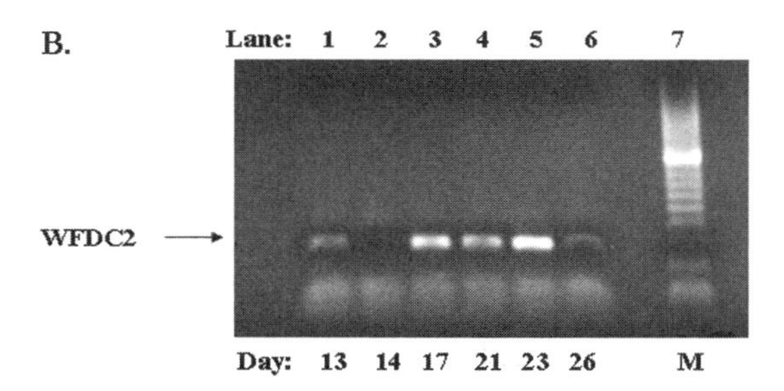

Fig. 6. Expression of selected gene fragments SLPI (218 bp, panel **A**) and WFDC2 (266 bp, panel **B**) in temporal cDNA populations of rhesus monkey endometrium by semiquantitative RT-PCR *(65)*. cDNA populations used in the microarray analysis were day 13 (proliferative phase, lane 1) and days 21 to 23 (secretory phase, lanes 4 and 5). Lane 7 is 100-bp marker (M).

proteins (whey acidic proteins) that have duplicated on chromosome 20 over evolution *(85)*.

Interestingly, WFDC2 was highly upregulated (27.5-fold) in our microarray studies. The expression profile of WFDC2 was confirmed by semiquantitative PCR in parallel with SLPI (**Fig. 6**). In addition, WFDC2 showed a restricted expression similar to that of SLPI during the window of receptivity (**Fig. 6**). Although there is limited literature on WFDC2, it is interesting to speculate that one or more of these uncharacterized family members could provide functional redundancy regarding endometrial development in the female reproductive tract. Our data on SLPI and WFDC2 confirm the microarray analysis and provide an important avenue for future studies on the potential importance of this gene family in endometrial function.

6. GENE EXPRESSION DURING INADEQUATE SECRETORY PHASES

As shown above, the regulation of syncytin, SLPI, BAT2, and WFDC2 displayed a restricted expression during the expected window of receptivity during adequate secretory phases. We extended our studies to compare both adequate and inadequate secretory phases. Our results clearly showed that the expression of both SLPI and WFDC2 are not temporally elevated during an inadequate secretory phase (2.8- and 4.8-fold decrease, respectively) (**Fig. 7**). In addition, we have also shown that both syncytin and BAT2 are also temporally underrepresented during an inadequate secretory phase (2.3- and 3.0-fold decrease, respectively) (**Fig. 8**). Together, these data suggest

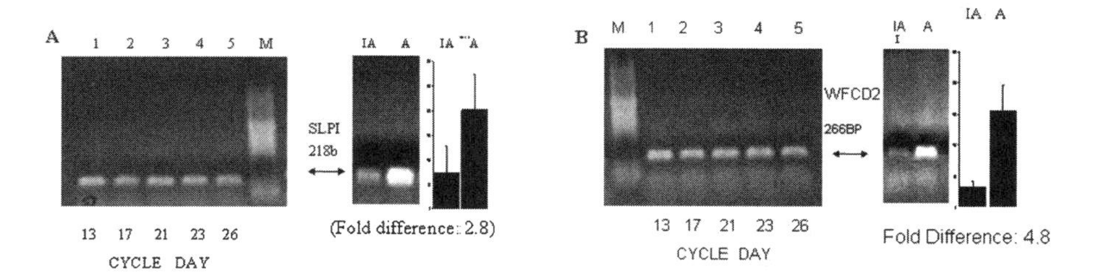

Fig. 7. Semiquantitative PCR analysis comparing inadequate (IA) and adequate (A) endometrial cDNA populations on cycle days 13, 17, 21, 23, and 26 for SLPI and WFDC2 (panels **A** and **B**, respectively) *(62)*. Lane M is a size marker (100-bp DNA ladder). The side panels compare IA and A cDNA populations on day 23 (midsecretory phase) for each gene, and the column charts represent average densitometric analysis of three experiments and margin of error (SD).

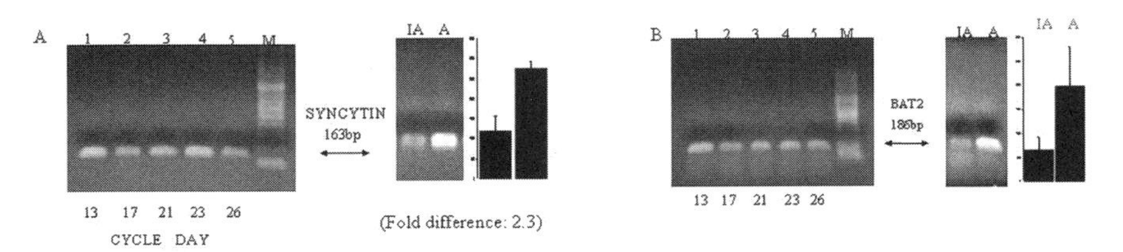

Fig. 8. Semiquantitative PCR analyses of endometrial cDNA populations for syncytin and BAT2 (panels **A** and **B**, respectively) prepared from inadequate (IA) cycle days 13, 17, 21, 23, and 26 ***(62)***. The side panels compare IA and A cDNA populations on day 23 for each gene. The column charts represent average densitometric analyses of three experiments with margins of error (SD).

that these four genes may play important roles in endometrial maturation and function during the receptive period of an adequate secretory phase

7. LCM AND DIFFERENTIAL DISPLAY

In order to couple LCM with differential display, we prepared cell-type–specific cDNA populations (glandular epithelia or stroma) from both the functionalis and basalis of adequate midsecretory endometria as described above. After differential display analysis, we selected six fragments that showed a putative cell-type–specific and/or a region-specific expression ***(27)***. The differential display gels and patterns of expression of these fragments are shown in **Fig. 9**. Although differential display reverse transcriptase-polymerase chain reaction (DDRT-PCR) is a potentially powerful and important approach, a drawback as noted above can be the appearance of false positives ***(86)***. After cloning and sequencing, specific primers for each of these fragments were designed and used to verify their expression patterns. Although three of these fragments were shown to be false positives with regard to their regional or cell-type specificity, they remain potentially important because of their elevated expression during an adequate secretory phase ***(27)***.

The three fragments identified in laser microdissected samples showing the expected regulatory pattern were F1 (highly expressed in the glands and stroma of the functionalis), BG-1 (highly expressed in the glands of the basalis), and FS-1 (highly expressed in the stroma of the functionalis). Although BG-1 and FS-1 are currently uncharacterized gene fragments, F1 showed a 94% homology to a known gene, namely, the human leukotriene B4 receptor ***(87,88)***.

Leukotriene B_4 (LTB_4) is one of the most potent chemoattractant mediators, acting mainly on neutrophils but also on related granulocytes,

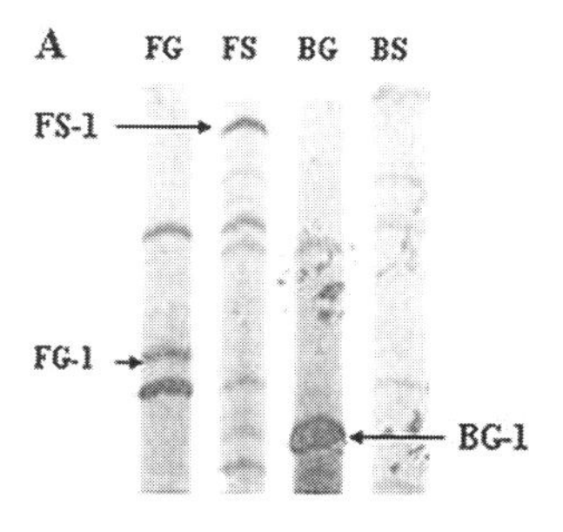

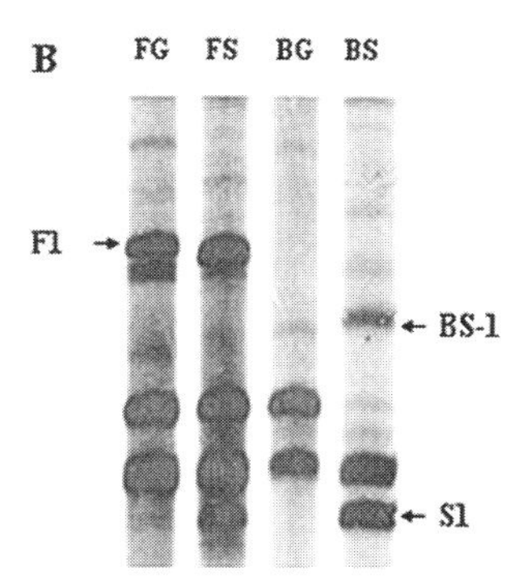

Fig. 9. A differential display gel. **(A)** A portion indicating putative functionalis/stromal-specific (FS-1), functionalis/glandular-specific (FG-1), and basalis/glandular-specific (BG-1) cDNA fragments. **(B)** A portion indicating basalis/stromal-specific (BS-1), functionalis-specific (F-1), and stromal-specific (S-1) cDNA fragments *(25)*.

macrophages, and endothelial cells. LTB_4 activates inflammatory cells by binding to its cell surface receptors BLTR1 and BLTR2 and has been implicated in a number of inflammatory diseases ***(87)***. Interestingly, levels of leukotrienes were elevated in the endometrium of women with primary dysmenorrhea and endometriosis ***(89)***. Levels of LTB_4 have also been shown to increase in the rat uterus during the peri-implantation phase implicating a role for this cytokine in uterine receptivity and implantation ***(90)***. Importantly, the expression of this gene was localized to the endometrial functionalis, the target for blastocyst invasion/implantation. To our knowledge, this is the first time expression of a BLTR2 receptor ortholog (F1) has been shown in the endometrium. Further studies will be required to identify the role of this receptor in proper maturation of the primate endometrium.

7.1. LCM and SLPI, WFDC2, BAT2, and Syncytin

We have also used LCM to study cell-type and regional differences in expression of SLPI, WFDC2, BAT2, and syncytin (see above) in the rhesus endometrium *(62)*. SLPI and BAT2 showed an increased expression primarily in the stromal compartment of the functionalis, whereas WFDC2 displayed elevated expression in both stroma and epithelia of the functionalis compared with the basalis (**Fig. 10**). No detectable differences for syncytin were observed with these LCM samples. These studies demonstrate a differential expression of several known genes during the window of receptivity that appears to be primarily localized in the functionalis of the primate endometrium, the primary target for blastocyst invasion. Preliminary studies (data not shown) showed that SLPI and WFDC2 are both upregulated in LCM harvested luminal epithelia during the receptive window of an adequate secretory phase. Further studies are necessary to understand the functions and roles of these genes in primate endometrial maturation and receptivity.

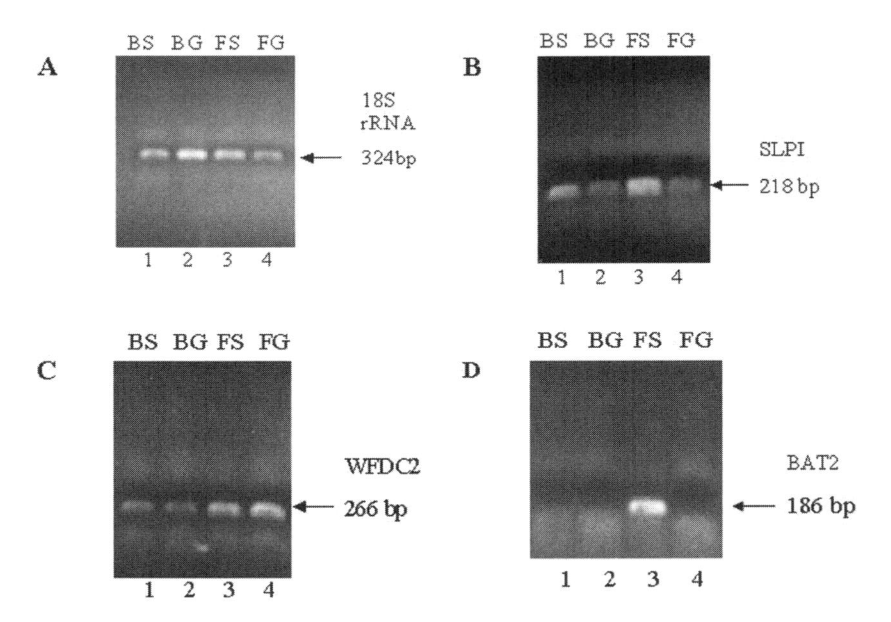

Fig. 10. Semiquantitative PCR analysis of cDNA populations prepared from laser capture microdissected rhesus monkey endometria from adequate secretory phases *(62)*. Basalis stroma (BS), basalis glands (BG), functionalis stroma (FS), and functionalis glands (FG) were analyzed for **(A)** 18S rRNA, **(B)** SLPI, **(C)** WFDC2, and **(D)** BAT2.

8. LCM AND MICROARRAY ANALYSIS OF ENDOMETRIAL CELLS

cDNA populations representing functionalis glandular epithelium (FG) and stroma (FS) isolated by LCM were constructed as previously described *(27)*. The two populations were used to prepare biotin-labeled probes by random primer extension and separately hybridized to Affymetrix (Santa Clara, CA) HGU95Av2 oligonucleotide microarrays. When compared, the results from our initial experiment showed that >50 genes were upregulated >5-fold in FG, while >100 genes were upregulated in FS. Likewise, these upregulated genes were downregulated in the comparative cDNA population. Examples of some of these data are presented in **Table 1** and **Table 2**. Genes upregulated in FG included transforming growth factor beta 1 (TGFB1), MAX interacting protein 1 (MXI1), fibroblast growth factor 9 (FGF9), and intercellular adhesion molecule 2 (ICAM2). TGFB1 has previously been shown by others expressed at elevated levels in secretory

Table 1
Genes Upregulated in the Functionalis Glands (FG), Secretory Phase

Accession no.	Fold-up	Gene	Biological function (NCBI)	Regulation in endometrium
L07648	10.3	Max interacting factor (MXI1)	DNA binding, transcription corepressor activity	
M38449	5.5	Transforming growth factor beta 1 (TGFB1)	Control of proliferation and differentiation	Upregulated in secretory phase glands ***(91)***
AC005175	5.1	Thromboxane A2 receptor (TBXA2R)	Muscle contraction	Expressed in glands ***(92)***
D14838	4.5	Fibroblast growth factor 9 (FGF9)	Embryonic development, cell growth, morphogenesis, tissue repair	
M32334	4.4	Intercellular adhesion molecule 2 (ICAM2)	Cell-cell adhesion, integrin binding	Expressed in vascular endothelium ***(93)***

Table 2
Genes Upregulated in Functionalis Stroma (FS), Secretory Phase

Accession no.	Fold-up	Gene	Biological function(NCBI)	Regulation in endometrium
AF084367	46	Inversin (INV)	Embryonic development, left-right axis determination	
D14043	39	Sialomucin (CD164 antigen)	Regulation of cell adhesion	
AA401397	28	Kallikrein 13 (KLK13)	Proteolysis and peptidolysis	
D83402	15	Prostacyclin (PTGIS)	Prostaglandin biosynthesis	Expressed in functionalis stroma ***(94)***
AB015051	12	Death-associated protein 6 (DAP6)	Regulation of transcription, DNA-dependent apoptosis	

phase glandular epithelium of the human endometrium ***(91)***. Genes upregulated in FS included inversin (INV), sialomucin (CD164), kallikrein 13 (KLK13), and death-associated protein 6 (DAP6). Expression of these genes in the endometrium has not previously been documented. These data provide support for the use of combined LCM and microarray analyses.

9. CONCLUSION

The use of LCM affords an important and powerful tool to characterize and analyze gene or protein expression patterns in tissues composed of multiple cell types within different regions. LCM coupled with techniques such as PCR analysis, differential display, and microarray analysis further augments the usefulness of this approach. Indeed, our data described above support the applicability of these approaches in combination with LCM. Our results demonstrate that regional and cell-type differences in gene expression are a property of a complex tissue such as the primate endometrium and that these approaches can be applicable to the analyses of other complex tissues. The application of the above tools to answer complex biological questions and other techniques yet to be developed will help lead us down new and exciting pathways of discovery and knowledge.

ACKNOWLEDGMENTS

The author thanks Dr. C. I. Ace, C. Franz, and F. A. Okulicz for their help and support of the studies described herein. The work described was supported in part by a grant from the NIH (NICHD, HD31620).

REFERENCES

1. Wynn RM. The human endometrium. Cyclic and gestational changes. In: Wynn RM, Jollie WP, eds. Biology of the Uterus. New York: Plenum Publishing Corp., 1989:289–331.
2. Maslar IA. The progestational endometrium. Semin Reprod Endocrinol 1988;6:115–128.
3. Bartelmez GW, Corner GW, Hartman CG. Cyclic changes in the endometrium of the rhesus monkey (Macaca mulatta). Contrib Embryol 1951;34:101–144.
4. Ferenczy A, Bergeron C. Histology of the human endometrium: From birth to senescence. Ann NY Acad Sci 1991;622:6–27.
5. Gurpide E, Tseng L. Induction of human endometrial oestradiol dehydrogenase by progestins. Endocrinology 1975;97:825–833.
6. Clarke CL, Sutherland RL. Progestin regulation of cellular proliferation. Endocrinol Rev 1990;11:266–301.
7. Fay TN, Grudzinskas JG. Human endometrial peptides: a review of their potential role in implantation and placentation. Hum Reprod 1991;6:1311–1326.
8. Tabibzadeh S, Mason JM, Shea M, Cai YQ, Murray MJ, Lessey B. Dysregulated expression of *ebaf*, a novel molecular defect in the endometria of patients with infertility. J Clin Endocr Metab 2000;85(7):2526–2536.
9. Salamonsen LA, Nie GY, Findlay JK. Newly identified endometrial genes of importance for implantation. J Reprod Immunol 2002;53(1–2):215–225.
10. Kao LC, Tulac S, Lobo S, Imani B, Yang JP, Germeyer A, et al. Global gene profiling in human endometrium during the window of implantation. Endocrinology 2002;143(6):2119–2138.
11. Lessey BA. Adhesion molecules and implantation. J Reprod Immunol 2002;55(1–2):101–112.
12. Nayak NR, Giudice LC. Comparative biology of the IGF system in endometrium, decidua, and placenta, and clinical implications for foetal growth and implantation disorders. Placenta 2003;24(4):281–296.
13. Daftary GS, Taylor HS. Pleiotropic effects of *Hoxa10* on the functional development of peri-implantation endometrium. Mol Reprod Dev 2004;67(1):8–14.
14. Okulicz WC. Regeneration. In: Glasser SR, Aplin JD, Giudice L, Tabibzadeh S, eds. The Endometrium. Reading, Bershire, UK: Harwood Academic Publishers, 2002:110–120.
15. Hartman CG. Regeneration of the monkey uterus after surgical removal of endometrium and accidental endometriosis. W J Surg Obstet Gynecol 1944;52:87–102.

16. Okulicz WC, Savasta AM, Hoberg LM, Longcope C. Biochemical and immunohistochemical analyses of estrogen and progesterone receptors in the rhesus monkey uterus during the proliferative and secretory phases of artificial menstrual cycles. Fertil Steril 1990;53:913–920.
17. Okulicz WC, Balsamo M, Tast J. Progesterone regulation of endometrial estrogen receptor and cell proliferation during the late proliferative and secretory phase in artificial menstrual cycles in the rhesus monkey. Biol Reprod 1993;49:24–32.
18. King WJ, Greene GL. Monoclonal antibodies localize oestrogen receptor in the nuclei of target cells. Nature 1984;307:745–747.
19. Press MF, Nousek-Goebl N, King WJ, Herbst AL, Greene GL. Immunohistochemical assessment of estrogen receptor distribution in the human endometrium throughout the menstrual cycle. Lab Invest 1984;51:495–503.
20. Press MF, Greene GL. Localization of progesterone receptor with monoclonal antibodies to the human progestin receptor. Endocrinology 1988;122:1165–1175.
21. McClellan MC, West NB, Tacha DE, Greene GL, Brenner RM. Immunocytochemical localization of estrogen receptors in the macaque reproductive tract with monoclonal antiestrophilins. Endocrinology 1984;114:2002–2014.
22. Snijders MPML, De Goeij AFPM, Debets-Te Baerts MJC, Rousch MJM, Koudstaal J, Bosman FT. Immunocytochemical analysis of oestrogen receptors and progesterone receptors in the human uterus throughout the menstrual cycle and after the menopause. J Reprod Fertil 1992;94:361–369.
23. Padykula HA, Coles LG, McCracken JA, King NW, Longcope C, Kaiserman-Abramof IR. A zonal pattern of cell proliferation and differentiation in the rhesus endometrium during the estrogen surge. Biol Reprod 1984;31:1103–1118.
24. Padykula HA, Coles LG, Okulicz WC, Rapaport SI, McCracken JA, King NW Jr, et al. The basalis of the primate endometrium: a bifunctional germinal compartment. Biol Reprod 1989;40:681–690.
25. Okulicz WC, Balsamo M, Tast J. Progesterone regulation of endometrial estrogen receptor and proliferation during the late proliferative and secretory phase in artificial menstrual cycles in the rhesus monkey. Biol Reprod 1993;49:24–32.
26. Okulicz WC, Ace CI, Scarrell R. Zonal changes in proliferation in the rhesus endometrium during the late secretory phase and menses. Proc Soc Exp Biol Med 1997;214(2):132–138.
27. Torres MST, Ace CI, Okulicz WC. Assessment and application of laser microdissection for analysis of gene expression in the Rhesus monkey endometrium. Biol Reprod 2002;67:1067–1072.
28. Okulicz WC, Ace CI, Torres MST. Gene expression in the rhesus monkey endometrium: differential display and laser capture microdissection. Frontiers Biosci 2003;8:d551–558.
29. Okulicz WC. Cellular and molecular regulation of the primate endometrium: A perspective. Reprod Biol Endocrinol 2006;4(Suppl 1):53:1–15.

30. Hodgen GD. Surrogate embryo transfer combined with estrogen-progesterone therapy in monkeys. Implantation, gestation, and delivery without ovaries. JAMA 1983;250:2167–2171.
31. Fritz MA. Inadequate luteal function and recurrent abortion: diagnosis and treatment of luteal phase deficiency. Semin Reprod Endocrinol 1988;6:129–143.
32. Jones GS. The luteal phase defect. Fertil Steril 1976;27:351–356.
33. Jones GS, Madrigal-Castro V. Hormonal findings in association with abnormal corpus luteum function in the human: the luteal phase defect. Fertil Steril 1970;21:1–13.
34. Jones GS. Luteal phase insufficiency. Clin Obstet Gynecol 1973;16:255–273.
35. Wilks JW, Hodgen GD, Ross GT. Luteal phase defects in the rhesus monkey: the significance of serum FSH:LH ratios. J Clin Endocrinol Metab 1976;43: 1261–1267.
36. Longcope C, Bourget C, Meciak PA, Okulicz WC, McCracken JA, Hoberg LM, et al. Estrogen dynamics in the female rhesus monkey. Biol Reprod 1988;39:561–565.
37. Okulicz WC, Balsamo M. A double immunofluorescent method for the simultaneous analysis of progesterone-dependent changes in proliferation (Ki-67) and the estrogen receptor in the rhesus endometrium. J Reprod Fertil 1993;99: 545–549.
38. Okulicz WC, Ace CI, Longcope C, Tast J. Analysis of differential gene regulation in adequate *versus* inadequate secretory-phase endometrial complementary deoxyribonucleic acid populations from the rhesus monkey. Endocrinology 1996;137(11):4844–4850.
39. Bonner RF, Emmert-Buck MR, Cole K, Pohida T, Chuaqui RF, Goldstein S, et al. Laser capture microdissection: molecular analysis of tissue. Science 1996;278:1482–1483.
40. Simone NL. Laser capture microdissection: opening the microscopic frontier to molecular analysis. Trends Genet 1998;14:272–276.
41. Fend F, Raffeld M. Laser capture microdissection in pathology. J Clin Pathol 2000;53(9):666–672.
42. Mills JC, Roth KA, Cagan RL, Gordon JI. DNA microarrays and beyond: completing the journey from tissue to cell. Nat Cell Biol 2001;3(8):E175–E178.
43. Jain KK. Application of laser capture microdissection to proteomics. Methods Enzymol 1903;356:157–167.
44. Craven RA, Banks RE. Use of laser capture microdissection to selectively obtain distinct populations of cells for proteomic analysis. Methods Enzymol 1903;356:33–49.
45. Ornstein DK, Gillespie JW, Paweletz CP, Duray PH, Herring J, Vocke CD, et al. Proteomic analysis of laser capture microdissected human prostate cancer and *in vitro* prostate cell lines. Electrophoresis 2000;21(11):2235–2242.
46. Lindeman N, Waltregny D, Signoretti S, Loda M. Gene transcript quantitation by real-time RT-PCR in cells selected by immunohistochemistry-laser capture microdissection. Diagn Mol Pathol 2002;11(4):187–192.

47. Lo YL, Shen CY. Laser capture microdissection in carcinoma analysis. Methods Enzymol 1903;356:137–144.
48. Paweletz CP, Charboneau L, Bichsel VE, Simone NL, Chen T, Gillespie JW, et al. Reverse phase protein microarrays which capture disease progression show activation of pro-survival pathways at the cancer invasion front. Oncogene 2001;20(16):1981–1989.
49. Rubin MA. Use of laser capture microdissection, cDNA microarrays, and tissue microarrays in advancing our understanding of prostate cancer. J Pathol 2001;195(1):80–86.
50. Luzzi V, Holtschlag V, Watson MA. Expression profiling of ductal carcinoma *in situ* by laser capture microdissection and high-density oligonucleotide arrays. Am J Pathol 2001;158(6):2005–2010.
51. Lawrie LC, Curran S, McLeod HL, Fothergill JE, Murray GI. Application of laser capture microdissection and proteomics in colon cancer. J Clin Pathol Mol Pathol 2001;54(4):253–258.
52. Miura K, Bowman ED, Simon R, Peng AC, Robles AI, Jones RT, et al. Laser capture microdissection and microarray expression analysis of lung adenocarcinoma reveals tobacco smoking- and prognosis-related molecular profiles. Cancer Res 2002;62(11):3244–3250.
53. Cong PJ, Raffeld M, Teruya-Feldstein J, Sorbara L, Pittaluga S, Jaffe ES. In situ localization of follicular lymphoma: description and analysis by laser capture microdissection. Blood 2002;99(9):3376–3382.
54. Suárez-Quian CA, Goldstein SR, Bonner RF. Laser capture microdissection: a new tool for the study of spermatogenesis. J Androl 2000;21(5):601–608.
55. Albrecht ED, Billiar RB, Aberdeen GW, Babischkin JS, Pepe GJ. Expression of estrogen receptors α and β in the fetal baboon testis and epididymis. Biol Reprod 2004;70(4):1106–1113.
56. Shimamura M, Garcia JM, Prough DS, Hellmich HL. Laser capture microdissection and analysis of amplified antisense RNA from distinct cell populations of the young and aged rat brain: effect of traumatic brain injury on hippocampal gene expression. Mol Brain Res 2004;122(1):47–61.
57. Kai N, Iwase K, Imai K, Nakahira E, Soma M, Ohtsuka S, et al. Altered gene expression in the subdivisions of the amygdala of Fyn-deficient mice as revealed by laser capture microdissection and mKIAA cDNA array analysis. Brain Res 2006;1073(1):60–70.
58. Kerman IA, Buck BJ, Evans SJ, Akil H, Watson SJ. Combining laser capture microdissection with quantitative real-time PCR: effects of tissue manipulation on RNA quality and gene expression. J Neurosci Methods 2006;153(1): 71–85.
59. Kerk NM, Ceserani T, Tausta SL, Sussex IM, Nelson TM. Laser capture microdissection of cells from plant tissues. Plant Physiol 2003;132(1):27–35.
60. Angeles G, Berrio-Sierra J, Joseleau JP, Lorimier P, Lefèbvre A, Ruel K. Preparative laser capture microdissection and single-pot cell wall material preparation: a novel method for tissue-specific analysis. Planta 2006;224(1):228–232.

61. Kreitmann-Gimbal B, Goodman AL, Bayard F, Hodgen DG. Characterization of estrogen and progesterone receptors in monkey endometrium: methodology and effects of estradiol and/or progesterone on endometrium of castrate monkeys. Steroids 1979;34:749–770.
62. Okulicz WC, Ace CI, Franz C. Temporal regulation of endometrial genes during inadequate secretory phases in the Rhesus monkey. Biol Reprod 2005;114 [Special Issue].
63. Ace CI, Balsamo M, Le LT, Okulicz WC. Isolation of progesterone-dependent complementary deoxyribonucleic acid fragments from rhesus monkey endometrium by sequential subtractive hybridization and polymerase chain reaction amplification. Endocrinology 1994;134:1305–1309.
64. Liang P, Pardee A. Differential display of eukaryotic messenger RNA by means of polymerase chain reaction. Science 1992;257:967–971.
65. Ace CI, Okulicz WC. Subtractive hybridization and differential display methods to identify changes in gene expression patterns. In: Rajendra Kumar T, ed. Molecular Endocrinology: Methods and Protocols. Totowa, NJ: Humana Press Inc., 2008 (in press).
66. Haag E, Raman V. Effects of primer choice and source of Taq DNA polymerase on the banding patterns of differential display RT-PCR. BioTechniques 1994;17:226–228.
67. Ace CI, Okulicz WC. Identification of progesterone-dependent mRNA regulatory patterns in the rhesus monkey endometrium by differential display reverse-transcriptase-polymerase chain reaction. Biol Reprod 1999;60: 1029–1035.
68. Okulicz WC, Ace CI. Progesterone-regulated gene expression in the primate endometrium. Semin Reprod Endocrinol 1999;17:241–255.
69. Okulicz WC, Ace CI. Temporal regulation of gene expression during the expected window of receptivity in the rhesus monkey endometrium. Biol Reprod 2003;69(5):1593–1599.
70. Yu C, Shen K, Lin M, Chen P, Lin C, Chang GD, et al. GCMa regulates the syncytin-mediated trophoblastic fusion. J Biol Chem 2002;277:50062–50068.
71. Mi S, Lee X, Li X, Veldman GM, Finnerty H, Racie L, et al. Syncytin is a captive retroviral envelope protein involved in human placental morphogenesis. Nature 2000;403:785–789.
72. Ono M, Kawakami M, Ushikubo H. Stimulation of expression of the human endogenous retrovirus genome by female steroid hormones in human breast cancer cell line T47D. J Virology 1987;61:2059–2069.
73. Zhang D, Simmen RC, Michel FJ, Zhao G, Vale-Cruz D, Simmen FA. Secretory leukocyte protease inhibitor mediates proliferation of human endometrial cells by positive and negative regulation of growth-associated genes. J Biol Chem 2002;277:29999–30009.
74. Ashcroft GS, Lei K, Jin W, Longenecker G, Kukarni AB, Greenwell-Wild T, et al. Secretory leukocyte protease inhibitor mediates non-redundant functions necessary for normal wound healing. Nat Med 2000;6:1147–1153.

75. King AE, Critchley HOD, Kelly RW. Presence of secretory leukocyte protease inhibitor in human endometrium and first trimester decidua suggests an antibacterial protective role. Mol Hum Reprod 2000;6:191–196.
76. Quayle AJ, Porter EM, Nussbaum AA, Wang YM, Brabec C, Yip KP, et al. Gene expression, immunolocalization, and secretion of human defensin-5 in human female reproductive tract. Am J Pathol 1998;152:1247–1258.
77. Jin FY, Nathan C, Radzioch D, Ding A. Secretory leukocyte protease inhibitor: a macrophage product induced by and antagonistic to bacterial lipopolysaccharide. Cell 1997;88:417–426.
78. Kikuno R, Nagase H, Ishikawa K, Irosawa M, Iyajima N, Anaka A, et al. Prediction of the coding sequences of unidentified human genes. XIV. The complete sequences of 100 new cDNA clones from brain which code for large proteins in vitro. DNA Res 1999;6:197–205.
79. Banerji J, Sands J, Strominger JL, Spies T. A gene pair from the human major histocompatibility complex encodes large proline-rich proteins with multiple repeated motifs and a single ubiquitin-like domain. Proc Natl Acad Sci USA 1990;87:2374–2378.
80. Carson DD, Lagow E, Thathiah A, Al-Shami R, Farach-Carson MC, Vernon M, et al. Changes in gene expression during the early to mid-luteal (receptive phase) transition in human endometrium detected by high-density microarray screening. Mol Hum Reprod 2002;8:871–879.
81. Borthwick JM, Charnock-Jones DS, Tom BD, Hull ML, Teirney R, Phillops SC, et al. Determination of the transcript profile of human endometrium. Mol Hum Reprod 2003;9:19–33.
82. Riesewijk A, Martin J, van Os R, Horcajadas JA, Polman J, Pellicer A, et al. Gene expression profiling of human endometrial receptivity on days LH+2 versus LH+7 by microarray technology. Mol Hum Reprod 2003;9:253–264.
83. Ace CI, Okulicz WC. Microarray profiling of progesterone-regulated endometrial genes during the rhesus monkey secretory phase. Reprod Biol Endocrinol 2004;2:54 (1–9).
84. Mirkin S, Nikas G, Hsiu JG, Díaz J, Oehninger S. Gene expression profiles and structural/functional features of the peri-implantation endometrium in natural and gonadotropin-stimulated cycles. J Clin Endocrinol Metab 2004;89(11):5742–5752.
85. Clauss A, Lilja H, Lundwall A. A locus on human chromosome 20 contains several genes expressing protease inhibitor domains with homology to whey acidic protein. Biochem J 2002;368(1):233–242.
86. Liang P, Pardee AB. Differential display. A general protocol. Methods Mol Biol 1997;85:3–11.
87. Tryselius Y, Nilsson NE, Kotarsky K, Olde B, Owman C. Cloning and characterization of cDNA encoding a novel human leukotriene B4 Receptor. Biochem Biophys Res Commun 2000;274:377–382.
88. Kamohara M, Takasaki JMM, Saito T, Ohishi T, Ishii H, Furuichi K. Molecular cloning and characterization of another leukotriene B4 receptor. J Biol Chem 2000;275:27000–27004.

89. Abu JI, Konje JC. Leukotrienes in gynaecology: the hypothetical value of anti-leukotriene therapy in dysmenorrhoea and endometriosis. Hum Reprod 2000;6(2):200–205.
90. Malathy PV, Cheng HC, Dey SK. Production of leukotrienes and prostaglandins in the rat uterus during peri-implantation period. Prostaglandins 1986;32: 605–614.
91. Chegini N, Zhao Y, Williams RS, Flanders KC. Human uterine tissue throughout the menstrual cycle expresses transforming growth factor-b1 (TGFb1), TGFb2, TGFb3, and TGFb type II receptor messenger ribonucleic acid and protein and contains [125I]TGFb1-binding sites. Endocrinology 1994;135:439–449.
92. Swanson ML, Lei ZM, Swanson PH, Rao CV, Narumiya S, Hirata M. The expression of thromboxane A_2 synthase and thromboxane A_2 receptor gene in human uterus. Biol Reprod 1992;47:105–117.
93. Thomson AJ, Greer MR, Young A, Boswell F, Tekfer JF, Cameron IT, et al. Expression of intercellular adhesion molecules ICAM-1 and ICAM-2 in human endometrium: regulation by interferon-gamma. Mol Hum Reprod 1999;5: 64–70.
94. Battersby S, Critchley HOD, de Brum-Fernandes AJ, Jabbour HN. Temporal expression and signalling of prostacyclin receptor in the human endometrium across the menstrual cycle. Reproduction 2004;127:79–86.
95. Okulicz WC, Ace CI, Tast J, Longcope C. Hormonal regulation of endometrial gene expression in the rhesus monkey. In: Carson DD, ed. Embryo Implantation: Molecular, Cellular and Clinical Aspects. Norwell, MA: Serono Symposium USA, Inc., 1999:185–201.

8

The Use of Two-Dimensional Gel Electrophoresis for Plasma Biomarker Discovery

Brad Jarrold, Alex Varbanov, and Feng Wang

Summary

Two-dimensional polyacrylamide gel electrophoresis (2-DE) is one of the most widely used and versatile methods of protein expression profiling among a rapidly growing arsenal of proteomics technologies. 2-DE combines two orthogonal and independent electrophoretic steps: isoelectric focusing (IEF) and sodium dodecyl sulfate–polyacrylamide gel electrophoresis (SDS-PAGE). At present, 2-DE is capable of simultaneously detecting and quantifying up to several thousand protein spots in the same gel image. As an example to illustrate the power of 2-DE, we provide comprehensive step-by-step instructions for the application of 2-DE to clinically collected human plasma samples to determine whether specific plasma proteins or protein expression patterns could serve as biomarkers for the development of a specific adverse side effect that arises upon treatment with an experimental drug. These instructions provide detailed information on how to deplete high-abundant proteins from human plasma, apply the depleted plasma proteins to immobilized pH gradient gels, perform IEF and SDS-PAGE separation, and perform silver staining to visualize proteins. The basic 2-DE protocol described here could easily be applied to other tissue types.

Key Words: biomarker; 2-DE; fractionation; plasma protein

1. INTRODUCTION

By coupling isoelectric focusing (IEF) in the first dimension with sodium dodecyl sulfate–polyacrylamide gel electrophoresis (SDS-PAGE) in the second dimension, two-dimensional polyacrylamide gel electrophoresis (2-DE) enables the separation of complex protein mixtures according to each protein's isoelectric point (p*I*) and molecular weight ***(1,2)***. Depending on the size of the immobilized pH gradient (IPG) and SDS-PAGE gel used,

From: *Methods in Pharmacology and Toxicology: Biomarker Methods in Drug Discovery and Development*
Edited by: F. Wang © Humana Press, Totowa, NJ

2-DE has the potential to resolve up to 10,000 proteins simultaneously ***(3,4)***, however this is unusual, and typically single 2-DE gel images generally contain 1000 to 2000 distinguishable protein spots ***(5)***. This typical proteome coverage can be increased if combined with additional prefractionation steps or spread over several narrow pH range IPG gels. Furthermore, 2-DE delivers a protein pattern map, which reflects changes in protein expression levels, isoforms, cleavage products, and posttranslational modifications.

2-DE is by no means the only platform for protein separations and there are inherent technical limitations. 2-DE provides good coverage of proteins with a p*I* between 3 and 11 and a molecular weight of 10 to 200 kDa; proteins possessing properties outside these ranges are typically underrepresented. In addition, hydrophobic proteins are rarely resolved and identified on 2-DE protein maps without resorting to specialized methods. Proteins found in low concentrations within the samples are also underrepresented, but through high-abundant protein depletion and prefractionation techniques, they can be brought above the detection threshold. 2-DE is labor intensive, difficult to automate, and, despite significant methodology improvements, a degree of technical proficiency is still required to routinely produce high-quality 2-DE protein maps.

Despite these limitations, the overall advantages and relative accessibility of 2-DE technology has made it an attractive methodology for biomarker discovery, and higher-throughput automated methods can be applied for biomarker validation and screening. Biomarkers are being used throughout the drug discovery and development processes to understand fundamental biological processes and relationships ***(6)***. Specific biomarkers for a disease state can be monitored for patient prognosis and response to therapy and serve as tools in the development of therapeutics. Biomarkers can also be used to stratify patients and to monitor adverse effects. Clinical biomarkers hold significant potential to make drug development process more effective and efficient ***(7)***. However, successful biomarker discovery through protein profiling is challenging because of the dynamic range of protein expression seen in eukaryotic tissues, ranging from 10^1 to 10^6 copies per cell. To illustrate this point, plasma biomarker discovery involves searching for extremely low abundance proteins, which comprise less than 1% of the total plasma proteome, whereas the 20 most abundant proteins represent more than 99% ***(8–10)***. The presence of high-abundant plasma proteins such as albumin and IgGs mask the detection and identification of low-abundant proteins in subsequent proteomic analysis. Removal of these abundant proteins is essential for the detection of less abundant proteins that may prove to be informative diagnostic or therapeutic markers. One technology developed to remove high-abundant plasma proteins is the Multiple Affinity Removal

System (MARS). MARS combines the specificity of antibody-antigen recognition and the efficiency of high-performance liquid chromatography (HPLC) for the simultaneous and reproducible removal of six abundant human plasma proteins (albumin, transferrin, IgG, IgA, haptoglobin, and antitrypsin) ***(11–15)***. As a result, the lower-abundant, more biologically relevant proteins can be brought into detectable levels by increasing their loading onto 2-DE gels for analysis.

Within this chapter, to illustrate the power of 2-DE, we provide comprehensive step-by-step instructions for the application of 2-DE to human plasma samples to determine whether specific plasma proteins or protein expression patterns could serve as biomarkers for the development of a specific adverse effect (AE) that arises upon treatment with an experimental drug.

2. MATERIALS

2.1. Equipment

Descriptions of specific equipment used in this experiment are given, however any model of comparable capability can easily be substituted.

1. Agilent 1100 HPLC system (Agilent Technologies, Wilmington, DE) and a 4.6 × 50 mm Multiple Affinity Removal LC Column-Human 6 (Agilent Technologies) for plasma fractionation. Sorvall RC-5C PLUS Superspeed Refrigerated Centrifuge equipped with a SH-3000 rotor (Thermo Electron Corporation, Asheville, NC) for the concentration and buffer exchange of fractionated plasma.
2. VMax microplate reader (Molecular Dynamics Corporation, Sunnyvale, CA) for determination of total protein concentration.
3. C75T Ultrasonic water bath sonicator (Misonix Inc., Farmingdale, NY) and a 5415D tabletop centrifuge (Eppendorf AG, Hamburg, Germany) for protein solubilization.
4. Milli-Q Synthesis A10 water purification system (Millipore, Billerica, MA) (**Note 1**).
5. DryStrip Reswelling Tray (Amersham Biosciences, Piscataway, NJ) for in-gel rehydration and equilibration of IPG strips.
6. Multiphor II Isoelectric Focusing (IEF) System equipped with a MultiTemp III Thermostatic Circulator and EPS 3501 XL power supply (Amersham Biosciences) for IPG strip isoelectric focusing.
7. Investigator 2-DE Electrophoresis System (Genomic Solutions, Ann Arbor, MI) for SDS-PAGE separation.
8. ImageScanner II (Amersham Biosciences) flat-bed scanner for gel image digitization (14 bits, 250 ppi).

9. Software: Z4000 Large-Scale 2-DE Gel Image Analysis System (Compugen, Jamesburg, NJ) and Statistical Analysis System (SAS) (SAS Institute, Cary, NC) for image and statistical analysis.

2.2. Reagents and Solutions

1. HPLC mobile phases: Buffers A and B (Agilent Technologies).
2. Plasma dilution buffer: 1 Complete EDTA-free Protease Inhibitor Cocktail tablet in 25 mL Buffer A.
3. Non-Interfering Protein Assay Kit (Genotech Biosciences, St. Louis, MO).
4. Immobiline DryStrip 18 cm pH 4–7 IPG gels (Amersham Biosciences).
5. Rehydration buffer: 5 M urea (Sigma, St. Louis, MO), 2 M Thiourea (Sigma), 2% CHAPS (Sigma), 2% SB3-10 (Sigma), 30 mM DTT (Sigma), and trace amount of bromophenol blue (Sigma).
6. Equilibration buffer 1: 50 mM Tris (Sigma) pH 8.8, 6 M urea, 30% glycerol (Sigma), 4% SDS (Sigma), 2% DTT, and trace bromophenol blue.
7. Equilibration buffer 2: 50 mM Tris (Sigma) pH 8.8, 6 M urea, 30% glycerol (Sigma), 2.5% iodoacetamide (Sigma), 2% DTT, and trace bromophenol blue.
8. SDS-PAGE lower tank buffer 10X stock: 2.1 M Tris pH 8.9.
9. SDS-PAGE upper tank buffer 10X stock: 1 M Tris, 1 M Tricine (Sigma), 1% SDS.
10. Investigator pre-cast homogenous 10% Tricine slab SDS-PAGE gels (Genomic Solutions).
11. Investigator Silver Stain Kits (Genomic Solutions).

3. METHODS

3.1. Plasma Fractionation and Concentration of Low-Abundant Proteins

High-abundant protein removal from crude human plasma was performed using Agilent Technologies' Human MARS. Briefly, crude human plasma was diluted with plasma dilution buffer (see **Section 2.2**) and filtered through 0.22-µm spin filters by centrifugation at 16,000 × *g* for 1 min. All chromatographic fractionations were performed at room temperature (22°C) on an Agilent 1100 HPLC system with automated sample injection. The flow-through and eluted fractions were manually collected.

1. Set up Buffer A and Buffer B as the only mobile phases.
2. Purge lines with Buffer A and Buffer B at a flow rate of 1.0 mL/min for 10 min without the MARS column attached.
3. Set up liquid chromatography (LC) method (**Table 1**) and run two blanks by injecting 100 µL Buffer A without the MARS column attached.
4. Attach the 4.6 × 50 mm MARS column and equilibrate with Buffer A for 4 min at a flow rate of 1 mL/min.

Table 1
LC Method for 4.6 × 50 mm Multiple Affinity Removal Column–Human 6

Line	Time (min)	%B	Flow rate (mL/min)	Max pressure (barr)	Buffer*	Description
1	0	0	0.25	120	A	Sample injection
2	7	0	0.25	120	A	End of wash
3	7.01	100	1	120	B	Start elution
4	12.5	100	1	120	B	Stop elution
5	12.6	0	1	120	A	Start regeneration
6	20	0	1	120	A	End run

*Proprietary buffers supplied with Multiple Affinity Removal System.

5. Dilute crude plasma sample 5X in plasma dilution buffer (see **Section 2.2**) (e.g., 50 μL human plasma with 200 μL of plasma dilution buffer).
6. Remove particulates from the sample by filtering through an 0.22-μm spin filter for 1 min at 16,000 × *g* at room temperature.
7. Run LC method (**Table 1**). In this experiment, four independent 50-μL injections were made for each AE and non-AE sample (**Note 2**).
   7.1. Inject 50 μL diluted plasma at a flow rate of 0.25 mL/min.
   7.2. Collect the flow-through fraction which appears between 3 and 9 min (**Fig. 1**) and store at 4°C until all four injections for that particular sample are completed.

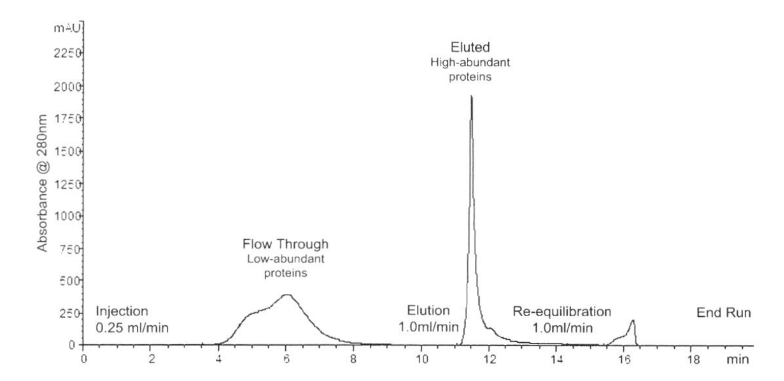

Fig. 1. Representative chromatogram of the affinity removal of high-abundant proteins from human plasma. A 50-μL injection of 5X diluted human plasma in Buffer A was made at a flow rate of 0.25 mL/min. The depleted protein fraction (flow-through) was collected from 3.5 to 9.0 min. The bound fraction was eluted with Buffer B at a flow rate of 1.0 mL/min for 3.5 min.

7.3. Collect eluted fraction which appears between 9 and 12 minutes (**Fig. 1**) and store at 4°C until all four injections for that particular sample are completed.
7.4. Regenerate column with Buffer A for 7.4 min at 1 mL/min.

8. Repeat **step 7** three more times (total of four injections per sample; **Fig. 2** illustrates the intra- and intersample reproducibility of LC fractionation).
9. Pool, concentrate, and buffer exchange flow-through fractions (**Note 3**).

9.1. Pool collected flow-through fractions from injections 1, 2, and 3 for each sample in a 4-mL, 5000 MWCO Spin concentrator.
9.2. Centrifuge at ~4000 × *g* for 6 min at 4°C.
9.3. Add flow-through fraction from injection 4 for each sample to the concentrator and centrifuge at ~4000 × *g* until approximately 500 μL remains in the filter compartment.
9.4. Add 3.5 mL of 2 M urea to concentrator and repeat centrifugation until approximately 500 μL remains in the filter compartment.
9.5. Repeat **step 9.4** three times to ensure Buffer A has been sufficiently replaced with 2 M urea.

10. Collect the concentrated/buffer exchanged samples and store at –20°C for later used.

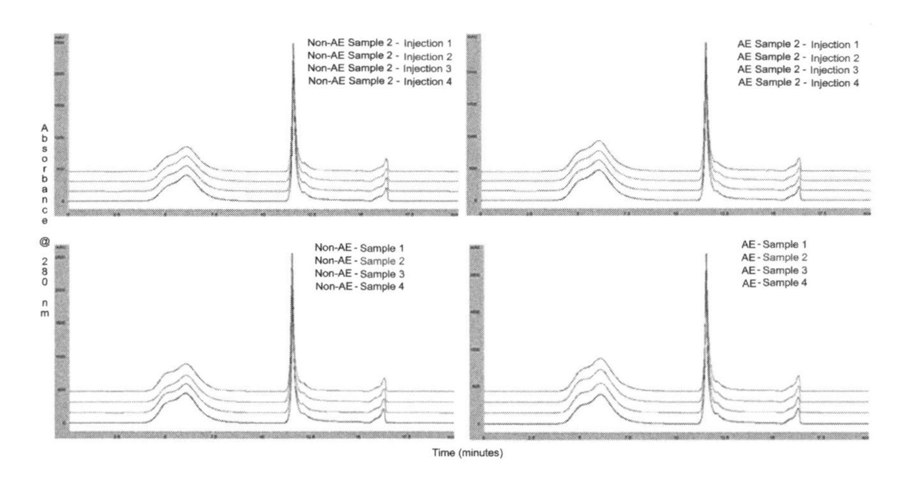

Fig. 2. Representative LC chromatographs demonstrating the reproducibility of affinity depletion of high-abundant plasma proteins from multiple injections of **(A, B)** the same sample and **(C, D)** multiple samples. **(A)** Four injections of non-AE sample 2: baseline sampling (prior to experimental drug administration). **(B)** Four injections of AE sample 2: termination sampling (end of drug administration). **(C)** Four injections of different non-AE samples: baseline sampling. **(D)** Four injections of AE samples: termination sampling.

3.2. Protein Quantification

The protein concentration of each pooled sample was estimated using the Non-Interfering Protein Assay Kit. This assay works in the presence of commonly used laboratory agents (reducing agents, detergents, amines, urea, etc.). The kit uses a proprietary Universal Protein Precipitating Agent (UPPA Genotech Biosciences, St. Louis, MO) for removing interfering agents, such as urea, from protein samples such that only clean protein remains in the tube to be assayed.

1. Thaw samples to be assayed on ice.
2. Prepare a standard curve (0 to 50 μg) by pipetting the appropriate volume of the BSA standard supplied with the kit into duplicate wells of a 96-deep well plate (VWR International, Westchester, PA) (**Note 4**).
3. Add 10 μL of each sample to duplicate wells of the plate.
4. Add 0.5 mL UPPA-I into each well, cover the plate with the sealing mat (VWR), and vortex.
5. Incubate the plate for 3 min at room temperature.
6. Add 0.5 mL UPPA-II into each well, cover the plate with the sealing mat, and vortex.
7. Centrifuge the plate at 8000 × *g* for 7 min to sediment precipitated protein.
8. Discard the supernatant by inverting the plate into an appropriate waste receptacle then softly blotting the plate on absorbent tissue paper. Allow the liquid to completely drain out of each well.
9. Add 100 μL Copper Solution and 0.4 mL water into each well, seal the plate, and vortex to completely dissolve the protein precipitate.
10. Add 1 mL Reagent II into each well—for simultaneous mixing, Reagent II should be added by rapidly shooting into the well.
11. Incubate the plate at room temperature for 20 min.
12. Remove 200 μL of solution and pipette into the corresponding wells of a 96-shallow well assay plate (VWR).
13. Read absorbance on a plate reader at 480 nm.
14. Plot absorbance against protein concentration of the standard curve, and determine concentration of unknowns.
15. Once protein concentrations have been determined, aliquot 150 μg of each sample into 1.5-mL Eppendorf tubes and store at –20°C.

3.3. In-gel Rehydration

In preparation for IEF, 150-μg aliquots of each sample were diluted to a total volume of 350 mL in rehydration buffer (see **Section 2.2**) (**Note 5**). The samples were loaded onto Immobiline DryStrip 18 cm pH 4–7 IPG gels during an overnight in-gel rehydration/sample loading step ***(18,19)***.

1. Thaw samples on ice.
2. Dilute each sample to a total volume of 350 mL in rehydration buffer.

3. Incubate at room temperature for 1 h with a 1-min water bath sonication every 15 min.
4. Clear any insoluble material through centrifugation at 14,000 rpm for 10 min at room temperature.
5. Level the Immobiline DryStrip Reswelling tray.
6. Pipette the entire 350 μL sample into the grooves of the Immobiline DryStrip Reswelling tray.
7. Peel off the protective cover sheets from the IPG strips, and position them such that the gel is in contact with the solution (**Note 6**).
8. Cover each IPG strip with ~2 mL of Immobiline DryStrip Cover Fluid (Amersham Biosciences) (low-viscosity paraffin oil).
9. Allow the strips to rehydrate overnight at room temperature.

3.4. Isoelectric Focusing

IEF was performed utilizing a Multiphor II IEF System at 20°C under the following voltage gradient: 0 to 500 V over 1 min; increased to 3500 V over 1.5 h and held at 3500 V for approximately 73,000 V·h.

3.4.1. Prepare the Multiphor II Unit

1. Set the thermostatic circulator to 20°C and begin cooling.
2. Pipette approximately 10 mL Cover Fluid on cooling plate. Position the DryStrip tray on the cooling plate so the anodic electrode connection of the tray is positioned at the top of the plate near the cooling tubes (**Note 7**).
3. Pipette 15 mL Cover Fluid into the tray. Place the DryStrip aligner (Amersham Biosciences), 12-groove-side up, into the tray on top of the Cover Fluid (**Note 8**).
4. Cut two IEF electrode strips (Amersham Biosciences) to a length of 110 mm. Place the electrode strips on a clean surface such as a pile of Kimwipes (Kimberly Clark Inc., Roswell, GA). Soak each electrode strip with 0.5 mL water. Blot with Kimwipes to remove excess water.

3.4.2. Load IPG Strips onto Multiphor II for IEF

1. Remove the rehydrated IPG strips from the reswelling tray and remove excess Cover Fluid by blotting on a Kimwipe, avoiding contact with the gel surface.
2. Immediately transfer the rehydrated IPG strips to adjacent grooves of the aligner. Place the strips with the acidic end at the top of the tray near the anode. Align the IPG strips so that the anodic gel edges are lined up.
3. Place the moistened electrode strips across the cathodic and anodic ends of the aligned IPG strips. The electrode strips must at least partially contact the gel surface of each IPG strip.
4. Align each electrode over an electrode strip, ensuring that the marked side corresponds with the side of the tray giving electrical contact. When the electrodes are properly aligned, press them down to contact the electrode strips.
5. Cover the IPG strips with 100 mL Cover Fluid.

6. Place the lid on the Multiphor II unit.
7. Connect the leads on the lid to the power supply. Ensure that the current check on the EPS 3501 XL power supply is switched off and begin IEF using the following gradient: 0 to 500 V over 1 min; increased to 3 500 V over 1.5 h and held at 3500 V for approximately 73,000 V·h.

3.5. Equilibration

The focused IPG strips were immediately equilibrated as follows: reduction through 15 min incubation in Equilibration buffer 1; followed by alkylation for 15 min in Equilibration buffer 2. This is a modification of the method described by Yan and co-workers *(20)* (see **Section 2.2**).

1. Remove the focused IPG strips and blot off excess Cover fluid on a Kimwipe.
2. Place in adjacent wells of the DryStrip Reswelling Tray and cover strips with 50 mL Equilibration buffer 1. Transfers tray to an orbital shaker and lightly agitate for 15 min at room temperature.
3. Aspirate off Equilibration buffer 1 and add 50 mL Equilibration buffer 2. Transfer tray to an orbital shaker and lightly agitate for 15 min at room temperature.

3.6. SDS-PAGE

Once equilibrated, strips were loaded onto Investigator Pre-cast 10% Tricine slab gels. Separation upon apparent molecular weight was conducted

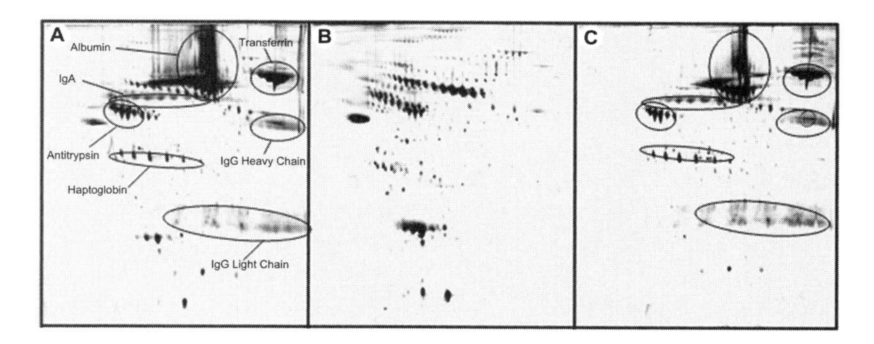

Fig. 3. Representative 2-DE protein profiles of unfractionated and fractionated human plasma proteins. **(A)** One hundred micrograms of unfractionated plasma, albumin, IgA, IgG, haptoglobin, transferrin, and antitrypsin account for 85% to 90% of the total protein mass of plasma and are targeted for removal. **(B)** One hundred micrograms of the flow-through fraction, low-abundance proteins previously masked by high-abundant proteins now become detectable. **(C)** One hundred micrograms of the eluted fraction; the six high-abundant plasma proteins are efficiently bound by the column and separated from other lower-abundant proteins.

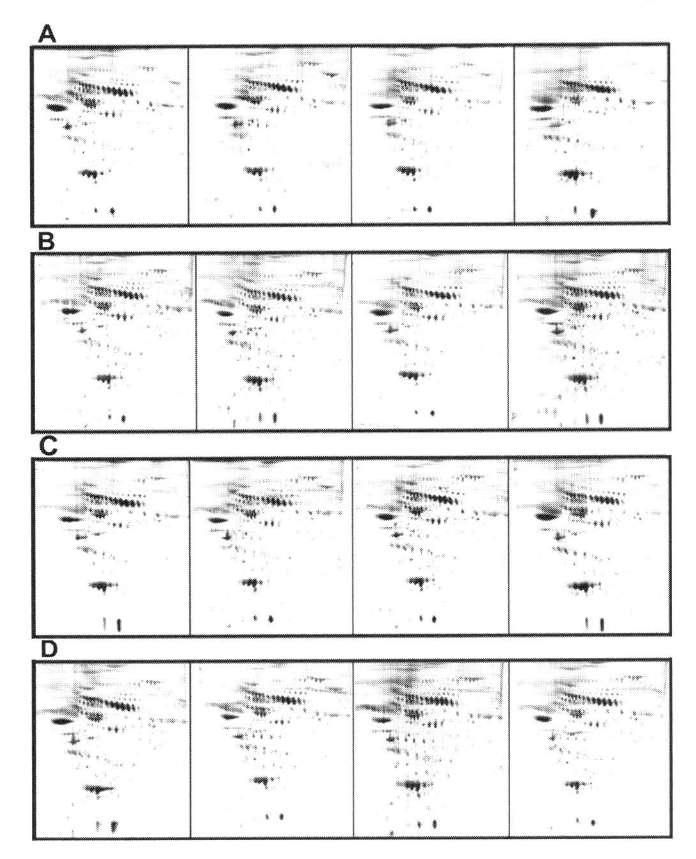

Fig. 4. Representative 2-DE gel protein profiles demonstrating the reproducibility of protein patterns after high-abundant plasma protein depletion. Each panel contains 2-DE gels from four different **(A, C)** non-AE or **(B, D)** AE patients from **(A, B)** baseline (prior to experimental drug administration) or **(C, D)** termination (end of drug administration) sampling time points. One hundred fifty micrograms of fractionated protein was loaded on each gel.

using the Investigator 2-DE Electrophoresis System at 15 W/gel for approximately 5 h.

3.6.1. Preparation of Gels and Buffers for SDS-PAGE (**Note 9**)

1. Prepare 10 L 210 mM Tris pH 8.9: dilute 1 L of 10X lower tank buffer with 9 L of water. Add to lower tank and allow chill to 4°C.
2. Prepare 3 L 100 mM Tris, 100 mM Tricine, 0.1% SDS: dilute 300 mL of 10X upper tank buffer with 2.7 L water. Place in the cold box and cool to 4°C.
3. Load the pre-cast gels into the gel rack and insert gaskets.
4. Add enough cooled upper tank buffer to cover the tops of the glass plates

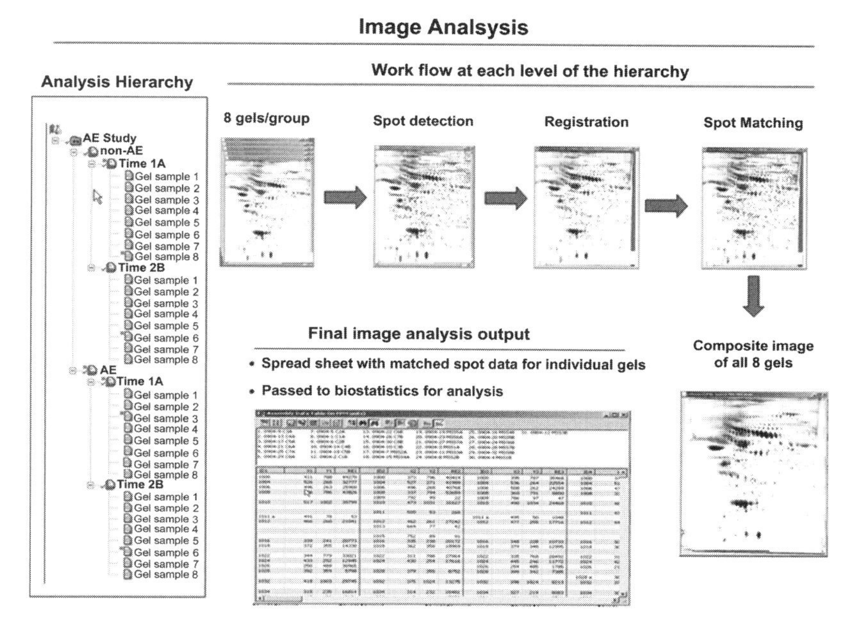

Fig. 5. Flow of 2-DE gel image analysis. The image analysis is performed in a hierarchical manner as illustrated in the boxed area to the left. The same procedure is followed for each group of eight gels (non-AE time 1A, non-AE time 2B, AE time 1A, and AE time 2B). The software detects spots, overlays and registers all gels in a group to match one specified reference gel from that group, matches spots on all the gels in the group, and then creates a single composite image of all the gels. This same procedure is then followed at the next level of the hierarchy: composite image of non-AE 1A is matched to the composite image of non-AE 2B; the same is done for the AE groups. The composite images from the non-AE and AE groups are then matched to one another, and the spot identification numbers are correlated between all the gel images in the experiment (i.e., the same spot on all the gel images will have the same spot identification). The final data output used for statistical analysis is a spreadsheet containing raw intensity data for each spot on each individual gel (the data from composite images are not used for analysis, these are simply a means to match and correlate spot identifications on all the gels in the experiment).

3.6.2. Load IPG Strips and Run SDS-PAGE

1. Remove strips from the equilibration buffer and blot off excess buffer.
2. Place the strip into the slits of the gel plates with the acidic end to the left.
3. Use the gel installer to push the strips down to make good contact with the SDS-PAGE gel.
4. Add the remaining upper tank buffer slowly to avoid disrupting the IPG strip position.

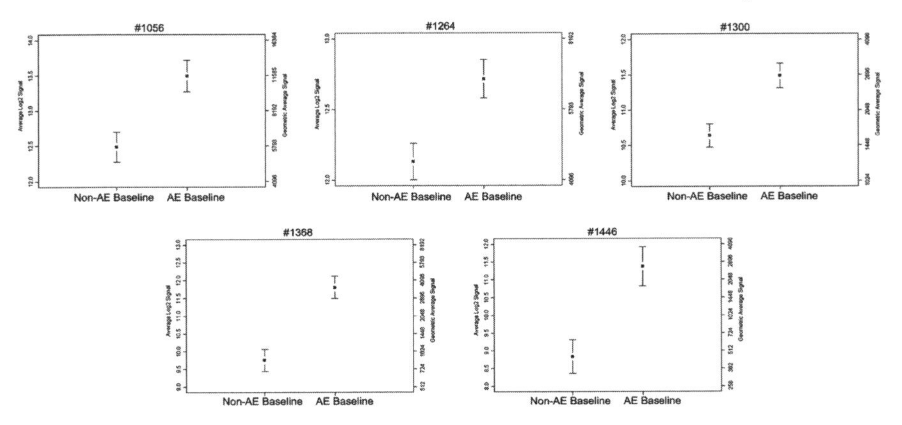

Fig. 6. Graphs depicting differential expression of the specified protein spots between non-AE and AE samples. These differences were present at the baseline sampling prior to administration of the experimental drug. Each filled square represents the average log2 spot expression level from the 2-DE gels of eight samples from each condition. The error bars represent the standard deviation of the mean. There were found to be statistically significant ($p < 0.005$) differences between non-AE and AE patients.

5. Place the lid on the tank and connect the lead to the power supply and start the 15 W/gel current. Run until the bromophenol dye front is approximately 1cm from the gel bottom.

3.7. Protein Pattern Visualization

Protein patterns were visualized through silver staining using the Investigator silver stain kit, which is a modification of the method described by Rabilloud *(21)*. The procedure was modified for nondestructive staining by omission of the glutaraldehyde in Fixative 2 solution.

3.7.1. Preparation of Staining Solutions

1. Fixative 1: Add 1.6 L ethanol to a 4-L beaker, then slowly add 400 mL acetic acid, bring up to 4 L with de-ionized water (final concentration of 40% ethanol and 10% acetic acid).
2. Fixative 2: Add contents of Fixative 2 bottle to a 4-L beaker and add 1.2 L ethanol, bring up to 4 L with de-ionized water, and mix (final concentration 8 mM potassium tetrathionate, 829 mM sodium acetate, and 30% ethanol).
3. Silver nitrate solution: Add content of silver bottle to a 4-L beaker and add 1 mL of formaldehyde solution, bring up to 4 L with de-ionized water, and mix (final concentration of 11.4 mM silver nitrate and 0.009% formaldehyde).

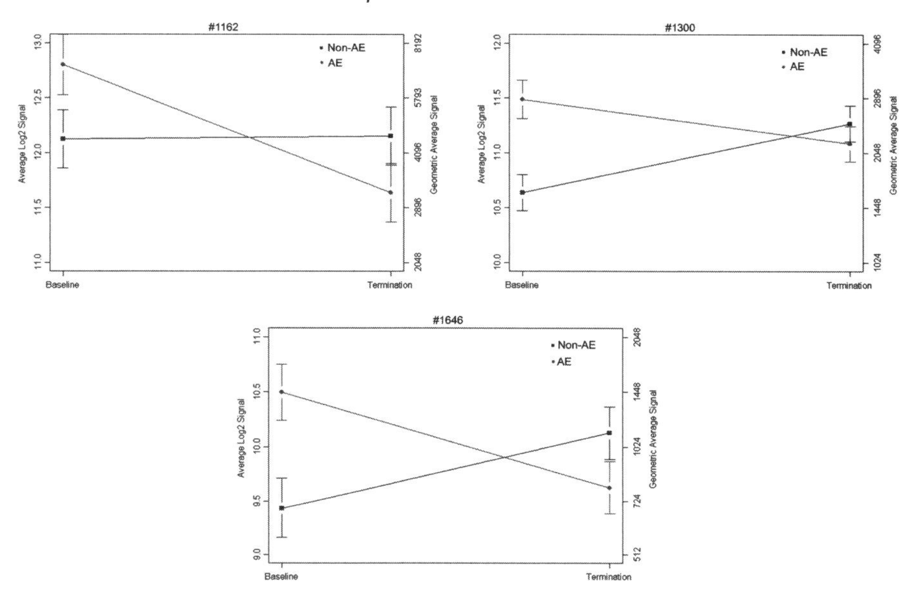

Fig. 7. Plots depicting "interaction" between condition and time. This means that the treatment group difference at one time point is significantly different ($p < 0.005$) from the treatment difference at the other time point. In other words, the time profiles for both treatment groups are statistically different and they are not parallel. Each filled square represents the average log2 spot expression level from the 2-DE gels of eight samples from each condition. The error bars represent the standard deviation of the mean.

4. Developer solution: add ~3.5 L de-ionized water to a 4-L beaker and stir briskly. While stirring, add contents of Developer bottle slowly until it is dissolved. Add 600 μL formaldehyde solution bring up to 4 L with de-ionized water, and mix (final concentration of 63 mM sodium thiosulfate, 0.2 M potassium carbonate, and 0.005% formaldehyde).
5. Stop solution: Add contents of Stop bottle to a 4-L beaker, add 80 mL acetic acid, bring up to 4 L with de-ionized water, and mix (final concentration of 0.4 M Tris and 2% acetic acid).

3.7.2. Silver Staining

1. Disconnect the gel unit from the power supply and disassemble the unit. Transfer the gels to trays containing 400 mL Fixative 1 solution. Gently agitate gels on a rotary shaker for 1 h at room temperature.
2. Decant off Fixative 1 solution and add 400 mL Fixative 2 solution. Gently agitate gels overnight.
3. Decant off Fixative 2 and add 400 mL de-ionized water. Gently agitate gels for 15 min. Repeat three times with fresh water each time.

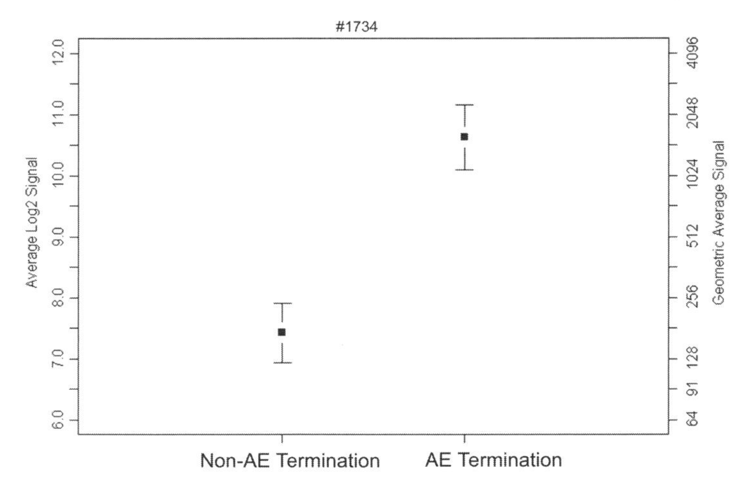

Fig. 8. Graph depicting differential expression of the specified protein spots between non-AE and AE samples. This difference was present at the termination (end of experimental drug administration) sampling. Each filled square represents the average log2 spot expression level from the 2-DE gels of eight samples from each condition. The error bars represent the standard deviation of the mean. There were found to be statistically significant ($p < 0.005$) differences between non-AE and AE patients.

4. Place gels in 400 mL silver nitrate solution and agitate for 1 h.
5. Decant off silver solution and wash gels in 400 mL de-ionized water for 1 min.
6. Decant off water and added 400 mL developer solution. Agitate gels until spots appear, ~8 min (**Note 10**).
7. Decant off developer solution and add 400 mL stop solution for 10 min, then add 400 mL de-ionized water. (**Figure 3** illustrates the protein pattern differences between crude and fractionated plasma samples, and **Fig. 4** illustrates the reproducibility of protein patterns after high-abundant plasma protein depletion).

3.8. Image Acquisition and Data Analysis

After electrophoresis and staining, gel image analysis is conducted to quantify protein spots, which allows gel-to-gel protein expression comparisons. This is usually accomplished with the help of commercially available 2-DE analysis software packages (**Note 11**). Although various software packages differ in price, degree of automation, analytical speed, and ease of use, they are generally labor intensive and require considerable time for familiarization. In general, 2-DE image analysis follows a series of steps: scanning the gel to digitize the protein pattern, filtering the image to remove background and artifacts, detecting spots, matching spots across gels, normalization, differential analysis, and statistical analysis ***(22)***.

Spot signals were measured by densitometric scanning. Image analysis was conducted using Z4000 Large-Scale 2D Gel Image Analysis System (according to the flowchart given in **Fig. 5**). The data with the raw signal values matched by spot was exported from Z4000 as a text file for further data analysis using SAS. The data was transformed to log (base 2) scale and normalized to have the same average log signal value across gels. ANOVA with mixed effects model (by spot) was used to perform statistical analysis of differential expression between conditions of interest allowing for correlation between observations from the same subject. For a pair of compared experimental groups, spots with p values less than 0.005 and at least two observations for both groups were identified as differentially expressed (**Figs. 6**, **7**, **8**).

4. NOTES

1. All solutions should be prepared in water of 18.2 MΩ cm and total organic content of less than five parts per billion.
2. Multiple injections of each sample were made in order to recover enough depleted protein to run several 2-DE gels. Consult the column certification of analysis to verify the columns capacity.
3. Eluted fractions were retained and stored at –80° C for future analysis. However, several were pooled, concentrated, and buffer exchange and resolved through 2-DE.
4. Our typical standard curve values are as follows: 0, 4, 8, 16, 24, 40, and 50 μg.
5. Sample preparation is by far the most important step for producing high-quality reproducible 2-DE gels. The goal of sample preparation is to have all proteins from the tissue of interest deactivated (proteases), denatured, disaggregated, reduced, and solubilized. In addition, the amount of interfering substances (i.e., RNA, DNA, etc.) should be reduced as much as possible; in particular, salt concentrations should be kept below 100 mM. Although a simple one-step extraction procedure would be ideal, no such procedure exists for universal application to the many types of tissue analyzed through 2-DE. Thus, sample preparation is generally tissue and protein class specific, but some guidelines should be followed; several reviews have provided detailed overviews of these ***(5,16,17)***.
6. Place IPG strip gel side down onto the rehydration buffer and avoid trapping air bubbles under the strip.
7. Remove any large bubbles between the tray and the cooling plate; small bubbles can be ignored.
8. The presence of air bubbles between the strip positions under the DryStrip aligner will not affect the experiment. Avoid getting IPG Cover Fluid on top of the aligner at this point, as it interferes with visualization of the grooves.

9. SDS-PAGE buffers and gel box assembly needs to be completed before IEF to allow time for the buffers to cool and reach 4°C. Typically, the buffers and gel unit are assembled shortly after IEF has been started and allowed to cool overnight.
10. This method is quantitative when the development time is kept to 10 min or less.
11. Commonly used software packages include ImageMaster 2D Platinum v6.0 (http://www.amershambiosciences.com), PDQuest (http://www.bio-rad.com), Progenesis and Phoretix (http://www.nonlinear.com), Z3 and Z4000 (http://www.2dgels.com), and Delta 2-D (http://www.decodon.com). Many of these packages are available in trial versions.

REFERENCES

1. O'Farrell PH. High resolution two-dimensional electrophoresis of proteins. J Biol Chem 1975;250(10):4007–4021.
2. Klose J. Protein mapping by combined isoelectric focusing and electrophoresis of mouse tissues. A novel approach to testing for induced point mutations in mammals. Humangenetik 1975;26:231–243.
3. Klose J. Genotypes and phenotypes. Electrophoresis 1999;20(4–5):643–652.
4. Hatzimanikatis V, Choe LH, Lee KH. Proteomics: theoretical and experimental considerations. Biotechnol Prog 1999;15(3):312–318.
5. Gorg A, Weiss W, Dunn MJ. Current two-dimensional electrophoresis technology for proteomics. Proteomics 2004;4(12):3665–3685.
6. Zolg JW, Langen H. How industry is approaching the search for new diagnostic markers and biomarkers. Mol Cell Proteomics 2004;3(4):345–354.
7. Scaros O, Fisler, R. Biomarker technology roundup: from discovery to clinical applications, a broad set of tools is required to translate from the lab to the clinic. BioTechniques 2005(Suppl);30–32.
8. Anderson NL, Anderson NG. The human plasma proteome: history, character, and diagnostic prospects. Mol Cell Proteomics 2002;1(11):845–867.
9. Tam SW, Pirro J, Hinerfeld D. Depletion and fractionation technologies in plasma proteomic analysis. Expert Rev Proteomics 2004;1(4):411–420.
10. Veenstra TD, Conrads TP, Hood BL, Avellino AM, Ellenbogen RG, Morrison RS. Biomarkers: mining the biofluid proteome. Mol Cell Proteomics 2005;4(4):409–420.
11. Cho Sy, Lee Ey, Lee JS, Kim HY, Park JM, Kwon MS, et al. Efficient prefractionation of low-abundance proteins in human plasma and construction of a two-dimensional map. Proteomics 2005;5(13):3386–3396.
12. Fang X, Huang L, Feitelson JS, Zhang WW. Affinity separation: divide and conquer the proteome. Drug Discovery Today Technol 2004;1:141–148.
13. Echan LA, Tang HY, Ali-Khan N, Lee K, Speicher DW. Depletion of multiple high-abundance proteins improves protein profiling capacities of human serum and plasma. Proteomics 2005;5(13):3292–3303.

14. Zolotarjova N, Martosella J, Nicol G, Bailey J, Boyes BE, Barrett WC. Differences among techniques for high-abundant protein depletion. Proteomics 2005;5(13):3304–3313.
15. Li X, Gong Y, Wang Y, Wu S, Cai Y, He P, et al. Comparison of alternative analytical techniques for the characterization of the human serum proteome in HUPO Plasma Proteome Project. Proteomics 2005;5(13):3423–3441.
16. Rabilloud T. Solubilization of proteins in 2-D electrophoresis. An outline. Methods Mol Biol 1999;112:9–19.
17. Shaw MM, Riederer BM. Sample preparation for two-dimensional gel electrophoresis. Proteomics 2003;3(8):1408–1417.
18. Rabilloud T, Valette C, Lawrence JJ. Sample application by in-gel rehydration improves the resolution of two-dimensional electrophoresis with immobilized pH gradients in the first dimension. Electrophoresis 1994;15(12):1552–1558.
19. Sanchez JC, Rouge V, Pisteur M, Ravier F, Tonella L, Moosmayer M, et al. Improved and simplified in-gel sample application using reswelling of dry immobilized pH gradients. Electrophoresis 1997;18(3–4):324–327.
20. Yan JX, Sanchez JC, Rouge V, Williams KL, Hochstrasser DF. Modified immobilized pH gradient gel strip equilibration procedure in SWISS-2DPAGE protocols. Electrophoresis 1999;20(4–5):723–726.
21. Rabilloud T. A comparison between low background silver diammine and silver nitrate protein stains. Electrophoresis 1992;13(7):429–439.
22. Marengo E, Robotti E, Antonucci F, Cecconi D, Campostrini N, Righetti PG. Numerical approaches for quantitative analysis of two-dimensional maps: a review of commercial software and home-made systems. Proteomics 2005;5(3):654–666.

9

Difference In-Gel Electrophoresis: *A High-Resolution Protein Biomarker Research Tool*

David S. Gibson, David Bramwell, and Caitriona Scaife

Summary

Difference in-gel electrophoresis (DIGE) is a recent adaptation of conventional two-dimensional gel electrophoresis (2-DE) that incorporates novel fluorescent labels, has multiplex attributes, and boasts software-assisted image analysis. Combined, these characteristics offer significant benefits in accuracy and reproducibility to quantify differential protein expression levels between biological samples. The DIGE technique and materials required to perform it are described in detail within. The principles behind consistent gel image acquisition and reliable image analysis are also considered. Within the context of biomarker and drug target discovery, this method simplifies analysis, increases sample throughput, and represents a reliable 2-DE platform.

Key Words: Biomarker; Cy dye; DIGE; fluorescent difference in-gel electrophoresis; proteomics; two-dimensional gel electrophoresis

1. INTRODUCTION TO DIFFERENCE IN-GEL ELECTROPHORESIS

Two-dimensional gel electrophoresis (2-DE) is an established platform that facilitates the analysis of complex protein mixtures. O'Farrell was first to introduce high-resolution two-dimensional electrophoresis by resolving proteins to individual isoelectric point and molecular weight coordinates *(1)*. The main asset of this method is that it provides a global view of the state of proteins within a sample. In theory, thousands of proteins can be visualized

From: *Methods in Pharmacology and Toxicology: Biomarker Methods in Drug Discovery and Development*
Edited by: F. Wang © Humana Press, Totowa, NJ

at once, giving a unique qualitative "map" or "fingerprint" of changes between given samples. Though many developments, such as standardized immobilized pH gradients, have led to vast improvement in inter-run consistency, deficiencies in sensitivity and spot matching have necessitated further adaptation using fluorescent stains. Comparison between large groups of conventionally (silver or Coomassie) stained gels is complicated by spot to spot warping, caused by variations in sodium dodecyl sulfate–polyacrylamide gel electrophoresis (SDS-PAGE) gel casting, electric and pH fields, and thermal fluctuations during electrophoresis. This leads to problems in spot matching and necessitates multiple gel replicates to prevent assumptions on mismatched proteins. In other words, gel to gel heterogeneity makes it difficult to distinguish with confidence between variations in the technique and those of genuine induced biological change, such as in disease states *(2)*. Difference in-gel electrophoresis (DIGE) addresses a number of these issues in that two to three samples can be subjected to exactly the same running conditions within a single gel. Unlu et al. developed DIGE to allow a more direct and reproducible comparison between protein samples, differentiated by prelabeling with spectrally resolvable fluorescent cyanine, or Cy, dyes *(3)*. The Cy dyes are charge matched with the residues they bind to within the proteins of a given sample and have similar molecular weights (0.5 kDa), thus result in only slight gel shifts. The Cy dyes are based on extended organic ring structures and hence are highly hydrophobic. Concerns with protein precipitation prior to electrophoresis have been surmounted by using a "minimal labeling" strategy, whereby binding is limited to only 1% to 2% of lysine residues available within a sample *(4)*.

Excitation of each fluor allows the creation of a digital image of each individually labeled sample. These dyes give additional validity to the two-dimensional technique in the form of higher sensitivity, wider dynamic range, and linearity of detection. Detection limits of 0.025 ng are possible, with a dynamic range around five orders of magnitude. One of the strongest features of the technique, however, is the ability to include an internal pooled standard, which is loaded on all gels within an experiment *(5)*. The internal standard permits the linking of all gels in an experiment, thus offering more reliable and intuitive software-assisted comparisons. The accuracy of protein quantification between samples is increased dramatically, and much smaller changes in protein expression can be studied with greater confidence. Evaluations of DIGE alongside traditional and more recent proteomic methods using isotope-coded or isobaric tags (cICAT and iTRAQ) reveal that it remains competitive in sensitivity and can be used with confidence as a platform for drug discovery and development *(4,6)*.

2. DIGE EXPERIMENT CONSIDERATIONS

Because three cyanine dyes are available, up to three separate protein samples can be labeled per gel. A pairwise analysis and organization could also be used (akin to gene chip analysis), where control versus drug-treated samples are labelled with Cy3 and Cy5 only. When normalization of expression levels is desired across a number of different experiments and within the one experiment, adequate quantities of each sample should be available to create a common pooled internal standard. The internal standard can be distinguished from experimental samples by labeling with Cy2 dye. Anomalies in spot intensity due to preferential labeling can be eliminated by randomized or reciprocal labeling, in which half of each experimental group is labeled with Cy3 and the other with Cy5 *(7)*. In order to distinguish intrinsic, interindividual biological variation from genuine changes in protein expression, biological replicates should be included in each experimental group. A recent study focused on the DIGE technique has shown that a minimum of four replicate gels is required to maintain a 95% chance of avoiding false negatives, when a twofold change in expression is considered significant *(8)*.

3. SAMPLE PREPARATION

Plasma and synovial fluid are used in this chapter to illustrate and describe the steps required for the purification and minimal fluorescent labeling of body fluid samples. For details of how to prepare cell lysates, with both minimal and saturation types of labeling, one can refer to the Ettan DIGE system user manual *(9)*. The following reagents and conditions have been used in our laboratory to produce reliable data with clinical relevance to patient outcome but could also be applied to prospective drug trial to monitor therapeutic response. Sample preparation should be consistent and kept as simple as possible to reduce inter-run inconsistencies. Protein modifications during sample preparation must be prevented, particularly degradation due to endogenous proteolytic enzymes. Such changes in samples analyzed by gel-based approaches can translate into misleading artifact spots with novel molecular weights.

3.1. Sample Purification and Assay

Cellular or particulate material should be removed from the body fluid by centrifugation prior to any further purification steps. This circumvents contamination by sub-proteomes other than that of the body fluid that is to be analyzed. Endogenous protease activity should be inactivated for reasons already eluded to above. A number of approaches are possible, with varying consequences to the resulting sample integrity. Protease inhibitor cocktails (such as the Complete Protease Inhibitor Cocktail Tablets; Roche, Diagnostics Ltd., Burgess Hill, U.K.) can be used to inactivate a wide variety of degradative enzyme classes including cysteine, matrix metallo and serine proteases. This remains our preferred method of body fluid stabilization, and an adaptation is now also available, in the form of a blood tube with proprietary inhibitors for immediate and convenient sample protection (BD P100, Becton Dickinson, Oxford, U.K.). Some authors, however, caution against their use in certain applications, as artifacts can result from modified protein charge, or peptide-based inhibitors such as leupeptin may interfere with mass spectrometry analysis ***(10,11)***. Proteases may also be inactivated by high or low pH extremes with Tris buffer or trichloroacetic acid (TCA), respectively, or alternatively total protein can be precipitated by TCA/acetone. In balance though, protein yield may be diminished by incomplete precipitation or resolubilization. Once stabilized, salts can be removed from protein samples (if higher than 10 mM) by dialysis with low-molecular-weight cut-off membranes, though if analysis of small peptides is desired, precipitation could be implemented. Other macromolecules such as lipids, polysaccharides, and nucleic acids should be removed by organic solvent, unless present at low concentrations (as with plasma). The sample can be lyophilized if concentration is necessary (5 to 10 mg/mL is an ideal protein concentration, though labeling of 1 mg/mL is possible) and resuspended in a minimal quantity of DIGE-compatible lysis buffer [DLB; 30 mM Tris, 7 M urea, 2 M thiourea, 4% (w/v) CHAPS, pH 8.5]. Ampholytes and dithiothreitol (DTT) are omitted from the lysis buffer prior to the labeling reaction as both primary amines and thiol groups will compete with the proteins for the available Cy dye. The pH of the sample to be labeled is also critical to the reaction, so check that the sample pH is 8.5 by spotting on a pH indicator strip and, if necessary, make drop-wise adjustments with dilute sodium hydroxide. The concentration of protein in each sample should be assayed either by Bradford reagent or using the proprietary Ettan 2D Quant kit (GE Healthcare, Lifesciences, Amersham, U.K.).

3.2. *Sample Labeling*

Aside from the pH and protein concentration recommendations already made, the efficiency of minimal dye labeling is dependent on the ratio of dye to protein (400 pmol Cy dye to 50 μg protein is cited in the DIGE user manual; GE Healthcare). The Cy dye fluors should be reconstituted in anhydrous dimethyl formamide under the manufacturer's guidelines to create a 1 mM stock solution. Each has a characteristic deep color as follows: Cy3, red; Cy5, blue; and Cy2, yellow (as shown in **Fig. 1**). The sample and dye quantities required for a six-gel, three-dye pilot experiment are shown

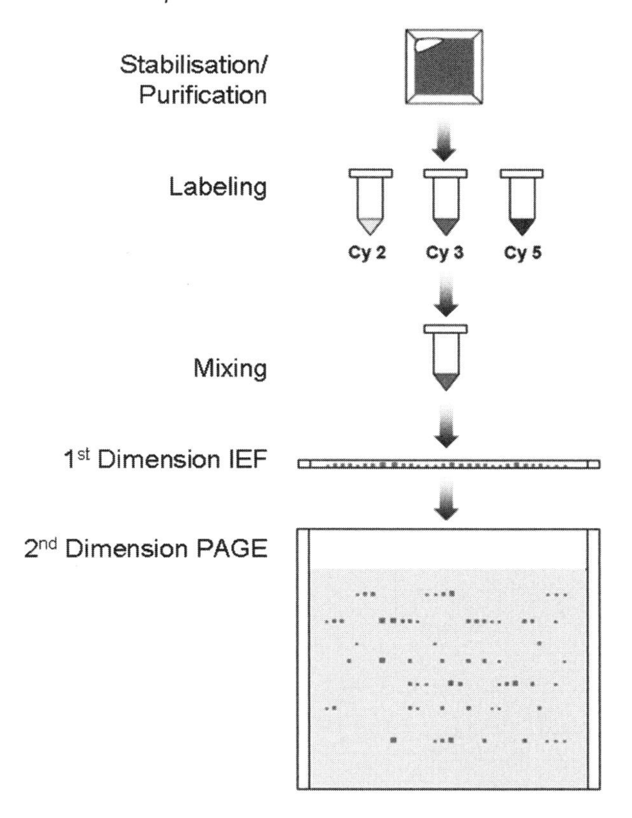

Fig. 1. Schematic representation of the laboratory procedures involved in a typical DIGE experiment.

in the following worked example (**Table 1**). Paired plasma and synovial fluid samples from six patients (A–F) are labeled with Cy3 and Cy5. An experimental design incorporating randomization of sample labeling and loading across gels is demonstrated to avoid systematic errors. An internal pooled standard is generated by combining equal amounts of all matched plasma and synovial fluid samples, followed by Cy2 dye labeling. Sufficient pooled internal standard is prepared to allow enough aliquots for each gel in the experiment. It is also prudent to create a slight excess (10% to 20%) of each dye reaction to ensure a complete aliquot is loaded on each gel. Thus for the individual plasma and samples, 60 μg is labeled with Cy3 or Cy5, but only 50 μg will be loaded of each. A single internal standard is therefore prepared, which comprises 30 μg of each of the 12 samples (6 plasma and 6 synovial fluid) and labeled with Cy2 dye. Before labeling, it is recommended that all sample concentrations are normalized to 10 μg/μL, to make subsequent pipetting easier (**Table 2**).

Table 1
The Sample and Dye Quantities Required for a Six-Gel, Three-Dye Pilot Experiment Are Shown for a Three-Dye Run Analyzing Plasma (PL) and Synovial Fluid (SF) from Six Patients (Anonymized as A, B, C, D, E, and F)

Gel	Cy2 pooled standard	Cy5	Cy3
1	50 μg (4.17 μg of each sample SF A–F and PL A–F)	50 μg of **PL** C	50 μg of **SF** D
2	50 μg (4.17 μg of each sample SF A–F and PL A–F)	50 μg of **SF** A	50 μg of **PL** F
3	50 μg (4.17 μg of each sample SF A–F and PL A–F)	50 μg of **PL** B	50 μg of **SF** E
4	50 μg (4.17 μg of each sample SF A–F and PL A–F)	50 μg of **SF** C	50 μg of **PL** A
5	50 μg (4.17 μg of each sample SF A–F and PL A–F)	50 μg of **PL** D	50 μg of **SF** B
6	50 μg (4.17 μg of each sample SF A–F and PL A–F)	50 μg of **SF** F	50 μg of **PL** E

Beforehand: Label Reaction Tubes

1. Label 12 Microfuge tubes (0.5 mL) as **Table 2** (*SF* A-F, *PL* A-F) for preparation of samples prior to labeling.
2. Label a second fresh set of 12 Microfuge tubes as above, for the labeling reaction.
3. Label one Microfuge tube as *PS* for the pooled standard.

Sample Preparation

4. Aliquot volumes of each sample equivalent to 100 μg into each of the first set of *SF* or *PL* individually labeled tubes. Adjust protein concentrations of all samples to 10 μg/μL by addition of DIGE compatible lysis buffer (DLB), as shown in **Table 2**.
5. Aliquot 6 μL of each normalized sample (60 μg) from the above, into each of the second fresh set of *SF* or *PL* individually labeled tubes and 3 μL (30 μg) of each sample into the one *PS* labeled tube. This gives a total of 36 μL (or 360 μg protein) in the pooled standard tube (*PS*).

Sample Labeling

6. The Cy dyes are diluted from 1 mM stock concentration to a working concentration of 400 pmol/μL with DMF, and 1.2 μL of the Cy3 and Cy5 dyes is added to individual samples in a randomized fashion as described in **Table 1**. An aliquot of 7.2 μL of 400 pmol/μL Cy2 is added to the pooled standard tube. (*Note*: Only reconstitute minimal quantities of Cy dye working dilutions for

Table 2
The Volumes of Samples and Diluent Required for Individual and Pooled Dye Reactions Are Shown*

Sample	[Protein] μg/μL	Volume required for 100 μg (μL)	Volume of DLB for 10 μg/μL (μL)	Volume add to PS labeling tube (μL)	Volume add to individual labeling tubes (μL)	Labeling tube
SFA	21.0	4.8	5.2	3	6	*SF* A
SF B	29.2	3.4	6.6	3	6	*SF* B
SF C	17.6	5.7	3.3	3	6	*SF* C
SF D	11.8	8.5	1.5	3	6	*SF* D
SF E	17.4	5.7	4.3	3	6	*SF* E
SF F	27.6	3.6	6.4	3	6	*SF* F
PL A	39.4	2.5	7.5	3	6	*PL* A
PL B	35.5	2.8	7.2	3	6	*PL* B
PL C	23.2	4.3	5.7	3	6	*PL* C
PL D	26.5	3.8	6.2	3	6	*PL* D
PL E	23.6	4.2	5.8	3	6	*PL* E
PL F	31.4	3.2	6.8	3	6	*PL* F
Total for pooled internal standard				36	PS	

*Sample concentrations are normalized to 10 μg/μL with DIGE compatable loading buffer (DLB).

the experiment as any remaining stock can be kept for future work at –20°C for 3 months.)

7. All labeling reaction tubes should be mixed thoroughly by pipette and vortex and pulse-centrifuged to collect the mixture at the bottom. Incubate the labeling reaction tubes on ice in the dark for 30 min. (*Note*: Subsequent exposure of all dye reactions to ambient light, whether in IPG strip or gel, should be minimized to prevent degradation/bleaching of the fluorophore.)
8. Add 1.2 μL of 10 mM lysine to each of the Cy3 and Cy5 dye reactions and 7.2 μL to the Cy2 dye reaction to stop the labeling. Again, mix and centrifuge briefly before incubating for a further 10 min on ice in the dark. The labeling reaction is now complete, and labeled samples can be stored for up to 3 months at –70°C in a light protected container, if not used immediately.

4. FIRST-DIMENSION ISOELECTRIC FOCUSING (IEF)

The ampholytes and DTT that had been omitted prior to the labeling are added at this point in the form of a 2X sample buffer. The sample is denatured with dithiothreitol (DTT) and a volume equivalent to 50 μg of each individual Cy3 and Cy5 labeled sample are then combined with 50 μg of the Cy2 pooled internal standard. The mixture is subsequently rehydrated onto 24-cm immobilized pH gradient (IPG) strips for highest resolution (sample in-gel rehydration). Samples with larger quantities of high molecular weight proteins, alkaline proteins, or hydrophobic proteins are likely to be poorly absorbed into the IPG strip gel matrix and would benefit substantially from cup loading detailed elsewhere ***(12)***.

Beforehand: Prepare Buffers

1. Prepare 2X sample buffer [8 M urea, 130 mM DTT, 4% (w/v) CHAPS, 2% (v/v) IEF ampholytes 4–7] and rehydration buffer (8 M urea, 13 mM DTT, 4% (w/v) CHAPS, 1% (v/v) IEF ampholytes 4–7].
2. Remove pH 4–7 IPG strips from freezer to thaw on bench and ensure rehydration tray is level. Note strip numbers and label six fresh microcentrifuge tubes as **Table 1** ***(1–6)***.

Pooling Samples, Strip Rehydration, and Isoelectric Focusing

1. Add equal volumes of 2X sample buffer to individual Cy3 and Cy5 labeled samples (8.4 μL each) and to the Cy2 labeled pooled standard (50.4 μL), mix and leave on ice for 10 min (each tube now has 50 μg labeled protein in 14 μL).
2. Aliquot 14 μL of each Cy3, Cy5, and Cy2 labeled samples to be focused on the same IPG strip into the tubes as indicated in **Table 1**. Add 408 μL of rehydration buffer to each tube, mix, and centrifuge briefly.
3. Pipette each mixture into a separate channel of the rehydration tray. Peel off the protective cover from the IPG strip and carefully lower it gel side down into the rehydration buffer-sample; remove any air bubbles with a pipette tip.

Table 3
Isoelectric Focusing Conditions Appropriate to Proteins Soluble Within the Acidic Range pH 4–7 for 24-cm IPG Strips

Voltage mode	Voltage(V)	Duration (h:min)	Volt-hours (kVh)
1. Step	3500	—	75
2. Gradient	1000	0:10	—
3. Step	8000	1:00	—
4. Step and hold	100	>24 h	—

4. Overlay each strip with ~2 mL of IPG strip cover fluid to prevent evaporation and slide on the plastic cover. Cover with aluminum foil and incubate overnight (or at least 10 h) at room temperature. Low voltage applied during rehydration can improve entry of high molecular weight protein ***(13)***.
5. After adequate rehydration, use clean forceps to remove the IPG strips from the rehydration tray. Blot off excess fluid and carefully position the IPG strips (gel side up) within the ceramic manifold of the IPGphor isoelectric focusing unit (GE Healthcare). Ensure that the positive (anodic) end of each strip is oriented toward the anode of unit.
6. Apply one filter paper electrode pad, which has been moistened with de-ionized water (remove excess fluid with blotting paper), to each end of the IPG strip in such a way as to overlap the gel by approximately 3 to 5 mm. Place the respective electrodes over the filter paper pads and clip firmly in place and overlay each strip with cover fluid to prevent dehydration (108 mL for whole manifold).
7. Close the IPGphor IEF unit safety lid and the instrument running conditions can be programmed. We suggest conditions appropriate to proteins soluble within the acidic range pH 4–7, for 24-cm IPG strips (**Table 3**). Current should be limited to 50 μA per IPG strip. Other run conditions for the various strip sizes and pH ranges are available [Ettan IPGphor manifold user manual (80-6499-52 edition AO); GE Healthcare].
8. Once the strips have been focused, they can be equilibrated straight away for the second dimension or stored for several months at –80°C between plastic sheets to prevent damage to the brittle frozen strips.

5. SECOND-DIMENSION SDS-PAGE (2-DE)

In order to improve protein migration from focused IPG strip to the second-dimension separation gel, it is important to equilibrate the strips in a buffer containing sodium dodecyl sulfate (SDS), urea, and glycerol. In the first equilibration step, DTT is added to completely reduce any remaining disulfide bonds, whereas iodoacetamide is added in the second step to alkylate the resultant sulfydryl groups thereby preventing reoxidation, which may complicate downstream mass spectrometry identification. In addition, the iodoacetamide "mops up" any free DTT, which may otherwise cause point-streaking artifacts, apparent in silver-stained gels. The Protean Plus Dodeca multiple gel unit (Biorad, Hemel Hempstead, U.K.) is used in the worked example, as it can accommodate up to 12 large-format slab gels, has an in-built buffer recirculation and cooling system and plate electrodes for consistent resolution. A recipe for a 12% homogenous polyacrylamide Tris-glycine gel is also given using stabilized, high-purity Protogel reagents (National Diagnostics, Atlanta, Georgia, U.S.A), though gradient gels and other buffer systems may give better resolution for select protein molecular weight ranges ***(14,15)***. It should be noted that the nature of the Cy dyes used in the DIGE technique necessitate the use of specialized low-fluorescence glass plates. The glass plates require thorough cleaning with dilute alcohol and de-ionized water using lint-free tissues to remove any dried salts or gel fragments prior to casting.

Beforehand: Prepare Equilibration Buffer and Gels

1. Prepare the stock SDS Equilibration buffer [50 mM Tris pH 8.8, 6 M urea, 30% (v/v) glycerol, 2% SDS (w/v)] (10 mL per strip will be required, excess can be aliquoted and stored at –20°C for future use).
2. Prepare the working-strength Gel Running buffer [25 mM Tris base, 192 mM glycine, 0.1% SDS (w/v)] (up to 25 L is required to fill the Dodeca gel tank).
3. Prepare 600 mL of gel casting solution (sufficient for six 1.0-mm, 12% gels; 100 mL per gel) [240 mL 30% (w/v) acrylamide/methylene bisacrylamide solution (37.5:1 ratio), 156 mL 4X Laemmli Resolving Gel buffer (0.375 Tris-HCl pH 8.8, 0.1% SDS final concentration, 197.4 mL de-ionized water, 240 μL tetramethylethylene-diamine (TEMED), 2.4 mL of 10% (w/v) ammonium persulfate (APS)]. Pass the solution through an 0.2-μm filter and degas under vacuum, prior to the addition of APS.
4. Apply waterproof adhesive tape to the sides of assembled glass plates and spacers, place into the gel caster with plastic sheet spacers, and gradually pour in the prepared mixture. Carefully layer ~2 mL of water-saturated butanol on top of each gel to remove bubbles and create a level surface. Flush the butanol off the gels with double de-ionized water after approximately 1 h and cover the casting unit with tin foil. Ideally, gels should be cast the day before use to ensure complete polymerization.
5. Prepare a 1.0% agarose solution in Gel Running buffer with 50 mg bromophenol blue incorporated per 100 mL. Dissolve the agarose by a short incubation in a microwave on low-medium power.

IPG Strip Equilibration

1. Prepare Equilibration buffer A by dissolving 100 mg of DTT per 10 mL of stock SDS Equilibration buffer (5 mL needed per IPG strip). Dispense a 5-mL

aliquot of buffer A into equilibration tubes and place the IPG strips carefully into each. (Disposable plastic 10-mL pipettes with the conical tip broken off are sufficiently long; reseal with parafilm between incubations.) Incubate the strips at room temperature with gentle rocking for 15 min, then decant the buffer.

2. Prepare buffer B by dissolving 250 mg of iodoacetamide per 10 mL of stock SDS Equilibration buffer (5 mL needed per IPG strip). Dispense a 5-mL aliquot of buffer B into each equilibration tube and reseal. Equilibrate the strips for a further 15 min at room temperature with gentle rocking, decant the buffer, and proceed to electrophoresis section.

SDS-PAGE

1. Rinse the equilibrated IPG strips in Gel Running buffer and using forceps place each strip across the top of a gel, such that the plastic backing of the strip makes contact with the back glass plate and the anodic end of the strip is at the top left of the gel. A thin spatula can then be used to maneuver the strip down into the well, with care so no bubbles are introduced between strip and gel. A good tip is to push one end of the IPG strip down, between the glass plates, so it makes contact with the top of the gel and gradually push the opposite end down so the strip sits level on the surface.
2. Layer approximately 2 mL of premelted 1% agarose sealing solution carefully across the strip, again try to eliminate any air bubbles trapped between the strip and gel. Allow to cool and solidify for 5 min and repeat the procedure for the outstanding IPG strips.
3. Insert the prepared gel cassettes into position in the electrophoresis unit (this can be made easier by prewetting each cassette by dipping it briefly into the Gel Running buffer). Ensure the buffer chamber is filled to the manufacturer's recommended level and place the safety lid on paying attention to the orientation of the electrodes.
4. Start the electrophoresis with 2 W per gel for 45 min followed by 17 W per gel for 4 h, both at 20°C. Alternately, for convenience, the gels can be run overnight at 0.75 W per gel. (Such an overnight run can take anything from 18 to 21 h.) The run should be terminated when the bromophenol blue dye front reaches the bottom of the gel.
5. In preparation for scanning the gels, the scanner instrument should be switched on and left to warm up for 30 min and the glass plate should be thoroughly cleaned with lint-free tissues. As DIGE gels are scanned while they are in the glass cassettes, these too should be cleaned carefully to remove any residual running buffer, gel smears, and so forth, from the surface of the plates. If not scanned immediately, gels can be individually wrapped in cling film and stored at 4°C but should be imaged as soon as possible to avoid signal loss and dissipation of focused spots. Gels stored at 4°C should also be allowed to come up to room temperature before scanning to avoid any condensation forming on the surface of the glass.

6. IMAGE ACQUISITION

The capture of gel image data is a critical stage in the whole analysis workflow that can often present a source of easily prevented experimental noise. It is possible to make careless choices at this stage that can result in a 200-fold drop in sensitivity. A variety of DIGE-compatible gel scanning instruments based on charge-coupled device (CCD) or photomultiplier tube (PMT) detection technologies are available (GE Healthcare, Lifesciences, Amersham, U.K., Typhoon; Syngene, Cambridge, U.K., ChemiGenius; Fuji flim Lifescience, Dusseldorf, Germany, FLA 3000; Biorad, Hemel Hempstead, U.K., FX Molecular Imager). Depending on the make of scanner, there are several parameters to optimize; the spatial resolution, the dynamic range, the scan content, and background offsets. These features are therefore briefly discussed and recommendations made in each case to improve the process of image acquisition.

6.1. Spatial Resolution

Image (or spatial) resolution relates to the number of pixels (picture elements) displayed per unit length of a digital image and is often measured in dots per inch (dpi) or in micrometers (the size of the area each pixel represents). Images with a higher spatial resolution are composed of a greater number of pixels and have more image detail than those of lower spatial resolution. It is important to be aware that variations in spatial resolution will not only affect the final appearance of the image but will also impinge on the quality of spot detection and the accuracy of any subsequent quantitative measurements.

At low resolutions, there are fewer pixels available to represent each spot, and as a result, spot detection and quantitative accuracy will be compromised. Image (spatial) resolution is illustrated in more detail in **Fig. 2**. Higher resolution means that more pixels, and hence more data, are available for the analysis: the spot indicated by arrow in **Fig. 2** is represented by 63 pixels at a relatively low 100 dpi resolution (**Fig. 2A**), compared with 485 pixels at a higher 300 dpi resolution (**Fig. 2B**).

There is, however, a maximum resolution, which once exceeded produces minimal additional information. Once resolution is sufficient to adequately represent the smallest features, any further increases in spatial resolution simply increase the ability to represent the system "noise." In addition, every doubling in spatial resolution quadruples the amount of data that has to be processed, which can cause problems in processing speed and file and memory management. For example, a typical 20 × 20 cm DIGE single dye image captured at 100 dpi versus 300 dpi would result in diverse file sizes of 1.2 Mb and 10.6 Mb, respectively (both at 16-bit depth).

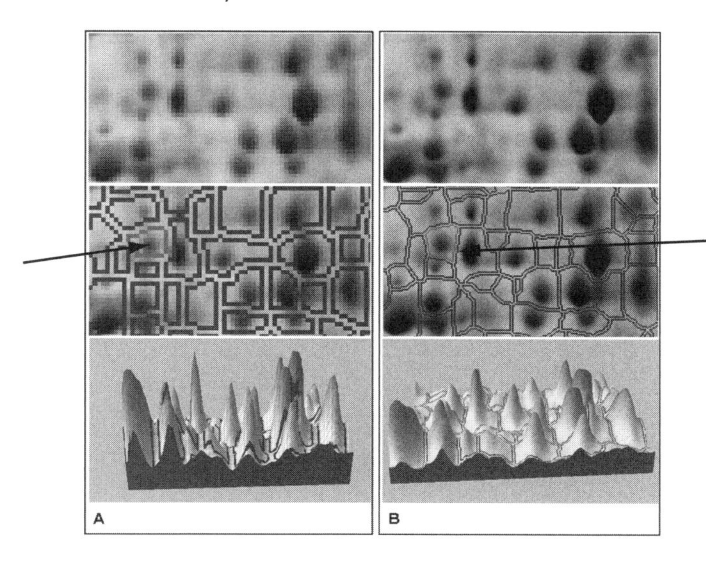

Fig. 2. Portions of two-dimensional gel image scans showing typical image quality, spot outline, and three-dimensional spot intensity (top to bottom) at 100 dpi (8-bit) and 300 dpi (16-bit) resolutions. The spot indicated by arrow is represented by 63 pixels at 100 dpi (A), compared with 485 pixels at 300 dpi (B) resolution.

To summarize, in most situations, 300 dpi or 100 µm will provide an image that is large enough for accurate analysis and small enough for efficient processing. However, if your gels are small (e.g., minigels), then you may need to increase the resolution to achieve this. As a rule of thumb, the active area of the gel (i.e., the area of spot material) should fall in the range 1000 to 1800 pixels in both horizontal and vertical directions. This range provides a good trade-off in information content and analysis performance.

6.2. Dynamic Range (Bit Depth)

Also referred to as color depth, bit depth or pixel depth is the number of bits used to represent the color (grayscale, intensity levels) of each pixel in an image. Greater bit depth allows a greater range of colors or shades of gray to be represented by a pixel. If possible, scan at 16-bit rather than 8-bit. The bit depth of a 16-bit image (65,536 levels of grayscale) compared with an 8-bit image (256 levels of grayscale) results in enhanced sensitivity and accuracy of quantification for less-abundant proteins. The possible grayscale levels available along with the resultant dynamic range (orders of magnitude) for the types of images commonly used in two-dimensional gel image analysis are indicated in **Table 4**. In reality, the images displayed on the computer

Table 4
The Resultant Grayscale Levels and Dynamic Range (Orders of Magnitude) Are Shown for a Range of Image Bit Depths Commonly Used to Analyze Two-dimensional Gels

Bit depth	Intensity (grayscale) levels	Orders of magnitude	Percent of 16-bit scan
8	256 (2^8)	2.4	0.39
10	1024 (2^{10})	3.01	1.56
12	4096 (2^{12})	3.6	6.25
16	65,536 (2^{16})	4.8	100

screen will only be represented in 256 shades of gray, and so an 8-bit image will look identical to a 16-bit image by eye. However, image analysis software can distinguish between the different levels of gray. As a rule, the more levels of gray represented in an image, the better the ability to differentiate low-abundance spots from background, and the greater the quantitative accuracy. This is further illustrated in **Fig. 2**, comparing spot detection in an identical area on the same two-dimensional gel, captured at 8-bit and 16-bit, respectively.

The dynamic range can be adjusted in CCD camera systems by altering the exposure time and in laser-based systems by fine tuning the voltage of the PMT detector. The dynamic range should be optimized to maximize the use of available grayscale values. Aim for the maximum gray levels in the image to be 5% to 10% less than those available. Scanner response curves can be nonlinear, and inconsistent settings can cause issues. For DIGE experiments without a Cy2 internal standard, chose settings that optimize the dynamic range for each stain and keep these consistent. When optimizing the dynamic range, it is important to avoid saturation effects. Saturation occurs when gray levels exceed the maximum available. When a spot becomes saturated, any differences in high pixel intensities cannot be resolved, and the spot appears truncated when viewed in three dimensions. No reliable quantitative data can be generated from a saturated spot, and saturated spots may also have an overall effect on normalization if included in later analysis.

6.3. Scan Content

Wherever possible, it is best to try to keep the area scanned as valid gel area. It is common to see scans where there are lots of scanner bed, labels, and so forth, in the captured image. Some scanners automatically adjust the scan settings based on what they "see" so a significant proportion of the dynamic range is lost by representing scanner bed or labels. It may be best to switch off auto gain control features or, at least, outline the areas you are interested in so the scanner optimizes only the region of interest. Extra "non gel" areas provide no useful information and should be cropped prior to image analysis. These scanning artifacts can skew the image statistics, "steal" dynamic range, increase storage requirements, and cause extra work in manual stages of analysis. Another good tip for consistency (and to save time later) is to always scan gel images using the same orientation and with the same settings.

Postscan processing of two-dimensional gel images using Adobe Photoshop or other general image processing software should be avoided, as these do not maintain the integrity of your original data, and any calibration information contained in the image file will be lost. The manipulations may make the images "look better" to the human eye but are simply transforming the original data. If the images look bad in the first place, then you should try to optimize the scanning not manipulate the images digitally. If possible, use GEL or IMG/INF files formats rather than generating TIFF files. The former often contain additional grayscale calibration information, which will not be included in the TIFF version. In any case, do not use JPEG file storage format as it is a standard for "lossy" image compression, which is optimized to allow the loss of information that is least noticeable to the human eye. This does not mean it does not affect measures made by computers. Lossy compression throws information away and manipulates the image data. Converting a JPEG image back to a TIFF is not a solution; once the image has been compressed in this way, the data has been lost and cannot be retrieved.

7. IMAGE ANALYSIS: SOFTWARE PRINCIPLES AND WORKFLOWS

Image analysis of conventional silver or Coomassie stained two-dimensional gels of individual samples can be highly subjective and very time consuming, due to inherent unpredictable distortions between gels. These inconsistencies prevent the perfect alignment and matching of spots between gels. On the other hand, DIGE-derived images from the same gel are precisely superimposable, and gel to gel spot matching is uniquely assisted by the internal standard (Cy2 labeled). So by virtue of the prelabeling and comigration of differentially labeled samples, variation in spot "coordinates" and intensity are accounted for, and gel images analysis is much more efficient ***(16)***. It is important to understand the principles

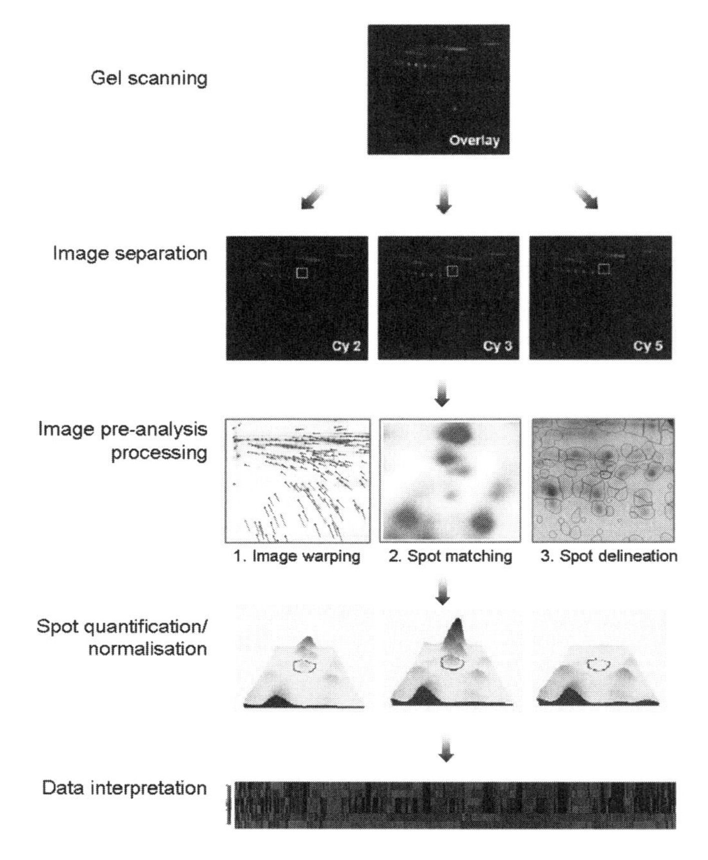

Fig. 3. Schematic representation of the processes involved in DIGE image analysis. (*see* Color Plate 4, following p. 230)

that traditional software applications apply, highlight their limitations, and provide a recommended analysis workflow (depicted in **Fig. 3** and *see* Color Plate 4, following p. 230).

7.1. Image Warping and Spot Matching

The biggest problem in the image analysis of a gel-based experiment is data alignment. This is easy to see by comparing the state of the art statistical analysis for array-based experiments and the sorts of analysis performed on gel-based systems. The key difference between the two workflows is that the exact locations of data points are known in the array scenario. In traditional two-dimensional gel analysis, the alignment issue is tackled by a between gel spot matching stage. Most traditional analysis strategies follow these key steps; detect spots on each gel individually, attempt to match the spots across the experiment (possibly with the assistance of whole image gel warping), measure spots, apply some form of normalization, apply statistical analysis ***(17)***. The recommended workflow (**Fig. 3**) is distinguished by initially warping gels, then matching and delineating spots, and imposing the same spot outlines across all gels prior to quantification.

The main problem with the traditional workflow is that it is not able to provide data of sufficient quality required to perform advanced statistical analysis. The core issue is focused on missing values; that is, a data point (usually a spot) that is not available across all samples. A recent study measured 42% missing values (i.e., not experimentally induced data omission) for a 16-gel experiment ***(18)***. It has also been shown that the number of missing values increases with the number of replicates ***(19)***. This produces the predicament of reducing the extent and quality of data, as the investigator endeavors to improve statistical power by increasing the number of replicates. In the traditional analysis workflow, missing values arise from two main sources: (1) the same measurements not being taken from each gel (usually due to spot detection) and (2) the measurements not being correctly matched between gels.

The same measurement not being taken from each gel has two main effects. The first is down to the fact that in traditional analysis scenarios, each gel has spots detected in isolation. This can lead to inconsistent results because essentially the pattern is determined from a single instance and as such is more prone to technical variance. The second issue is usually attributed to experimental conditions where a spot has zero expression in one or more of the groups. In this case, a spot will not be detected on an individual gel basis, and we are left with a hole in the data. Strangely, the proteomics community tends to differ with standard scientific practice with zero expression spots preferring to have no measurement or "unmatched" rather than measuring the value "zero." This is analogous to measuring air temperature and saying "there was no temperature" when it hits zero instead of measuring and recording zero. This particular stance also forces a multiple instead of a unified statistical framework approach to analysis, which is laborious, prone to bias, and still may be suboptimal.

7.2. Spot Delineation/Outline

Geometric correction alone does not solve all of the issues as we would still be prone to threshold of detection issues on a per gel basis and also not matching spots that were not expressed in certain groups. The obvious next step is to stabilize and standardize the spot detection across all of the images within an experiment. The geometric correction combined with an "ensemble," experiment-wide, spot pattern would go a large way to not only removing the bulk of missing values but also removing a large amount of manual intervention. We would also gain most in the low-expression spots that are harder to detect consistently on a per gel basis. If we apply geometric correction and derive a suitable outline for spots that are not undergoing experimentally induced variation, then the software can handle more than half (usually more) of the spots in an experiment fully automatically. This is essentially reverse logic where it is more efficient to bias the analysis to discount what is not changing rather than to optimize for what may be. This also removes the situation where you have to spend a lot of time editing a series of spots to find "they don't change." Considering the efficiency of overall analysis, applying the same spot outlines across geometrically corrected images is advantageous, and the benefits increase as the number of spots subject to experimentally induced expression changes decreases. A restriction imposed by the same spot outlines is that the area considered to contain a spot is consistent across images. This means that there are only going to be issues if the expression change is so large that it cannot be represented by the same area across the gels. For spots in isolation, this is not an issue as a larger area can be used (any extra pixels included in the larger boundary will be zero in the smaller spots). Tight clusters of spots that vary slightly in location and potentially patterns of spots that overlap when considered across all groupings may however prove problematic and require additional editing. Interestingly, posttranslational modification shifts of proteins are better handled by the same outlines workflow than the traditional approach, as one can apply a robust statistical framework to discovering them. Therefore, contrary to current practice and "gut feel," it appears to be more favorable to apply the same spot outline pattern to geometrically corrected gels and robustly analyze the bulk of the spots within the experiment in a "first pass" with minimal manual intervention and hence bias.

Once the same outlines analysis has been completed, one will have a list of "interesting" areas. These will either be completely satisfactory with the automatic outlines or it may be deemed that editing should be considered. It is advisable that the initial analysis be completed fully without intervention and this analysis saved. From this, a list of areas for further examination should be made and these explored in a mode that adds to the information and not replaces it. At this point, we may consider editing outlines. The editing process is less biased when one uses the same outline across all gels within the experiment. This is because the volume of a spot is affected by the number of pixels chosen to include in its boundary; if imposed inconsistently between groups, it is easy to introduce unintentional bias. Because there is a much reduced requirement on manual intervention, one can spend a lot more time on the areas that matter. Feedback from multiple labs has shown that differential editing is rarely necessary at all and, if it is, tends not to be a major factor in the results of the experiment. Areas where this is deemed necessary should always be treated with caution and may be a target area for future experiments where zoom gels are used or alterations to the sample preparation improve the data quality in these areas.

In summary, the suggested workflow is

1. Geometrically correct your gels.
2. Create a spot pattern representative of all of the spots within your experiment.
3. Analyze the spots completely.
4. Complete statistical analysis.
5. Create a copy of the experiment and edit/apply differential outlines on areas where it is deemed strictly necessary.
6. Complete statistical analysis.

8. PREPARATIVE GELS FOR SPOT PICKING

Because DIGE offers superlative sensitivity to quantify minute differences in protein expression levels, relative to conventional staining techniques, sample load is reduced accordingly (from as low as 75 μg to 200 μg per gel). Once spots of interest have been identified, it is then necessary to run a "preparative" gel with a higher sample loading (0.5 mg to 2.0 mg) ensuring sufficient protein in these spots to obtain reliable identifications by mass spectrometry ***(20)***. A spot "pick list" can be generated, such that isoelectric point and molecular weight coordinate data derived from the DIGE "analytical" gels can be transposed onto a Coomassie or silver stained gel.

9. CONCLUSION

Whereas the concept of DIGE is relatively novel, it has unique attributes that make it particularly suitable in the drug and biomarker discovery process. Because the technique is not dependent on antibody-protein or oligonucleotide-RNA/DNA affinity, it does not preclude the target identity, therefore represents changes in the sample without preemptive bias. Although extended multiplex capability and an increased dynamic range are desirable, it remains the most sensitive validated gel-based proteomic tool with direct relevance to innovation in clinical and pharmaceutical research. Improvements in dedicated analytical software have greatly increased the throughput and confidence in data derived from DIGE images. DIGE holds much promise in the challenge to uncover new molecular targets, screen putative drug efficacy, and monitor therapeutic response for a wide range of debilitating diseases.

REFERENCES

1. O'Farrell PH. High resolution two-dimensional electrophoresis of proteins. J Biol Chem 1975;250(10):4007–4021.
2. Marouga R, David S, Hawkins E. The development of the DIGE system: 2D fluorescence difference gel analysis technology. Anal Bioanal Chem 2005;382(3):669–678.
3. Unlu M, Morgan ME, Minden JS. Difference gel electrophoresis: a single gel method for detecting changes in protein extracts. Electrophoresis 1997;18(11):2071–2077.
4. Tonge R, Shaw J, Middleton B, et al. Validation and development of fluorescence two-dimensional differential gel electrophoresis proteomics technology. Proteomics 2001;1(3):377–396.
5. Alban A, David SO, Bjorkesten L, et al. A novel experimental design for comparative two-dimensional gel analysis: two-dimensional difference gel electrophoresis incorporating a pooled internal standard. Proteomics 2003;3(1):36–44.
6. Wu WW, Wang G, Baek SJ, Shen RF. Comparative study of three proteomic quantitative methods, DIGE, cICAT, and iTRAQ, using 2D gel- or LC-MALDI TOF/TOF. J Proteome Res 2006;5(3):651–658.
7. Lilley KS, Friedman DB. All about DIGE: quantification technology for differential-display 2D-gel proteomics. Expert Rev Proteomics 2004;1(4): 401–409.
8. Karp NA, Lilley KS. Maximising sensitivity for detecting changes in protein expression: experimental design using minimal CyDyes. Proteomics 2005;5(12):3105–3115.
9. GE Healthcare. Ettan DIGE System User Manual. AB(18–1173–17). GE Healthcare, Lifesciences, Amersham, U.K., 2006:1–112.
10. Gorg A, Drewes O, Weiss W. Separation of proteins using two-dimensional gel electrophoresis. In: Simpson RJ, ed. Purifying Proteins for Proteomics—A Laboratory Manual. New York: CSHL Press, 2003:391–430.
11. Gibson DS, Blelock S, Brockbank S, et al. Proteomic analysis of recurrent joint inflammation in juvenile idiopathic arthritis. J Proteome Res 2006;5(8): 1988–1995.
12. Gorg A, Obermaier C, Boguth G, et al. The current state of two-dimensional electrophoresis with immobilized pH gradients. Electrophoresis 2000;21(6):1037–1053.
13. Gorg A, Obermaier C, Boguth G, Weiss W. Recent developments in two-dimensional gel electrophoresis with immobilized pH gradients: wide pH gradients up to pH 12, longer separation distances and simplified procedures. Electrophoresis 1999;20(4–5):712–717.

14. Laemmli UK. Cleavage of structural proteins during the assembly of the head of bacteriophage T4. Nature 1970;227(5259):680–685.
15. Anderson NL, Anderson NG. Analytical techniques for cell fractions. XXII. Two-dimensional analysis of serum and tissue proteins: multiple gradient-slab gel electrophoresis. Anal Biochem 1978;85(2):341–354.
16. Corzett TH, Fodor IK, Choi MW, et al. Statistical analysis of the experimental variation in the proteomic characterization of human plasma by two-dimensional difference gel electrophoresis. J Proteome Res 2006;5(10):2611–2619.
17. Fodor IK, Nelson DO, Alegria-Hartman M, et al. Statistical challenges in the analysis of two-dimensional difference gel electrophoresis experiments using DeCyder. Bioinformatics 2005;21(19):3733–3740.
18. Grove H, Hollung K, Uhlen AK, Martens H, Faergestad EM. Challenges related to analysis of protein spot volumes from two-dimensional gel electrophoresis as revealed by replicate gels. J Proteome Res 2006;5(12):3399–3410.
19. Houtman R, Krijgsveld J, Kool M, et al. Lung proteome alterations in a mouse model for nonallergic asthma. Proteomics 2003;3(10):2008–2018.
20. Mahnke RC, Corzett TH, McCutchen-Maloney SL, Chromy BA. An integrated proteomic workflow for two-dimensional differential gel electrophoresis and robotic spot picking. J Proteome Res 2006;5(9):2093–2097.

10

Label-Free Mass Spectrometry-Based Protein Quantification Technologies in Protein Biomarker Discovery

Mu Wang, Jin-Sam You, Kerry G. Bemis, and Dawn P.G. Fitzpatrick

Summary

Major technological advances have made proteomics an extremely active field for biomarker discovery in recent years due primarily to the development of newer mass spectrometric technologies and the explosion in genomic and protein bioinformatics. This leads to an increased emphasis on larger scale, faster, and more efficient methods for detecting protein biomarkers in human tissues, cells, and biofluids. Most current proteomic methodologies for biomarker discovery, however, are not highly automated and are generally labor-intensive and expensive. More automation and improved software programs capable of handling a large amount of data are essential in order to reduce the cost of discovery and increase the throughput. In this chapter, we will discuss and describe a case study of a unique proteomic method that uses a non-gel, label-free LC/MS-based protein quantification technology.

Key Words: Biomarkers; label-free quantitative analysis; mass spectrometry; proteomics

1. INTRODUCTION

Quantitative proteomics has become a widely applied analytical tool for protein identification and quantification in complex biological samples *(1–6)*. One of the goals of proteomics is to measure protein expression and characterize the protein expression profile in specific tissues and biofluids for potential biomarker discovery. Even though a tremendous effort has been made to improve the proteomics technologies, there are still numerous challenges associated with even the most advanced technologies in analyzing global protein expression due to the inherent complexity of biological samples. These challenges include (1) sensitivity of the instrument

From: *Methods in Pharmacology and Toxicology: Biomarker Methods in Drug Discovery and Development*
Edited by: F. Wang © Humana Press, Totowa, NJ

and ability to identify novel proteins; (2) low to moderate throughput of the system from sample preparation to data analysis; (3) wide coverage of protein mass and abundance (dynamic range); and (4) the ability to quantitatively analyze protein expression. At the present time, there is no consensus within the field of proteomics that any one technology can attain a complete and quantitative protein coverage of all proteins in a given tissue or biofluid. The most commonly used proteomic approach is accomplished by a combination of either two-dimensional gel electrophoresis (2-DE) or liquid chromatography (LC) to separate and visualize proteins/peptides and mass spectrometry (MS) to identify, characterize, and quantify them. Although 2-DE has been the workhorse in proteomics research efforts in the past decade, its lack of ability to widen the protein dynamic range and to analyze hydrophobic proteins or those with very high or low molecular weight is still its biggest disadvantage. One alternative approach to 2-DE is the non–gel-based "bottom up" LC/MS-based shotgun proteomic technology ***(7–10)***. Although some successes using stable isotopic labeling technology for protein quantification have been reported ***(11)***, it remains technically difficult to comprehensively characterize the global proteome due to the high costs of the labeling reagents and the nature of the methodology (i.e., in isobaric tagging of peptide (iTRAQ™) or isotope-coded affinity tag (ICAT®), it depends on labeling efficiency). In addition, chromatographic shifts can make quantitative analysis of differentially labeled peptides computationally difficult and error-prone. Furthermore, simultaneous quantification of proteins from multiple samples is problematic due to limitation of available labeling reagents. Recently, the ion intensity–based label-free quantitative approach has gradually gained more popularity as mass spectrometer performance has improved significantly ***(12–15)***. It provides a powerful tool to resolve and identify thousands of proteins from a complex biological sample. In this method, proteins are first digested with a protease into a peptide mixture that is then analyzed by LC/MS and LC/tandem mass spectrometry (MS/MS) and subsequently identified by database searching and relative protein abundance determined by chromatographic peak intensity measurement. This approach is rapid and more sensitive, and it increases the protein dynamic range three- to four-fold compared with 2-DE. This method can also be automated and has the ability for large-scale proteome analysis. Along with all these advantages compared with other proteomic technologies, however, significant bioinformatics and biostatistical resources for data analysis are required.

2. LABEL-FREE PROTEIN QUANTIFICATION

Label-free and signal intensity–based relative protein quantification shotgun proteomics approaches are promising alternatives to stable isotope labeling approaches and have been applied to quantify differentially expressed proteins from complex biological samples ***(16–19)***. One approach is to directly compare peptide peak areas between LC/MS runs. This approach is simple and cost effective and has demonstrated high reproducibility and linearity comparing peptide level or protein abundance. Inclusion of statistical analysis with this method allows detection of small significant changes that are biologically meaningful. Several studies have demonstrated that extracted ion chromatograms (XIC) of selected peptide ions correlate well with protein abundances in complex biological samples ***(17–19)***. Another label-free method, referred to as spectral counting, compares the number of MS/MS spectra assigned to each protein ***(20)***. One advantage of spectral counting method is its ability to potentially measure relative abundances of different proteins in the same biological system as shown by significant correlations between spectral counts and independent estimates of protein copy number in yeast ***(21)***. However, the application of these methods, especially to mammalian systems, requires more robust computing power and algorithm capable of handling chromatographic peak alignment and peptide ion intensity measurements for analyzing changes in protein abundances in complex biological samples.

3. A PEAK INTENSITY–BASED LABEL-FREE PROTEIN QUANTIFICATION METHOD

All label-free LC/MS-based protein quantification methods discussed previously include four fundamental steps as depicted in **Fig. 1**: (1) sample preparation (protein extraction, reduction, alkylation, and digestion); (2) sample separation and analysis (LC and MS/MS); (3) data analysis (identification, quantification, and statistical analysis); and (4) data interpretation (protein-protein interaction network and pathway analysis). In this chapter, we will focus on the peak intensity–based method for protein quantification followed by a case study.

3.1. Protein Extraction, Reduction, Alkylation, and Digestion

In general, proteins are extracted from tissues, cultured cells, or biofluids in freshly made lysis buffer containing 8 M urea and 10 mM dithiothreitol (DTT). For cells grown in 6-well plates, culture media are first removed by aspiration. Then, 100 μL of lysis buffer is added to each well, and cells are lysed by pipetting. Lysed cells are transferred to microcentrifuge tubes and incubated at room temperature with gentle agitation. Unlysed cells and insoluble particles are removed by centrifugation (12,000 × g for 5 min) prior to protein assay. In order to take the same amount of proteins

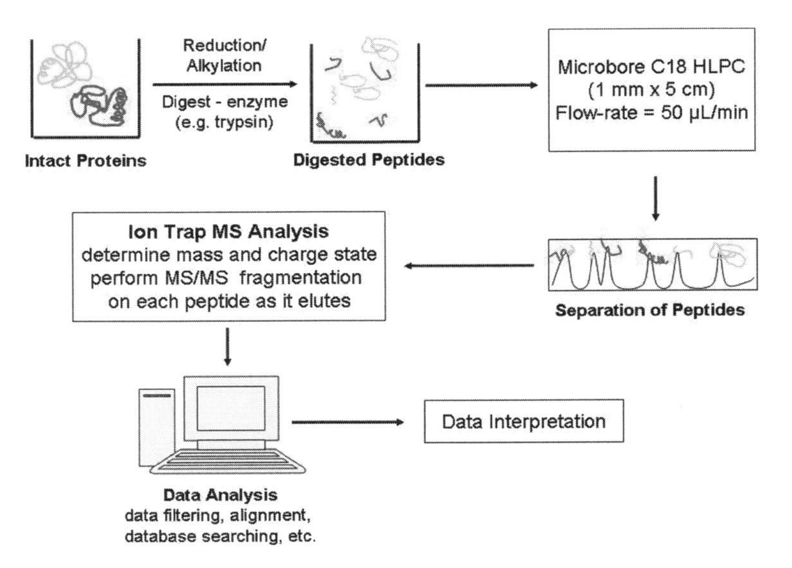

Fig. 1. A schematic flowchart describing the general steps involved in the MS-based label-free protein quantification technology.

from each sample, protein concentrations are measured by Bradford assay *(22)*. The same lysis buffer should be used as the background reference for protein assay in order to obtain a relatively accurate measurement among all samples (due to the presence of urea in lysis buffer).

Resulting protein extracts are subsequently reduced and alkylated with DTT and iodoacetamide to block sulfhydryl groups in proteins. However, in the method described here, the volatile reagents triethylphosphine and iodoethanol are used instead. This volatile reduction and alkylation protocol has been described previously *(23)*. It allows all sample preparation steps to be carried out in one tube without washing, filtering, or sample transferring, which minimizes sample preparation variations. Briefly, 100 µg protein is taken from each sample, and pH values are raised by adding 100 µL 100 mM ammonium carbonate (pH 11). Two hundred microliters of reduction/alkylation cocktail (97.5% acetonitrile, 2% iodoethanol, and 0.5% triethylphosphine) is added to protein samples and incubated for 1 h at 37°C. After incubation, protein samples are dried under vacuum overnight to remove residual chemicals.

Protein mixtures are then digested by trypsin. Dried pellets are resuspended in 500 µL 100 mM ammonium bicarbonate buffer containing 2 µg trypsin and incubated at 37°C overnight. Protein to trypsin ratio is 50:1, and urea concentration should be below 1.6 M. Tryptic digests are filtered with 0.45-μm spin filters.

3.2. Mass Spectrometric Analysis

All digested samples should be randomized for injection order to remove systematic bias from data acquisition. Typically, up to 20 μg of the tryptic peptides are required for injection onto a C18 microbore column (i.d. = 1 mm, length = 5 cm). Peptides are eluted with a linear gradient from 5% to 45% acetonitrile developed over 120 min at a flow rate of 50 μL/min, and effluent is electro-sprayed into a LTQ linear ion-trap mass spectrometer (San Jose, CA, USA). The electron-spr ionization source is operated with 4.8-kV potential, a sheath gas flow of 20 arbitrary units, and a capillary temperature of 225°C. The instrument is tuned using an angiotensin I peptide. Maximum ion time is set to 50 ms for parent ion scan and 500 ms for zoom scan and MS/MS scan. This method requires all the MS data be collected in the data-dependent "Triple-Play" mode as shown in **Fig. 2** (MS scan, Zoom scan, and MS/MS scan). Parent ion scans and MS/MS scans are collected in centroid mode, and zoom scans are collected in profile mode. Dynamic exclusion is set to a repeat count of one, exclusion duration of 120 s, and

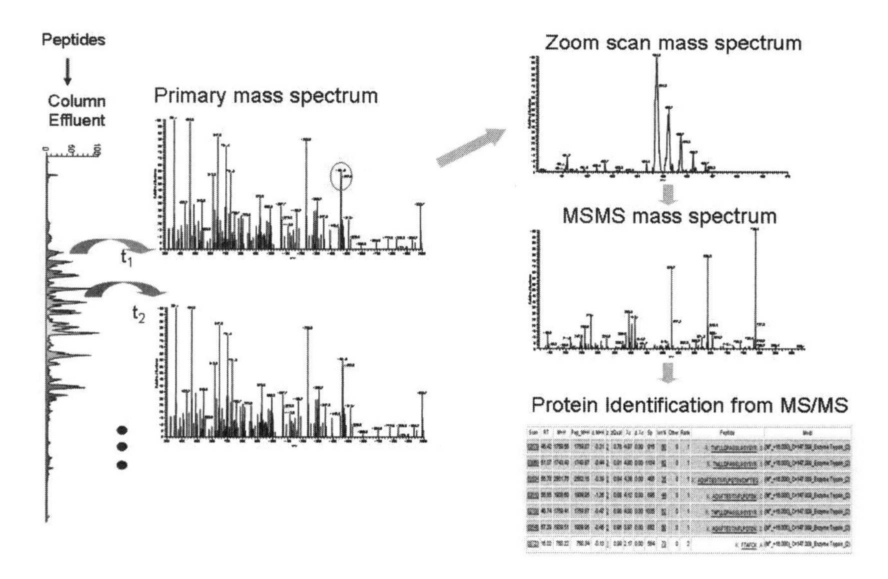

Fig. 2. Triple-play MS method. After the effluent is electro-sprayed into a mass spectrometer, primary MS scans are acquired for peptide mass determination, followed by zoom scans for charge state determination, and MS/MS scans for peptide sequence determination.

rejection widths of –0.75 m/z and + 2.0 m/z. The acquired data are filtered and analyzed by an algorithm developed by Higgs et al. and described in detail in the published article *(24)*. Database searches against the IPI (International Protein Index) and the Non-Redundant (NCBI) databases are carried out using both the SEQUEST and X!Tandem algorithms. Protein quantification is also carried out using the same algorithm described above *(24)*. Briefly, once the raw files are acquired from the LTQ, all extracted ion chromatograms (XIC) are aligned by retention time (**Fig. 3**). To be used in the protein quantification procedure, each aligned peak must match parent ion, charge state, daughter ions (MS/MS data), and retention time. After alignment, the area under the curves (AUCs) for individually aligned peaks from identified peptides from each sample are computed; the AUCs are then compared for relative protein abundance.

3.3. Protein Identification

Proteins identified by SEQUEST and X!Tandem are categorized into priority groups based on the quality of the protein identification as shown in **Table 1**. The Peptide ID Confidence assigns a protein to a "HIGH" or "LOW" classification based on the peptide with the highest peptide ID Confidence (the best peptide). Proteins whose best peptide has a confidence between 90% and 100% are assigned to the "HIGH" category regardless of whether there are other peptides having low confidence or not. Proteins whose best peptide has a confidence between 75% and 89% are assigned to the "LOW" category. All peptides with confidence less than 75% are filtered

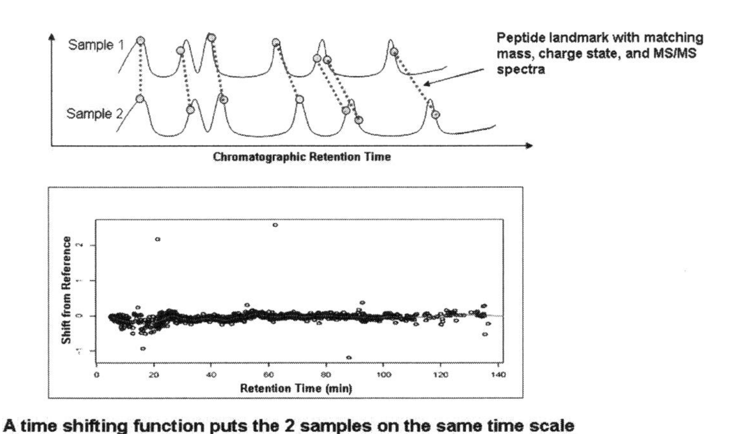

Fig. 3. Chromatographic peak alignment requires all landmark peaks matching peptide mass, charge state, MS/MS spectra, and retention time within 1-min window.

Table 1
Prioritization of Protein Identification

Priority	Protein ID confidence	Multiple sequences
1	High (90–100%)	Yes ($\geq$2 unique sequences)
2	High (90–100%)	No (single sequence)
3	Low (75–89%)	Yes ($\geq$2 unique sequences)
4	Low (75–89%)	No (single sequence)

out by the software before further analysis. SEQUEST and X!Tandem database search algorithms are used for peptide sequence identification. Each algorithm compares the observed peptide MS/MS spectrum and a theoretically derived spectra from the database to assign quality scores (*XCorr* in SEQUEST and *E-Score* in X!Tandem). These quality scores and other important predictors are combined in the algorithm that assigns an overall score, %ID Confidence, to each peptide. The assignment is based on a random forest recursive partition supervised learning algorithm ***(24)***. The %ID Confidence score is calibrated so that approximately X% of the peptides with %ID Confidence > X% are correctly identified.

The confidence in protein identification is increased with the number of distinct amino acid sequences identified. Therefore, proteins are also categorized depending on whether they have only one or multiple unique sequences of the required confidence. A protein is classified as "YES" in the "Multiple Sequences" column if it has at least two distinct amino acid sequences with the required ID confidence; otherwise it is classified as "NO." Priority assignments reflect the level of confidence in the protein identification. Priority 1 proteins would have the highest likelihood of correct identification and Priority 4 proteins the lowest. This priority system is based on the quality of the amino acid sequence identification (Peptide ID Confidence) and whether one or more sequences are identified (Multiple Sequences). Many would view any protein identification outside of priority 1 as questionable ***(25)***. All data processing is carried out on a Linux cluster using highly parallel processing and data qualification and filtering software.

3.4. Protein Quantification

One of the key features of the algorithm described here for protein quantification is the chromatographic peak alignment because large biomarker studies can produce chromatographic shifts due to multiple injections of the samples onto the same HPLC column. Unaligned peak comparison will result in larger variability and inaccuracy in peptide quantification *(24)*. A graphical example of a comparison of peptide quantities across a complex biological sample is shown in **Fig. 4**. All peak intensities are transformed to a $\log_2$ scale before quantile normalization *(26)*. Quantile normalization is a method of normalization that essentially ensures that every sample has a peptide intensity histogram of the same scale, location, and shape. This normalization procedure removes trends introduced by sample handling, sample preparation, total protein differences, and changes in instrument sensitivity while running multiple samples. If multiple peptides have the same protein identification, then their quantile normalized $\log_2$ intensities are averaged to obtain $\log_2$ protein intensities. The average of the normalized log peptide intensities is a weighted average. A peptide is weighted proportionally to the peptide ID Confidence for its protein category and receives a weight of zero if it is outside that category. For example, peptides with <90% confidence contribute zero quantitative weight to a "HIGH" category protein. The log transformation serves two purposes. First, relative changes in protein expression are best described by ratios. However, ratios are difficult to statistically model, and the log transformation converts a ratio to a difference, which is easier to model.

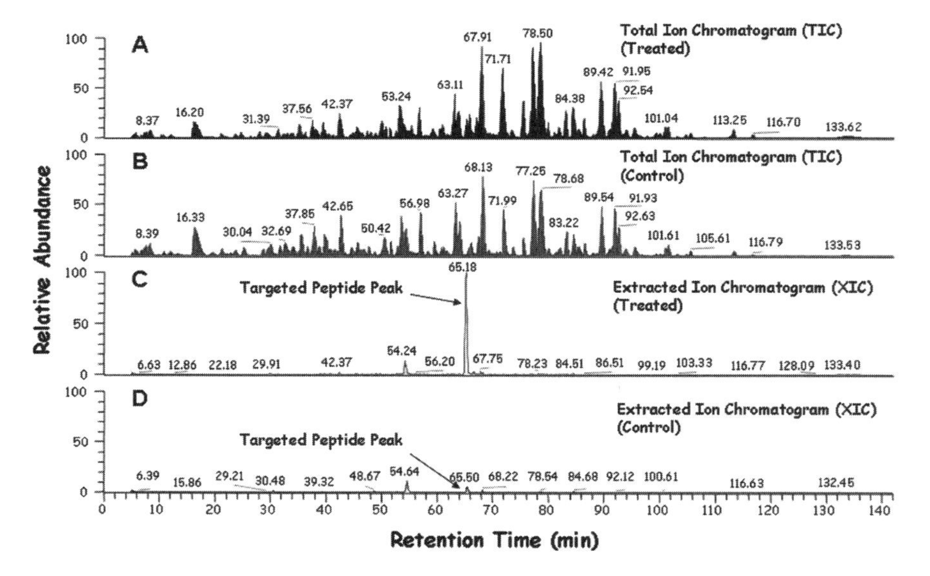

Fig. 4. Peptide quantification by extracted ion chromatograms. **(A, B)** Total ion chromatogram (TIC) from treated and control samples. **(C, D)** Extracted ion chromatogram (XIC) from treated and control samples. By integrating the AUC, the relative quantity of the peptide of interest (indicated by arrows) can be calculated and compared.

Second, as is frequently the case in biology, the data better approximate the normal distribution on a log scale ***(27)***. This is important because normality is an assumption of the analysis of variance (ANOVA) statistical model used to analyze the data. The base of the log transform is arbitrary with base 2 the most common with genomic data. Base 2 is popular because a two-fold change (or doubling, or 100% increase) yielding an expression ratio of 2 is transformed to 1 on a log base 2 scale (i.e., a two-fold change is a unit change on the log base 2 scale). The log protein intensity is the final quantity that is fit by a separate ANOVA model for each protein shown below:

$$[\log_2 \text{(Intensity)} = \text{overall mean} + \text{group effect (fixed)} + \text{sample effect (random)} + \text{replicate effect (random)}]$$

In this model, group effect refers to the effect caused by the experimental conditions or treatments being evaluated. Sample effect represents the random effects from individual biological samples. It also includes random effects from sample preparation. The replicate effect refers to the random effects from replicate injections of the same sample. All of the injections should be therefore randomized and the instrument be operated by the same operator for a particular study. The inverse log of each sample mean is calculated to determine the fold change between samples.

3.5. Quality Assurance and Quality Control

To assess the stability of the HPLC system and MS instrument, a known purified standard protein is commonly spiked into every sample at a constant amount as an internal reference for assessment of technical variations before tryptic digestion. Several considerations should be given for the selection of the standard: (1) the protein must not come from the same species as the sample of interest; (2) a series of signature peptides should be easily detected and identified by the instrument; and (3) the amount of the standard protein spiked into each sample should be comparable with the amount of median abundant proteins in the sample. After global protein identification and quantification, these peptides and their relative quantities should be inspected for quality assurance and quality control purposes.

3.6. Statistical Analysis

The number of significant changes between groups, the fold changes, and the variability (coefficient of variation) for each priority level can be determined from the ANOVA. The threshold for significance is set to control the false-discovery-rate (FDR) for each comparison at an investigator-desired percentile, normally 5% ***(28)***. The FDR is estimated by the q value, which is an adjusted *p* value. The FDR is the proportion of significant changes that are false positives. If proteins with a *q* value ≤ 0.05 are declared significant, it is expected that 5% of the declared changes will be false positives. In this method, the *p* value to *q* value adjustment is done separately for Priority 1, Priority 2, and the "LOW" confidence categories.

Fold change (FC) is computed from the means on the AUC scale (antilog) as follows:

When Mean Treated Group ≥ Mean Control Group (up regulation):

$$FC = \text{Mean Treated Group} / \text{Mean Control Group}$$

When Mean Control Group > Mean Treated Group (down regulation):

$$FC = -\text{Mean Control Group} / \text{Mean Treated Group}$$

and

$$\text{Absolute FC} = |FC| = \text{absolute or positive value of the FC}$$

A fold change of 1 means there is no change.

Also, the median % coefficient of variation (%CV) for each priority level is determined. The %CV is the standard deviation/mean on a % scale. The %CV is given both for the replicate variation as well as the combined replicate plus sample variation. An example of a data set is shown in **Table 2**.

3.7. Pathway Analysis and Protein Classification

To understand the biological significance of the protein expression changes, the results including protein IDs and FC from the LC/MS analysis can then be analyzed using protein-protein interaction and/or pathway analysis software. This software allows for creation of protein-protein interaction networks, biological pathways, and gene regulation networks from a data set, which will help better understand specific biological processes that are involved in a particular study.

4. A PRACTICAL EXAMPLE: SEARCHING FOR BIOMARKERS OF CISPLATIN RESISTANCE IN HUMAN OVARIAN CANCER CELLS

To illustrate the utility of this LC/MS-based method described previously, a practical example is provided that outlines how a proteomic experiment is designed and what steps are involved in data interpretation. Prior to any experiment, study design must be completed with a statistician in

Table 2
Summary of the Study Using LC/MS-Based Label-Free Protein Quantification Method

Protein priority	Peptide ID confidence	Multiple sequences	Number of proteins	Number of significant changes	Maximum absolute fold change	Median%CV replicate	Median %CV replicate + sample
1	HIGH	YES	855	95	5.51	11.71	12.87
2	HIGH	NO	583	25	18.80	21.58	23.65
3	LOW	YES	27	0	5.12	22.03	24.77
4	LOW	NO	652	16	78.77	28.70	31.95
Overall			2117	136	78.77	17.96	19.55

order to ensure that the study answers the questions of interest and has sufficient technical and biological replicates to detect small but significant changes using appropriate statistical methods. A technical replicate is a replicate sample from the same biological sample. For example, split a single biological sample into two parts and run both replicates in the experiment. This will allow for assessment of instrument errors. Biological replicates are samples from independent experimental units (e.g., each of 10 human plasma samples from different individuals). Whereas a technical replicate estimates the precision for the assay itself, biological replicates provide an estimation of biological variation ***(29,30)***. In general, biological replicates are more informative than technical replicates.

Group size determination depends on the size of effect to be detected (FC) and the sample-to-sample biological variation expected (CV), and which error rates to be controlled. It is best to control the FDR instead of the false-positive-rate (FPR) when hundreds of proteins are analyzed. The FDR can be large (e.g., >0.05) even if the FPR is small (e.g., <0.05). If control of FDR is chosen, then the proportion of proteins that will change (the prevalence) has to be estimated. With this information, the group size required for given power (probability of determining a true change, i.e., the sensitivity) can be computed. **Table 3** shows a suggested group size with given FC and %CV. As the percent of proteins expected to change varies, the group size required should be matched accordingly.

4.1. Cell Lines and Study Design

Two pairs of cell lines, A2780 and 2008 cisplatin-sensitive human ovarian cancer cell lines and their resistant counterparts, A2780/CP and

Table 3
Group Size Determination*

Fold change	%CV (%error)	%Proteins changed	Group size
2	20	5	5
2	20	10	4
2	15	5	4
2	15	10	4
1.5	20	5	10
1.5	20	10	9
1.5	15	5	7
1.5	15	10	6

*The power is fixed at 95%, the FDR is 5%.

2008/C13*5.25, were used in this study. These cell lines have been used as model systems for drug-resistance studies ***(31–34)***. All cell lines were handled under identical conditions and maintained at 37°C in a humidified incubator containing 5% CO_2 in RPMI-1640 supplemented with 15% fetal bovine serum. Proteins were prepared and subjected to LC/MS/MS analysis as previously described ***(24)***. There were eight groups in this study: two pairs of cell lines each including a sensitive cell line and a resistant cell line with no drug treatment (**Table 4**, groups 1 & 2 and 5 & 6), the resistant cell lines (A2780-CP and 2008-C13*5.25) with 10 μM and 20 μM of cisplatin treatment, respectively (**Table 4**, groups 3 & 4 and 7 & 8). There were six samples per group except for group 3, yielding a total of 93 randomized HPLC injections. Samples were run on a Surveyor HPLC system (Thermo-Finnigan) with a C18 microbore column (Zorbax 300SB-C18, 1 mm × 5 cm).

4.2. Data Normalization and Quality Assessment

In this study, all injections were performed using the same C18 microbore column. Chicken lysozyme was spiked into every sample at a constant amount before tryptic digestion. After tryptic digestion, nine chicken lysozyme peptides were quantified (all peptides with ID confidence >75%). In the plot shown in **Fig. 5** the individual protein quantities (as presented by the peak intensities) are displayed for each injection. The overall mean for each group is displayed by the line across the plot. This plot presents a visual quality control assessment of the ability of the instrument to measure a constant amount of protein over sequential injections. Because a constant amount of chicken lysozyme was spiked into all the samples, it should show no significant change between groups. If there is a significant group effect (i.e., if q value <0.05), then one should interpret significant changes in other proteins with smaller fold changes as possibly due to spurious effects. In this experiment, the largest absolute fold change for chicken lysozyme was 1.16 with a q value of 0.0667. Even though this is not a significant change based on the q-value threshold of <0.05, any significant fold changes of absolute magnitude less than 1.16 in this study should be interpreted with caution.

4.3. Ranking and Prioritization of Data

After database search, protein identification and quantification, and statistical analysis, a total of 5282 distinct amino acid sequences corresponding with 2117 proteins were identified and quantified from all samples (summarized in **Table 2**). Among them, 855 proteins were in the Priority 1

Table 4
Experimental Design

Group number	Group name	Group abbreviation	Cell type	Cisplatin treatment (μM)	Number of samples per group	Total number of replicate injections per group
1	A2780	C1S	Sensitive	0	6	12
2	A2780CP	C1R00	Resistant	0	6	12
3	A2780CP-10	C1R10	Resistant	10	5	10
4	A2780CP-20	C1R20	Resistant	20	6	12
5	2008	C2S	Sensitive	0	6	11
6	2008C13	C2R00	Resistant	0	6	12
7	2008C13-10	C2R10	Resistant	10	6	12
8	2008C13-20	C2R20	Resistant	20	6	12

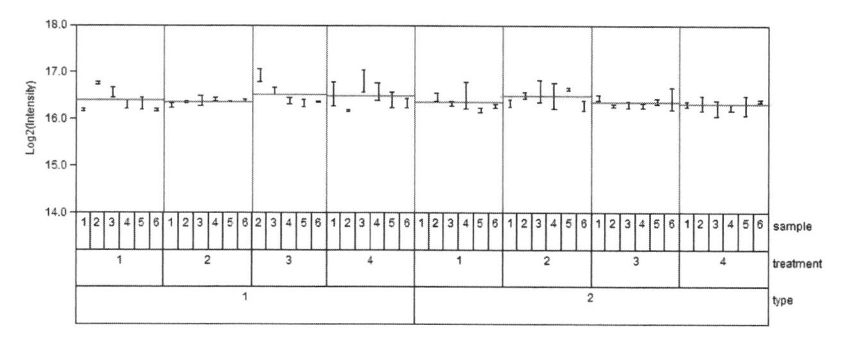

Fig. 5. The individual protein intensities for chicken lysozyme are plotted on a $\log_2$ scale. The horizontal gray line is the group mean. Intensities for the two replicates are joined by a vertical line, and each sample is plotted separately. This plot allows the replicate, sample, and group variation to be visually assessed. Type 1 is the cell line A2780/A2780-CP, type 2 is the cell line 2008/2008-C13*5.25; treatment groups 1 through 4 correspond respectively with sensitive cell line and resistant cell line with 0, 10, and 20 μM cisplatin treatment; samples 1 through 6 correspond with individual samples within the group; please note that treatment group 3 in type 1 has only five samples, and sample no. 5 of treatment group 2 from type 1 was analyzed by single injection.

group and 95 of them showed significant changes that are potentially associated with cisplatin resistance. The significance threshold is set to control the FDR at less than 5%. The replicate median %CV (technical variation) for the Priority 1 proteins was 11.71% and the combined replicate plus sample median %CV was 12.87% (technical plus biological variations). There were also 41 proteins that had significant changes among the 1262 proteins that were less confidently identified (Priorities 2 to 4) (136 total proteins as shown in **Fig. 6** The cisplatin resistance–specific protein changes were extracted when the two types of cell lines were directly compared (A2780/A2780-CP vs. 2008/2008-C13*5.25), suggesting that other protein expression level changes in the two cell lines may be cell-type–specific.

4.4. Data Interpretation

The results from the LC/MS were then analyzed by PathwayStudio 4.0 (Ariadne Genomics, Rockville, MD). This software allows for creation of biological pathways, gene regulation networks, and protein interaction diagrams from a data set. With a list of gene names derived from the proteomic results, more than 1700 connections were made from the protein

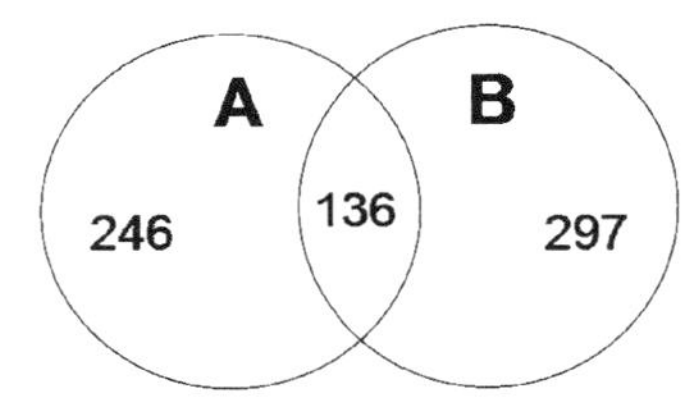

Fig. 6. Venn diagram of combined differentially expressed proteins in the two cell lines. All priority group proteins are compared between cell lines **(A)** A2780/A2780-CP and **(B)** 2008/2008-C13*5.25. Only sensitive and resistant cells without cisplatin treatment are compared (groups 1 & 2 vs. groups 5 & 6).

list. A "Classification Table" was thus created based on these results. **Table 5** shows the biological processes that these proteins were found to be involved in when they were run against the software's ResNet database. These processes included proliferation, DNA repair, apoptosis, and more. Based on these experimental data and the information obtained from the pathway analysis and other literature searches, a hypothesis can be generated involving several specific pathways that may contribute to cisplatin drug resistance (**Fig. 7**).

Table 5
Biological Processes in Which the Differentially Expressed Proteins are Involved*

Biological process	Number of proteins involved
Amplification	6
Angiogenesis	2
Apoptosis	24
Differentiation	16
Mitosis	9
Regulation of signal transduction	13
Cell survival	16
Damage	6
DNA damage recognition	2
DNA recombination	35
Transformation	8

*Protein classification based on the analysis by PathwayStudio 4.0. All proteins (Priority 1 to 4) with significant changes ($q < 0.05$) are included.

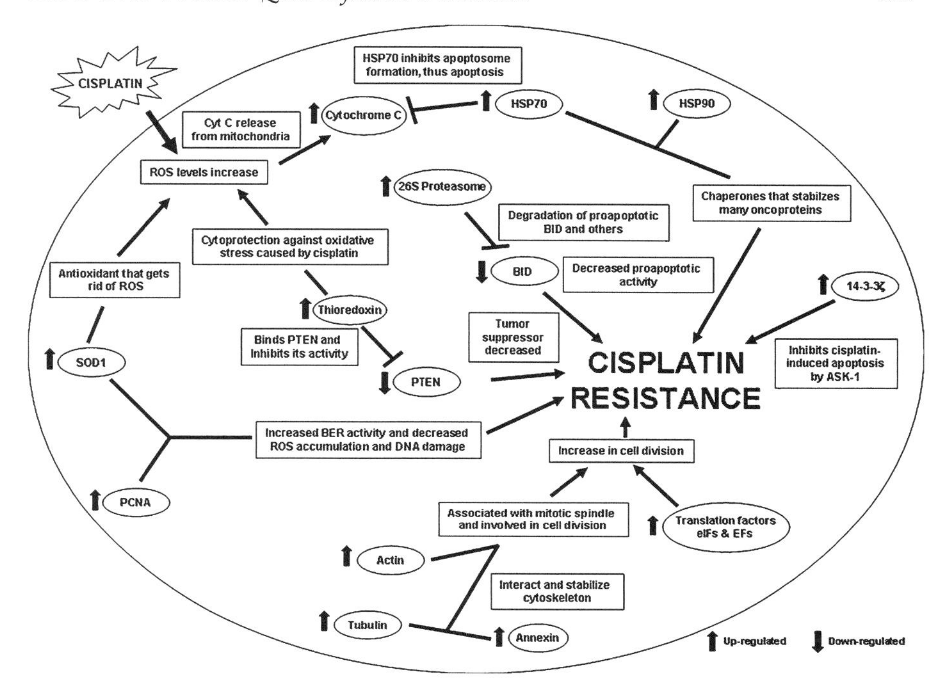

Fig. 7. A proposed biological pathway involved in cisplatin resistance in ovarian cancer.

5. CONCLUSION

Although the need for the development of high-throughput, sensitive, and cost-effective MS-based quantitative protein analysis methods for biomarker discovery remains a significant challenge in proteomics, the promise of the proteomic technologies to enhance the biomarker discovery has been shown in this chapter. To reduce false-positive discoveries, significant development in bioinformatics and robust validation methods will be required. As the future of proteomics unfolds, it will certainly create many more opportunities for the development of novel drug targets, improved understanding of proteomic data, and underlying biology. The development in label-free protein quantification technologies will no doubt meet the demands of both proteomics and clinical applications.

REFERENCES

1. Blackstock WP, Weir MP. Proteomics: quantitative and physical mapping of cellular proteins. Trends Biotechnol 1999;17:121–127.

2. Gygi SP, Rist B, Aebersold R. Measuring gene expression by quantitative proteome analysis. Curr Opin Biotechnol 2000:11:396–401.
3. Rabilloud T. Two-dimensional gel electrophoresis in proteomics: old, old fashioned, but it still climbs up the mountains. Proteomics 2002:2:3–10.
4. Conrads TP, Issaq HJ, Veenstra TD. New tools for quantitative phosphoproteome analysis. Biochem Biophys Res Commun 2002;290:885–890.
5. Ong SE, Foster LJ, Mann M. Mass spectrometric-based approaches in quantitative proteomics. Methods 2003;29:124–130.
6. Tao WA, Aebersold R. Advances in quantitative proteomics via stable isotope tagging and mass spectrometry. Curr Opin Biotechnol 2003;14:110–118.
7. McDonald WH, Yates JR 3rd. Shotgun proteomics and biomarker discovery. Dis Markers 2002;18:99–105.
8. Wu CC, MacCoss MJ, Howell KE, Yates JR 3rd. A method for the comprehensive proteomic analysis of membrane proteins. Nat Biotechnol 2003;21:532–538.
9. Washburn MP, Ulaszek R, Deciu C, Schieltz DM, Yates JR 3rd. Analysis of quantitative proteomic data generated via multidimensional protein identification technology. Anal Chem 2002;74:1650–1657.
10. Washburn MP, Wolters D, Yates JR 3rd. Large-scale analysis of the yeast proteome by multidimensional protein identification technology. Nat Biotechnol 2001;19:242–247.
11. Gygi SP, Rist B, Gerber SA, Turecek F, Gelb MH, Aebersold R. Quantitative analysis of complex protein mixtures using isotope-coded affinity tags. Nat Biotechnol 1999;17:994–999.
12. Yan W, Chen SS. Mass spectrometry-based quantitative proteomic profiling. Brief Funct Genomic Proteomics 2005;4:27–38.
13. Zhang B, VerBerkmoes NC, Langston MA, Uberbacher E, Hettich RL, Samatova NF. Detecting differential and correlated protein expression in label-free shotgun proteomics. J Proteome Res 2006;5:2909–2918.
14. Wang G, Wu WW, Zeng W, Chou C-L, Shen R-F. Label-free protein quantification using LC-coupled ion trap or FT mass spectrometry: reproducibility, linearity, and application with complex proteomes. J Proteome Res 2006;5:1214–1223.
15. Ono M, Shitashige M, Honda K, Isobe T, Kuwabara H, Matsuzuki H, Hirohashi S, Yamada T. Label-free quantitative proteomics using large peptide data sets generated by nanoflow liquid chromatography and mass spectrometry. Mol Cell Proteomics 2006;5:1338–1347.
16. Griffin TJ, Lock CM, Li XJ, Patel A, Chervetsova I, Lee H, Wright ME, Ranish JA, Chen SS, Aebersold R. Abundance ratio-dependent proteomic analysis by mass spectrometry. Anal Chem 2003;75:867–874.
17. Bondarenko PV, Chelius D, Shaler TA. Identification and relative quantitation of protein mixtures by enzymatic digestion followed by capillary reversed-phase liquid chromatography-tandem mass spectrometry. Anal Chem 2002;74:4741–4749.

18. Chelius D, Bondarenko PV. Quantitative profiling of proteins in complex mixtures using liquid chromatography and mass spectrometry. J Proteome Res 2002;1:317–323.
19. Wang W, Zhou H, Lin H, Roy S, Shaler T, Hill L, et al. Quantification of proteins and metabolites by mass spectrometry without isotopic labeling or spiked standards. Anal Chem 2003;75:4818–4826.
20. Liu H, Sadygov RG, Yates JR III. A model for random sampling and estimation of relative protein abundance in shotgun proteomics. Anal Chem 2004;76:4193–4201.
21. Ghaemmaghami S, Huh WK, Bower K, Howson RW, Belle A, Dephoure N, et al. Global analysis of protein expression in yeast. Nature 2003;425: 737–741.
22. Bradford MM. A rapid and sensitive method for the quantitation of microgram quantities of protein utilizing the principle of protein-dye binding. Anal Biochem 1976;72:248–254.
23. Hale JE, Butler JP, Gelfanova V, You JS, Knierman MD. A simplified procedure for the reduction and alkylation of cysteine residues in proteins prior to proteolytic digestion and mass spectral analysis. Anal Biochem 2004;333:174–181.
24. Higgs RE, Knierman MD, Gelfanova V, Butler JP, Hale JE. Comprehensive label-free method for the relative quantification of proteins from biological samples. J Proteome Res 2005;4:1442–1450.
25. Carr S, Aebersold R, Baldwin M, Burlingame A, Clauser K, Nesvizhskii A. The need for guidelines in publication of peptide and protein identification data: Working Group on Publication Guidelines for Peptide and Protein Identification Data. Mol Cell Proteomics 2004;3:531–533.
26. Bolstad BM, Irizarry RA, Astrand M, Speed TP. A comparison of normalization methods for high density oligonucleotide array data based on variance and bias. Bioinformatics 2003;19:185–193.
27. Limpert E, Stahel WA, Abbt M. Log-normal distributions across the sciences: keys and clues. BioScience 2001;51:341–352.
28. Reiner A, Yekutieli D, Benjamini Y. Identifying differentially expressed genes using false discovery rate controlling procedures. Bioinformatics 2003;19:368–375.
29. Simon R, Radmacher MD, Dobbin K. Design of studies using DNA microarrays. Genet Epidemiol 2002;23:21–36.
30. Yang YH, Speed T. Design issues for cDNA microarray experiments. Nat Rev Genet 2003;19:649–659.
31. Holzer AK, Samimi G, Katano K, Naerdemann W, Lin X, Safaei R, Howell SB. The copper influx transporter human copper transport protein 1 regulates the uptake of cisplatin in human ovarian carcinoma cells. Mol Pharmacol 2004;66:817–823.
32. Holzer AK, Katano K, Klomp LW, Howell SB. Cisplatin rapidly down-regulates its own influx transporter hCTR1 in cultured human ovarian carcinoma cells Clin Cancer Res 2004;10:6744–6749.

33. Samimi G, Katano K, Holzer AK, Safaei R, Howell SB. Modulation of the cellular pharmacology of cisplatin and its analogs by the copper exporters ATP7A and ATP7B. Mol Pharmacol 2004;66:25–32.
34. Cheng TC, Manorek G, Samimi G, Lin X, Berry CC, Howell SB. Identification of genes whose expression is associated with cisplatin resistance in human ovarian carcinoma cells. Cancer Chemother Pharmacol 2006;58:384–395.

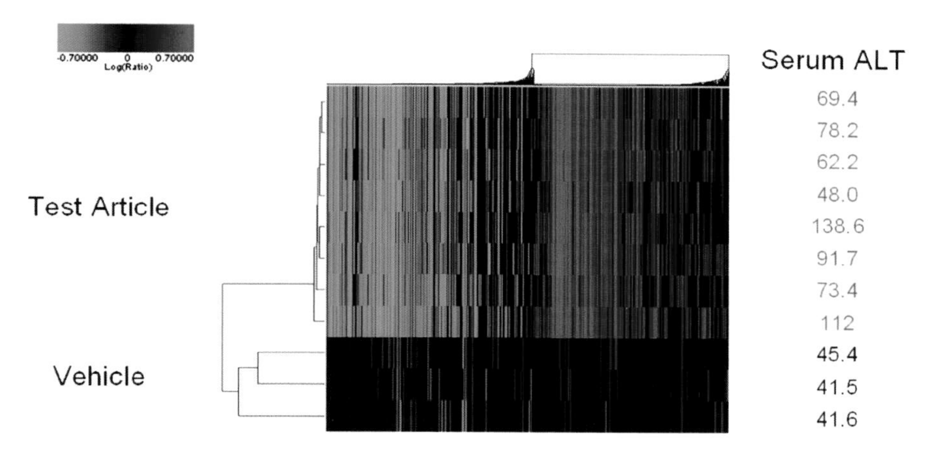

Color Plate 1. Fig. 1, Chapter 2: Comparison of serum alanine aminotransferase (ALT activity) and liver gene expression profiles induced by a hepatotoxicant. Rats were treated for 3 days with either the vehicle or the test article at daily doses causing moderate hepatotoxicity. Serum was collected and ALT activity levels quantified. Livers were sampled, and gene expression profiles were generated with microarrays. Whereas significant interindividual variability can be seen in serum ALT levels, the expression profiles are strikingly consistent among the eight test article–treated rats. Hierarchical cluster analysis was performed using Rosetta Resolver version 6.0 (Rosetta Inpharmatics, Seattle, WA). Genes shown include genes that were upregulated or downregulated by at least twofold with a p value less than 0.01. Green indicates downregulation, and red indicates upregulation (*see* discussion on p. 31).

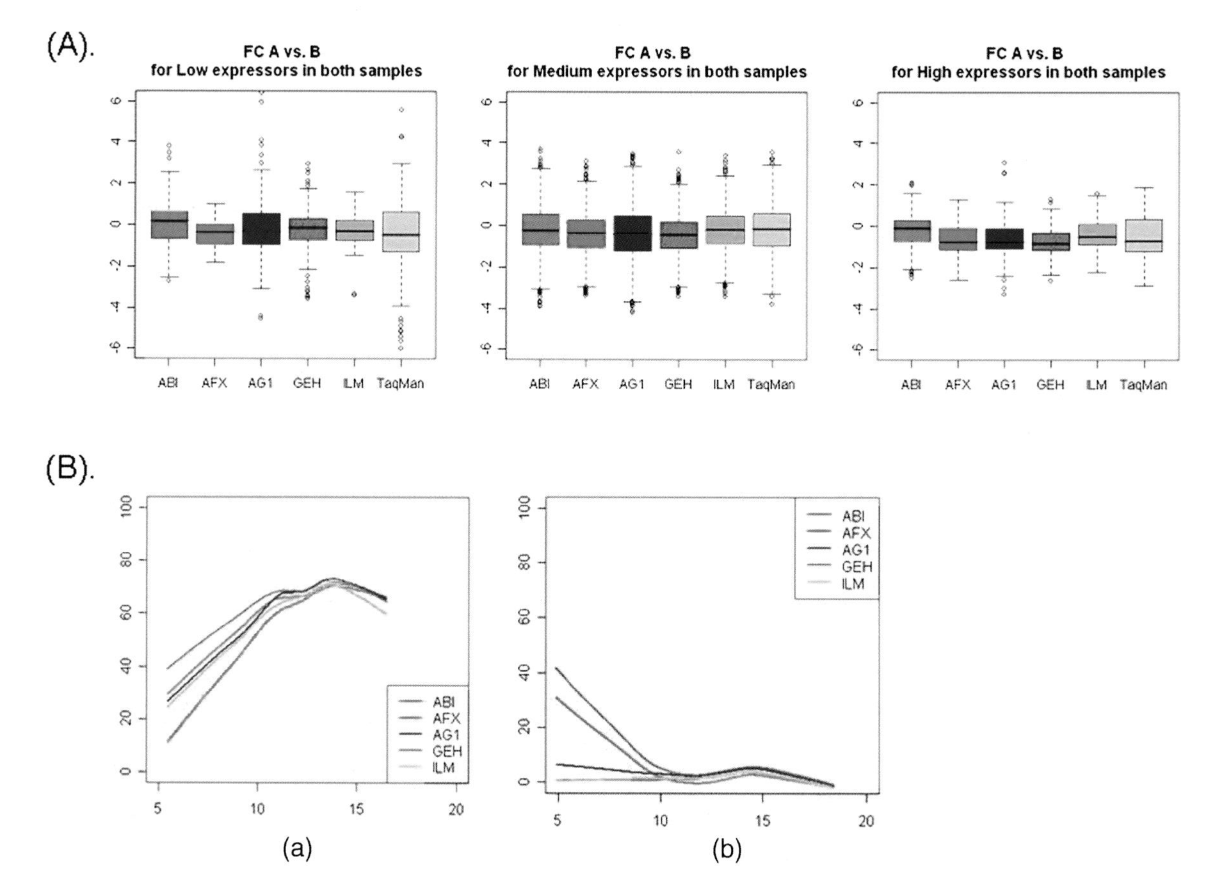

Color Plate 2. Fig. 6a, b, Chapter 4: Validation of microarray results using TaqMan® Gene Expression Assay data set as reference (*see* complete caption on p. 76 and discussion on p. 74).

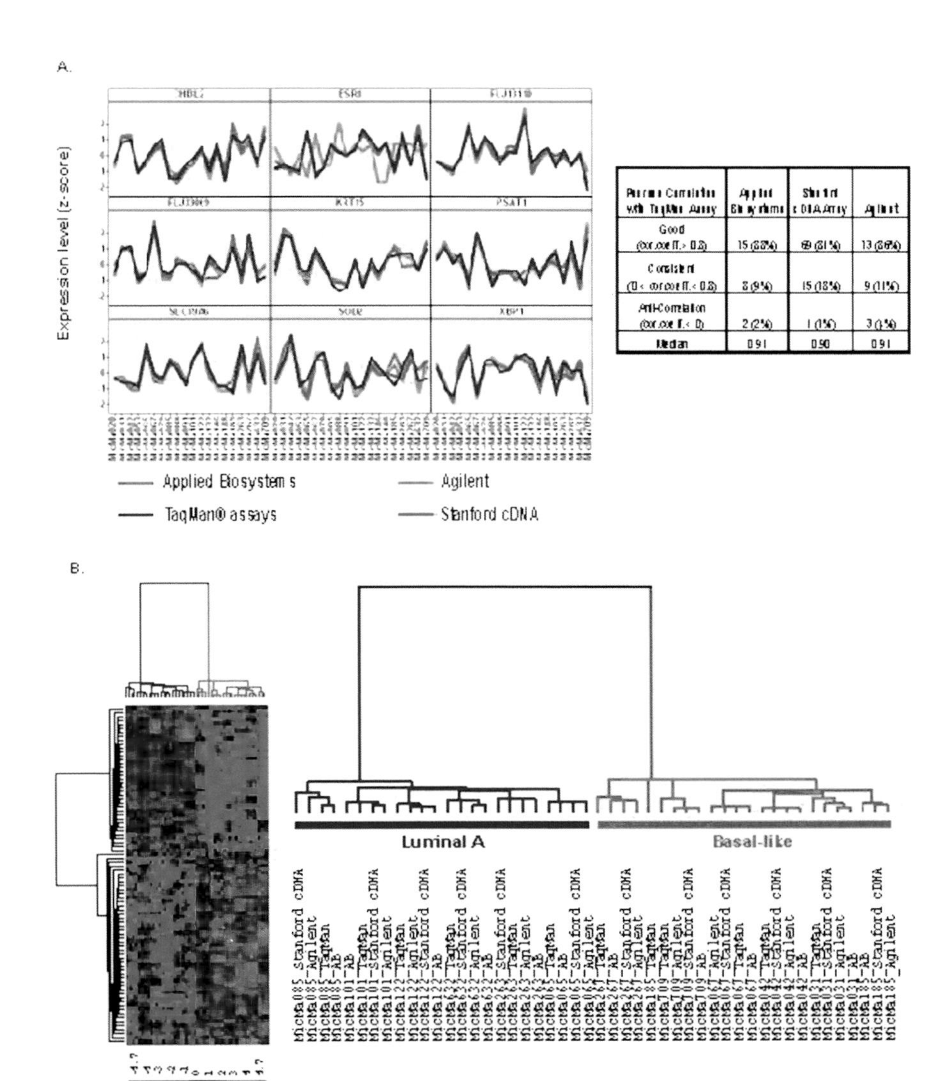

Color Plate 3. Fig. 7, Chapter 4: Validation of potential prognostic markers by TaqMan® assay–based real-time PCR. Eighty-five marker genes including the minimal set of 54 genes identified to best distinguish the luminal A and the basal-like subtypes were validated by TaqMan® Gene Expression assays (*see* complete caption on p. 78, 79 and discussion on p. 77).

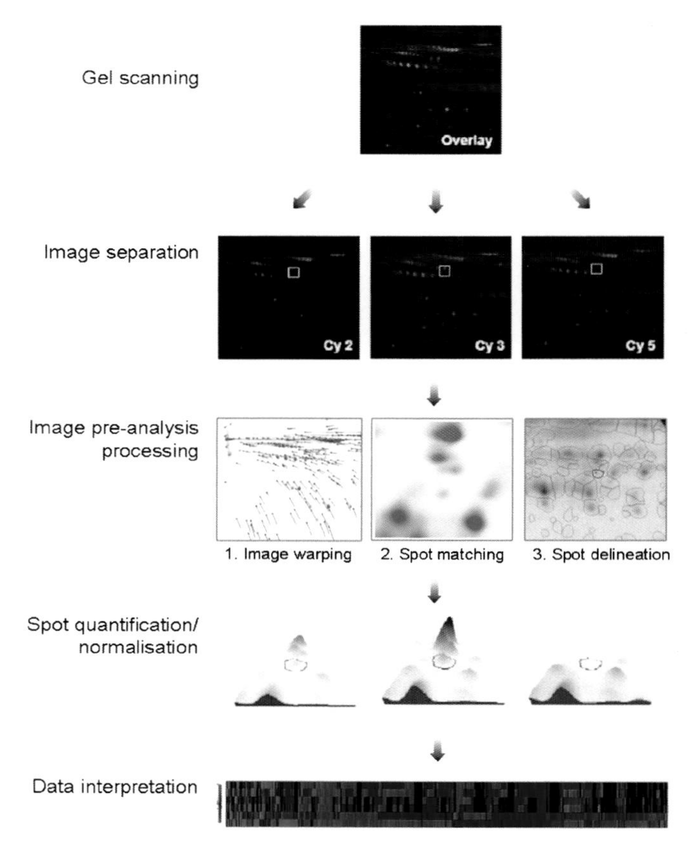

Color Plate 4. Fig. 3, Chapter 9: Schematic representation of the processes involved in DIGE image analysis (*see* discussion on p. 204).

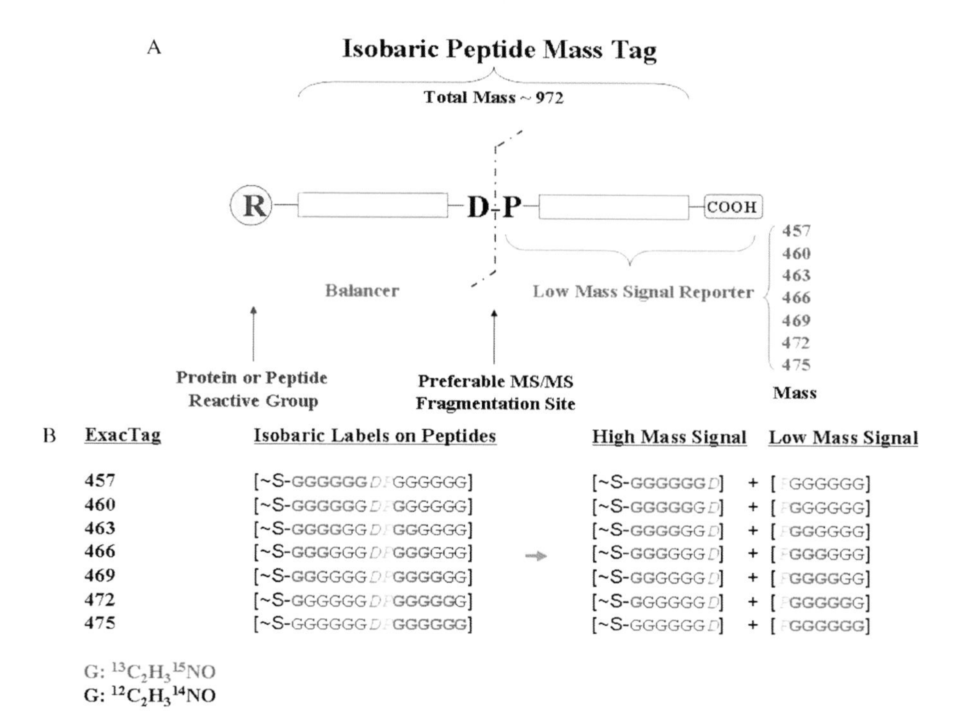

Color Plate 5. Fig. 1a, b, Chapter 11: Composition of isobaric mass tags. (**A**) Schematic structure of the isobaric mass tag described in this chapter. The tag contains four elements: ***(1)*** a reactive group with specificity toward a thiol (cysteine) or an amine (lysine, α-amino) group; ***(2)*** a cleavage enhancement moiety, which is an aspartic acid (Asp, D) proline (Pro, P) scissile bond group. Distributed around the DP sequence are the ***(3)*** low mass signal reporter and its ***(4)*** corresponding balancer sequence. (**B**) Detailed composition of seven isobaric mass tags. The tag contains a common amino acid composition. G represents glycine. The red bold "G" is heavy glycine and plain "G" is normal glycine. In MS/MS, the tags generate two sets of signals: low mass signals and high mass signals (*see* discussion on p. 234).

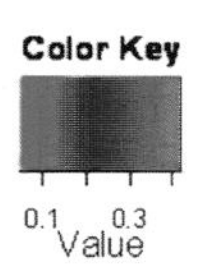

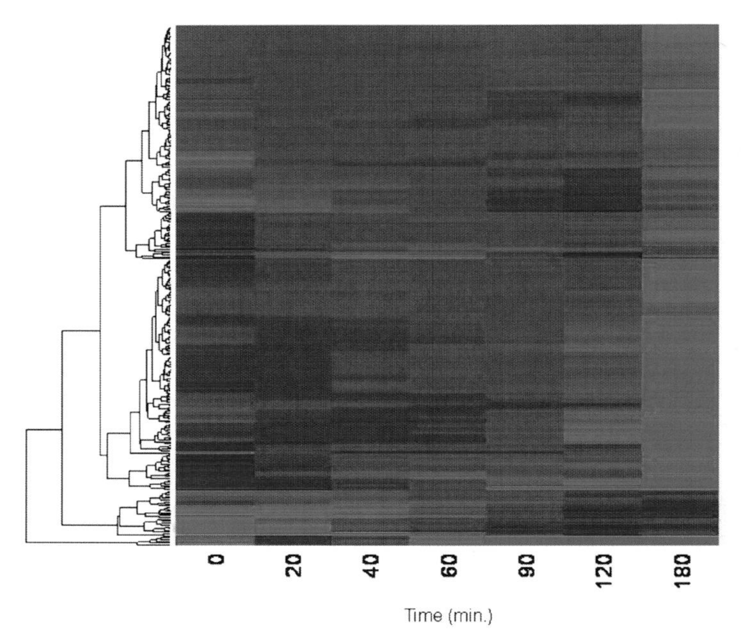

Color Plate 6. Fig. 5, Chapter 11: Hierarchical clustering of 440 proteins. The protein list was clustered by dynamic changes in expression levels (the most downregulated in dark-green and the most upregulated in dark-red) (*see* discussion on p. 242).

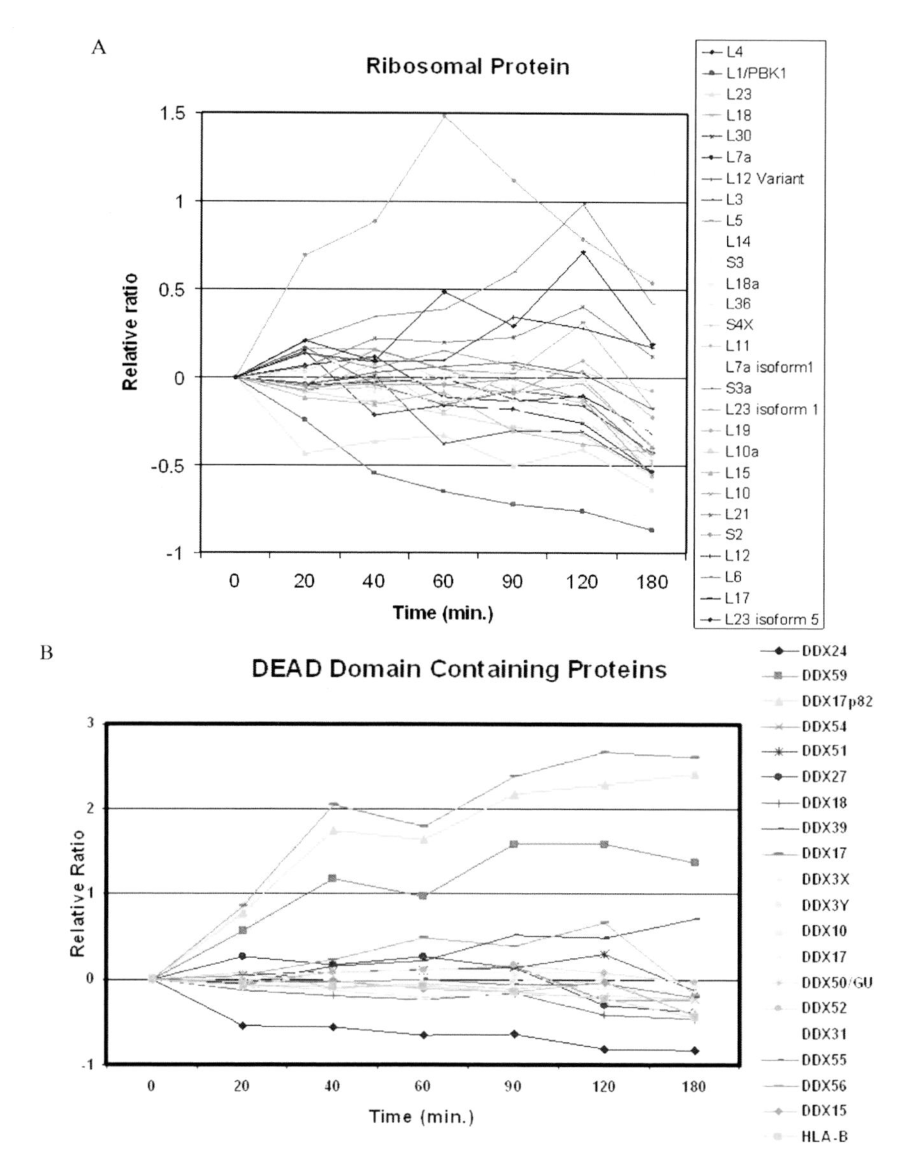

Color Plate 7. Fig. 6a, b, Chapter 11: Nucleolar protein profiling by the described isobaric mass tagging technology. **(A)** Changes in the abundance of various ribosomal protein subunits after actinomycin D treatment. Relative ratios of quantified ribosomal proteins are shown. **(B)** Changes in the abundances of DEAD box domain containing proteins after actinomycin D treatment. Relative fold changes of DEAD/H domain containing proteins are shown. HLA-B, HLA-B associated transcript (*see* discussion on p. 243).

11

Top-Down Quantitative Proteomic Analysis Using a Highly Multiplexed Isobaric Mass Tagging Strategy

Bing Xie, Wayne F. Patton, and Craig E. Parman

Summary

Proteomic analysis has proved to be key to determining drug mechanisms and assessing toxicologic potential during preclinical screening studies. A major goal in proteomics is to accurately measure changes in the relative abundance of large sets of proteins in complex biological systems as a function of experimental parameters, such as drug dose or exposure time. Until recently, top-down quantitative proteomics has been restricted to two-dimensional gel analyses or two-plex mass tagging. A new top-down approach based on isobaric mass tagging (ExacTag) for highly multiplexing (up to 10-plex) protein quantification is presented, involving chemically tagging cysteine or lysine residues of intact proteins isolated from cells, tissues, or biological fluids. As many as 10 labeled samples are then combined, fractionated, proteolytically digested, and analyzed by gel electrophoresis or liquid chromatography–tandem mass spectrometry. Proteins are identified using public domain search engines, such as Mascot (Matrix Science Ltd., London, UK) and quantified using an in-house developed software package. During the fragmentation, the tag-labeled peptides generate a set of low mass reporters that are unique to each sample. Measurement of the intensity of these reporters allows the relative quantification of the peptides and consequently the proteins from which they originated. The capabilities of the approach are demonstrated by analysis of the HeLa cell nucleolar proteome after treatment with the metabolic inhibitor actinomycin D for various time periods. A total of 440 proteins are qualitatively identified and quantified. The quantification data demonstrate that the nucleolar proteome changes significantly over time in response to differences in growth conditions, which is consistent with previous observations from several groups. The highly multiplexed and quantitative nature of the new technology should herald in new opportunities to provide diagnostic and functional insights into the proteomics discovery process.

Key Words: biomarkers; isotope-coded protein labels; mass tagging; quantitative proteomics; stable isotope labeling; top-down proteomics.

From: *Methods in Pharmacology and Toxicology: Biomarker Methods in Drug Discovery and Development*
Edited by: F. Wang © Humana Press, Totowa, NJ

1. INTRODUCTION

In the field of proteomics, an important task is to identify and quantify changes in protein content among multiple cellular states or disease stages. These changes may reflect differences among disease stages relative to a normal state and thus help direct researchers to potential targets for therapeutic intervention *(1)*. Proteomic differential display analysis has proved indispensable to determining drug mechanisms and for assessing toxicologic potential during preclinical screening studies. Early proteomics analysis was conducted using a combination of two-dimensional gel electrophoresis (2-DE) and mass spectrometry *(2–4)*. Recently, proteomics has expanded to include profiling, functional, and structural relationships based on a broader range of technologies. Techniques have been developed to simultaneously compare the changes in protein by the isotope coded affinity tag (ICAT) method *(5)*, solid phase mass tagging *(6)*, acetylation *(7)*, esterification *(8)*, and other heavy isotope approaches *(9)*. These technologies take advantage of the mass spectrometer's capability for sensitive peptide differentiation and identification, but the binary nature of these methods limits the scope of their application.

It is often observed that physiologic effects are manifested in the proteome over time through changes in the concentration of an effector or signaling molecule. Only by comparison with a common reference standard can most existing technologies track changes in a proteome as a function of time or concentration. Often, negative and positive controls are also required to clearly establish significant biological trends. The use of a two-state method for these applications requires a large number of experiments with the resultant loss in data precision *(10)*. Furthermore, multiple pairwise comparisons produce coverage gaps, as different peptides and proteins will invariably be identified in each experiment *(11)*. Newer technologies that can simultaneously measure changes in the proteome among multiple states and with higher sensitivity have greatly increased the useful information gleaned in proteomics experiments. Ideally, a multiplexed quantitative detection technology, not dependent on residue-specific chemistry, would be advantageous for differential profiling of large numbers of proteins. Multiplex protein quantification technologies, such as stable isotope labeling with amino acids in cell culture (SILAC; a three-state approach), which incorporates modified isotopic amino acids into proteins synthesized in cultured cells *(12)*, and isobaric tagging for relative and absolute quantitation (iTRAQ), which labels all of the tryptic peptides, have been developed *(13)*. However, a technology with easy-handling features and higher multiplexing power would have significant advantages for high-throughput protein quantification applications ***(14,15)***.

Typically, mass spectrometry (MS)-based proteomics methods have been based on a bottom-up strategy that depends upon proteolytic digestion of proteins into constituent peptides, followed by separation and detection using liquid chromatography–tandem mass spectrometry (LC-MS/MS). Most MS-based investigations to date have focused on qualitative identification, quantification, separation strategies, sample preparation, and bioinformatics approaches for the bottom-up methodology. In recent years, however, top-down proteomics approaches, involving the analysis of intact proteins, have gained increasing popularity. The ability to characterize posttranslational modifications, reduce false-positive identifications, and study protein-protein interactions make this an ideal approach for solving many biological problems in proteomics.

2DE/MS studies demonstrate that one gene results in an average of 10 to 15 different protein isoforms, most of this microheterogeneity being attributed to posttranslational modifications. Additionally, roughly 40% to 60% of human genes have alternative splice forms, indicating that alternative mRNA splicing is a very common mechanism for the generation of protein isoform diversity. Bottom-up proteomics is not well suited to the analysis of protein posttranslational modifications or the analysis of protein isoforms generated by alternative mRNA splicing. Bottom-up proteomics procedures, by the very nature of reducing all this protein diversity to constituent peptide fragments, are unable to distinguish among the various protein isoforms exhibited in a typical proteome. They simply cannot reassemble the pieces back together into meaningful biological data.

A new isobaric mass tagging technology, ExacTag Labeling reagents (PerkinElmer, Waltham, MA), are targeted to whole proteins and thus suitable for elucidating the rich protein diversity in living organisms. The technology provides a superior approach for determining the relative expression level of individual proteins among multiple samples by MS, with high precision (low coefficients of variation). Because multiple samples are tagged and mixed before processing steps, sample-to-sample recovery differences are completely eliminated. Unlike conventional immunoassays, the MS-based approach allows unambiguous detection and identification of the multitude of proteins and their variants, which improves the biological relevance of the assay. The highly multiplexed and quantitative nature of the ExacTag Labeling technology should herald in new opportunities to provide functional insights into the proteomics discovery process.

2. MATERIALS

2.1. Cell Culture, Serum, and Reagents

1. Dulbecco's Modified Eagle's Medium (DMEM) (Invitrogen Corp., Carlsbad, CA) supplemented with 10% fetal calf serum (Invitrogen), 100 units/mL penicillin (Invitrogen), and 100 μg/mL streptomycin (Invitrogen).
2. Actinomycin D (Sigma Chemical Co., St. Louis, MO) dissolved in DMSO to 1 mg/mL. The final concentration for treatment is 1 μg/mL.
3. HeLa cells (ATCC, Manassas, VA).

2.2. ExacTag Reagents

Whereas the composition of the ExacTag labels can vary according to the specific application, their general structure and use are similar for all applications of the technology. An ExacTag label set is composed of a family of reactive peptides with the identical amino acid composition (*see* **Fig. 1** and Color Plate 5, following p. 230). Each label contains an aspartate-proline (DP) dipeptide linkage positioned within an otherwise common peptide sequence. The DP linkage is a highly scissile bond that fragments in a mass spectrometer at lower collision energies than other dipeptide linkages ***(16–19)***. The common amino acid structure can fragment into a balancer and a low mass reporter region (**Fig. 1A**). Therefore, once peptides labeled with ExacTag labels are filtered away from a complex background in an MS/MS experiment, they can be selected and altered so that each labeled peptide generates a set of unique masses whose intensities can be used for quantification. This procedure generates a very low background for the highly sensitive detection and quantification of the multiplexed labels in the second stage of the MS. The detailed composition of labels is shown in **Fig. 1B**. The labels are peptide-based tags that have the combination of six heavy and six light glycines in the low mass reporter and balancer regions. By switching heavy and light glycines in these regions, a series of low mass reporters and corresponding high mass signals will be generated. ExacTag reagents are available in kit form from PerkinElmer (Waltham, MA). The kits contain the following components:

1. Dissolution Buffer: 0.5 M ammonium bicarbonate buffer (AMB), pH ~8.5.
2. Denaturant Buffer: 5% sodium dodecylsulfate (SDS).
3. Reducing Buffer: 50 mM Tris (2-carboxyethyl) phosphine hydrochloride (TCEP).
4. ExacTag Labels: *For the 7-plex kit*, HG-457, HG-460, HG-463, HG-466, HG-469, HG-472, and HG-475 labels, which can be used for labeling up to seven

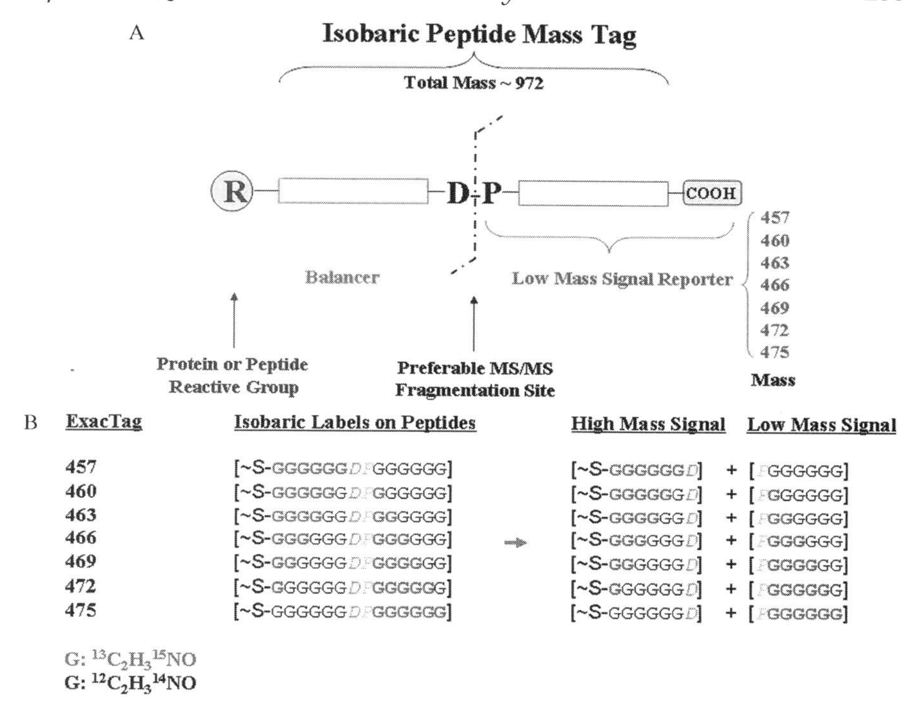

Fig. 1. Composition of isobaric mass tags. **(A)** Schematic structure of the isobaric mass tag described in this chapter. The tag contains four elements: (1) a reactive group with specificity toward a thiol (cysteine) or an amine (lysine, α-amino) group; (2) a cleavage enhancement moiety, which is an aspartic acid (Asp, D) proline (Pro, P) scissile bond group. Distributed around the DP sequence are the (3) low mass signal reporter and its (4) corresponding balancer sequence. **(B)** Detailed composition of seven isobaric mass tags. The tag contains a common amino acid composition. G represents glycine. The red bold "G" is heavy glycine and plain "G" is normal glycine. In MS/MS, the tags generate two sets of signals: low mass signals and high mass signals. (*see* Color Plate 5, following p. 230.)

different samples. The low mass signal reporters produce at a normal m/z of 457, 460, 463, 466, 469, 472, and 475, and are thus readily distinguishable based on their 3 Da mass differences.

2.3. Protein Separation by Sodium Dodecylsulfate–Polyacrylamide Gel Electrophoresis (SDS-PAGE) and Gel Staining

1. Tris/glycine/SDS Buffer (10X): 250 mM Tris, 1920 mM glycine, 1% SDS, pH 8.3 (Bio-Rad Laboratories, Inc., Hercules, CA).

2. Protein sample loading buffer (Laemmli sample buffer, 62.5 mM Tris-HCl, pH 6.8, 2% SDS, 25% glycerol, 0.01% Bromophenol Blue, Bio-Rad Laboratories, Inc.).
3. Prestained SDS-PAGE Standards, broad range (Bio-Rad).
4. Bio-Rad 4–20% Tris-HCl Criterion Precast Gel (Bio-Rad Laboratories, Inc.).
5. Coomassie Blue staining solution (Coomassie Brilliant Blue R-250; EM Science, Darmstadt, Germany): 0.2% Coomassie Blue, 10% acetic acid, 45% methanol.
6. Coomassie Blue Destaining Solution I: 50% methanol, 10% acetic acid.
7. Coomassie Blue Destaining Solution II: 10% methanol, 5% acetic acid.
8: Storage Solution: 5% acetic acid.

2.4. In-Gel Digestion

1. 50 mM ammonium bicarbonate (AMB).
2. Trypsin kit (Promega Corporation, Madison, WI), 20 μg/vial, sequencing grade.
3. Trypsin Reaction Buffer (50 mM AMB and 9% acetonitrile (ACN; EM Sciences, Gibbstown, NJ).
4. ZipTip C18 tips (Millipore Corporation, Bedford, MA).
5. ZipTip Re-hydration Buffer: 100% ACN.
6. ZipTip Equilibration and Washing Buffer: 0.1% trifluoroacetic acid (TFA; EM Sciences, Gibbstown, NJ).
7. ZipTip Elution Buffer: 50% ACN/0.1% TFA

2.5. Liquid Chromatography and Mass Spectrometry

1. Injection Sample Buffer: 0.1% Formic Acid
2. Standard LC gradient Buffers

3. METHODS

The isobaric mass tagging reagent is covalently affixed, via the reactive motif (R), to proteins or peptides, depending upon the targeted amino acids by standard chemistry. In this study, a given mass tag label was covalently attached to the cysteine residues of nucleolar proteins from HeLa cells treated with actinomycin D for a given time point using reactive labels containing a reactive iodoacetamide group. After labeling, the samples were combined, trypsin digested, and fractionated so that differentially labeled peptides could be isolated together. Thus, a peptide labeled with a unique isobaric mass tagging reagent represents the same protein from a different sample.

The labels from the ExacTag set are isobaric meaning they have same mass but a different amino acid sequence. When tags from the same isobaric set are separately attached to an identical peptide, in different reactions, the labeled peptide counterparts from different samples have identical masses. The labeled target peptides can ionize and be filtered from other molecules based on the differences in mass-to-charge ratio (m/z) in the MS/MS mode. The

DP scissile bond is preferentially fragmented or fragmented under collision-induced dissociation (CID), which gives rise to two quantifiable groups of signals: low mass signal reporters containing the portion after the Pro residue and high mass signals consisting of the target peptide affixed to the rest of tag (**Fig. 1B**). Here, only low mass reporters are used for quantitation.

Unlike many conventional protein-reactive isotope mass tag labeling technologies, such as isotope coded affinity tag (ICAT) and isotope coded protein label (ICPL), ExacTag isobaric mass tag–conjugated proteins or peptides co-elute in chromatographic separations due to their identical hydrophobicity and are indistinguishable in the first stage of MS, allowing for more efficient enrichment and accurate quantification (data not shown). In addition, two sets of signals (low and high mass signals) may be used in quantification separately or in combination to generate correlating ratios, making quantification more precise.

3.1. General Workflow of Top-Down Multiplexed Protein Labeling

In the ExacTag Thiol labeling protocol, proteins from each sample are reduced and labeled via their cysteine residues. Then, all samples are combined in a single tube for further processing and analysis. This workflow eliminates potential variation typically encountered in serial sample preparation that may cause inaccuracies in quantification.

Combined samples are fractionated by different processes depending upon the user's experiments (e.g., SDS-PAGE, HPLC, or other separation methods). Each fractionated protein sample (gel-slice, if using SDS-PAGE) is digested to its constituent peptides (using trypsin or other protein cleavage agent) and subjected to LC-MS/MS analysis. If sample losses occur during preparation, each sample experiences the same loss and the ratios of the components are preserved. **Figure 2** summarizes the ExacTag Thiol reagent general workflow for a standard 7-plex experiment. Currently, as many as 10 different samples can be prepared and analyzed in a single experiment, though fewer samples may be compared or replicates of fewer experimental conditions may be evaluated (e.g., three control samples, three experimental samples, and a simple set of standard proteins could be evaluated).

3.2. Sample Preparation

1. HeLa cells (ATCC, Manassas, VA) are grown in DMEM media supplemented with 10% fetal calf serum, 100 units/mL penicillin, and 100 μg/mL streptomycin. When cells are 90% confluent (1.5×10^7 per 150-mm dish), they are used for passage or harvested for biochemical studies.

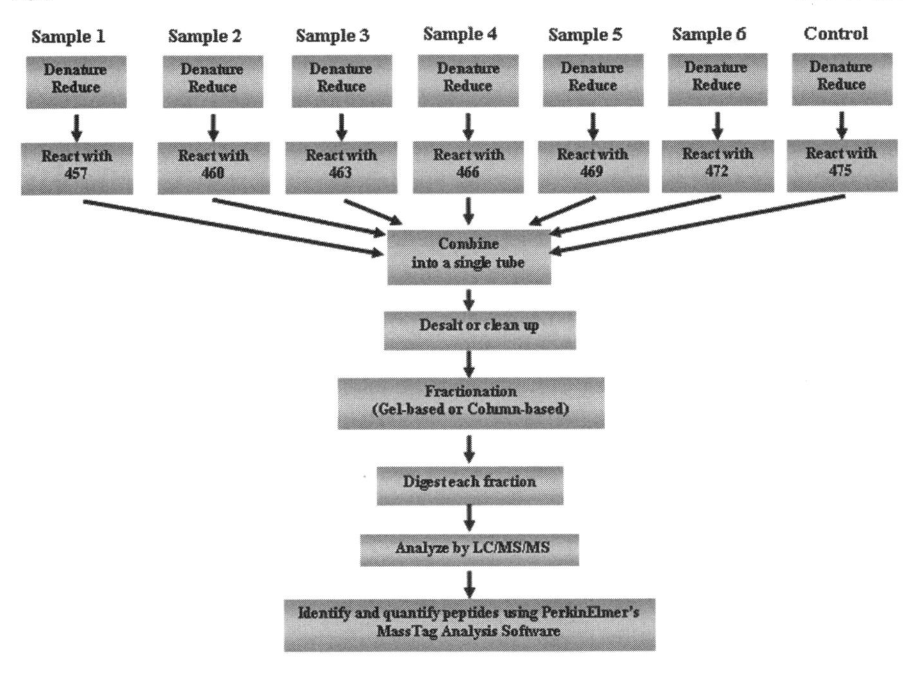

Fig. 2. Top-down workflow for the ExacTag labeling procedure.

2. Actinomycin D is added to seven sets of cells (each set contained five 150-mm plates) to a final concentration of 1 µg/mL to the HeLa cells and incubated for 0, 20, 40, 60, 90, 120, and 180 min, respectively.
3. Nucleoli are isolated from the HeLa cells for each time point as previously described (http://www.lamondlab.com/f7protocols.htm) ***(20)***.
4. Nucleolar proteins are released by boiling nucleoli fractions in 200 µL denaturing buffer (50 mM NH_4HCO_3, 0.5% SDS). After centrifugation at 14,000 rpm for 5 min, the protein concentration is determined by solution-phase Bradford assay ***(21)***.

3.3. ExacTag Labeling and Desalting

1. Nucleolar proteins (10 µL per time point) are reduced in 2 mM TCEP for 15 min by boiling. (The total amount of proteins from sample 1 to 7 was 83.3, 70.1, 70.1, 54.3, 71.8, 51.0, and 55.3 µg, respectively.)
2. After cooling, a unique isobaric mass tag (100.1, 84.2, 84.2, 65.2, 86.2, 61.3, and 66.5 µg, respectively) is added to each of the seven samples at a molar ratio of 10 to 1 (label to cysteine). The amount of material added was based on the purity of the tags.
3. The reactions are carried out at room temperature in the dark for 3 h.
4. The seven labeling reactions are combined and the protein mix is desalted by passing it through a Tricorn Superdex Peptide 10/300 GL column (GE

Healthcare, Amersham, UK) using a BioLogic DuoFlow 40 System with a BioFrac Fraction Collector (Bio-Rad Laboratories, Inc.). Ammonium bicarbonate (50 mM) is used as the buffer. The fractions that contain proteins (detected on-line with BioLogic DuoFlow System) are combined and dried down in a Speed-Vac Concentrator System (Thermo Fisher Scientific, Inc., Waltham, MA).
5. The desalting step can be conducted by acetone precipitation.
   (1) Place the sample tube with the sample on ice for 1 h.
   (2) Add six volumes of cold acetone (chilled at –20°C overnight) to the sample tube.
   (3) Replace the cap and invert the tube three times.
   (4) Incubate the tube at –20°C until a cloudy precipitate forms (30 min to 4 h).
   (5) Briefly centrifuge the sample (10,000 × *g* for 1 to 2 min) and carefully decant the acetone. *Do not dry the pellet.*
   (6) Use the precipitated material as the sample for further processing.

3.4. Protein Separation by SDS-PAGE and Coomassie Blue Staining

1. The labeled proteins from desalting step are resuspended in 150 μL 50 mM AMB.
2. About 50 μg proteins (10 μL) are mixed with equal volume of 2X protein sample loading buffer and are electrophoretically separated using a Bio-Rad 4–20% Tris-HCl Criterion Precast Gel.
3. The gel is stained by Coomassie Blue staining solution for 1 h.
4. Wash briefly with Coomassie Blue Destaining Solution I.
5. Destain with Coomassie Blue Destaining Solution I for 2 h. To speed up the destain process, a small piece of foam rubber can be placed in the container with the gel.
6. Destain with Coomassie Blue Destaining Solution II until background is very low.
7. Equilibrate the gel in the 500 mL of the storage solution for at least 1 h. The gel should return to its original dimensions during this process.
8. Store the gel in the storage solution as needed. It might be convenient to carefully transfer the gel to a heat-sealable bag for longer-term storage.
9. A typical image of unlabeled sample and labeled sample is shown in **Fig. 3**.

3.5. In-Gel Digestion

1. The whole lane is excised into 27 contiguous slices.
2. Wash the gel slices twice with 50 mM AMB.
3. Incubate the gel slices with 50% (v/v) ACN in 50 mM AMB, 30 min at 37 °C.
4. Repeat **step 3** two to three times until the blue dye in gel disappears.
5. Wash the gel plugs once with 50% (v/v) ACN in 50 mM AMB.
6. Speed-vac to dryness for 20 – 30 min.

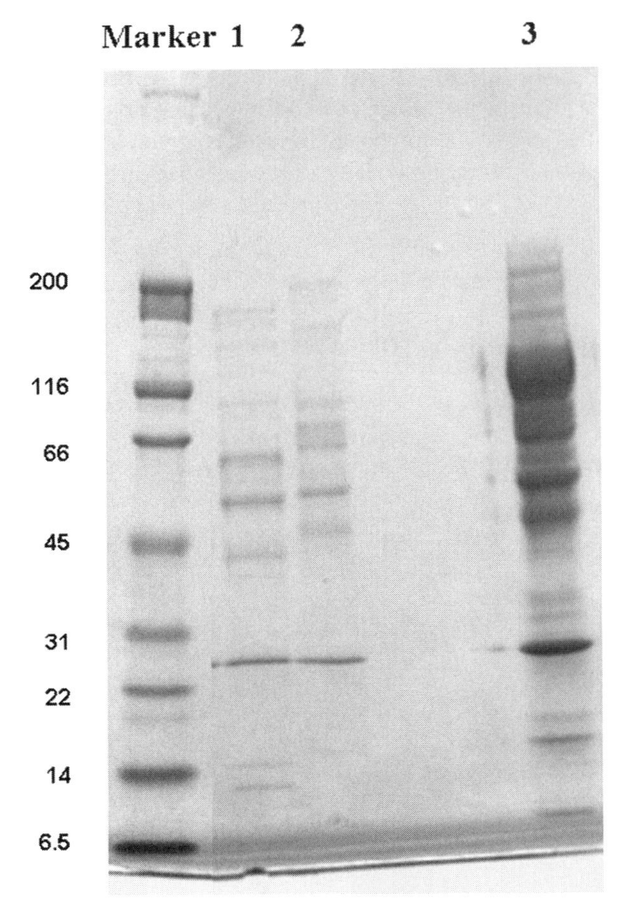

Fig. 3. Labeled proteins fractionation by SDS-PAGE. Labeled nucleolar proteins from 90-min time point are separated by SDS-PAGE. **Lane 1**, unlabeled nucleolar proteins; **lane 2**, labeled nucleolar proteins; **lane 3**, mixture of labeled nucleolar proteins from the seven time points.

7. Add 5 μL of the Trypsin Solubilization Reagent (50 mM acetic acid) from trypsin kit to one vial of trypsin. Mix the vial briefly to ensure the trypsin is dissolved.
8. Add 995 μL of the Trypsin Reaction Buffer to the vial and mix. The final concentration of trypsin is 20 μg/mL.
9. Add 20 μL (0.4 μg of trypsin) of the prepared Trypsin to each tube to swell the gel slices.
10. Add 50 μL of the Trypsin Reaction Buffer to the gel slices. Confirm that the gel plugs are at the bottom of the wells and covered with liquid.
11. Incubate overnight (10 to 16 h) at 37°C.

12. After the incubation, remove the liquid from the gel plugs and transfer the liquid to a new labeled tube. There is about 40 to 50 μL from each tube. This solution contains the extracted tryptic peptides.
13. Samples are desalted and concentrated using the reverse-phase C18 zip-tips.
14. For electrospray ionization (ESI) acquisition, samples are eluted with elution buffer [50% acetonitrile (ACN), 0.1% formic acid] into 96-well plates for automated mass spectrometry analysis.
15. For MALDI acquisition, samples are zip-tipped and eluted with 50% ACN containing 0.1% TFA.

3.6. Mass Spectrometry Acquisition

For ESI acquisition, samples are run in duplicate using LC-MS/MS.

1. Analysis is performed on a ThermoFinnigan LTQ (Thermo Fisher Scientific, Inc., Waltham, MA) ion trap mass spectrometer operating in a data-dependent mode. The six most intense ions were sequentially analyzed by MS/MS.
2. Parameter setting: The normalized collision energy setting is 35 and a full MS target value of 3×10^4 and an MSn target value of 1×10^4 are used. All other parameters for data-dependent analysis are at factory settings provided with the Xcalibur version 1.4 software (Thermo Fisher Scientific, Inc.).

3.7. Protein Identification and Quantitation by Mascot and ExacTag Analysis Software

3.7.1. Protein Identification

1. MS/MS spectrum from LTQ is extracted and converted into Mascot generic format (MGF) files using utility program extract_msn.exe supplied with the Xcalibur software. Any charge state information is removed from the MGF files to allow the protein search engine to control which charge states are searched.
2. The replicate data files from each of the gel slices are combined for searching. Proteins in each gel slice are identified by searching each combined peak-list against the IPI_Human 3.26 database using the Mascot (v2.2.1, Matrix Science, London, UK) algorithm. The enzyme is specified as trypsin with a potential of one missed cleavage. The peptide tolerance is limited to 2 Da, MS/MS fragment tolerance is limited to 0.8 Da, and variable modifications are oxidation of methionine and the ExacTag labeling of cysteines. The database is searched against the precursor charge states of 2+ and 3+.
3. Only proteins with a protein score above 40 are used for further quantification.

3.7.2. Protein Quantitation

The protein quantitation results were generated using the using ExacTag Analysis Software V 3.0 (PerkinElmer). This package can automatically parse the Mascot search results and the associated peak lists to generate quantitation data.

1. All spectra in the peak lists are searched for the report signals of the ExacTag labels. All spectra that contain signals for these reporters above background are then combined with the protein search results. The quantification results for a spectrum can be associated with a protein in two ways. First, any spectra that is identified in the protein search results as having the ExacTag label modification and has significant reporters signals is added to the quantitation results. More than half of the proteins quantified have labeled peptides identified.
2. Additionally, the sequences of identified proteins are subjected to *in silico* digestion, and a list of potential labeled peptides is generated. The masses of potential *in silico* labeled peptides are compared with the list of the m/z values of quantified peptides observed in the MS/MS analysis. Again, the data are searched using both the 2+ and 3+ precursor charges states. The measured mass of the precursor is required to be within 0.6 Da of the theoretical mass to be considered a match, and the data are added to the quantitation results.
3. Once all peaks from each gel slice are compared and a list of proteins with corresponding labeled peptides is generated, a complete list containing all proteins and their labeled peptides that are quantified is consolidated. The quantified peptides for the same protein from different gel slices are pooled for final quantitation under single protein assignment. For a protein to be included in the final result, there had to be a minimum three peptides. The average ratio and coefficient of variation (CV) of a protein are calculated based on the consolidated information. The final ratios are normalized by subtracting control values (ratio of time point 0 = 0).

4. RESULTS AND DISCUSSION

4.1. Typical Spectrum for ExacTag-Labeled Peptide

Using an ESI source, the labeled peptide may be represented with different mass-to-charge ratios (m/z) because ionization can produce multiple charge states. When these different peaks are assessed by MS/MS, the resultant spectra will show multiple peaks with different high mass signals and the same low mass signals. **Figure 4** shows a typical spectrum of a labeled peptide that is used for protein identification and quantitation. The Mascot search result shows this peptide sequence is CPQVEEAIVQSGQK and it is from Isoform 2 of Guanine nucleotide-binding protein-like 3.

4.2. The Kinetics of Nucleolar Proteome Changes After Actinomycin D Treatment

The relative levels of 440 cysteine-containing proteins are quantified over seven time points after inhibiting cellular transcription with actinomycin D. The heat map in **Fig. 5** (see Color Plate 6, following p. 230) shows the kinetics of the protein level change upon the treatment. A wide range of responses to actinomycin D treatment for different proteins was observed

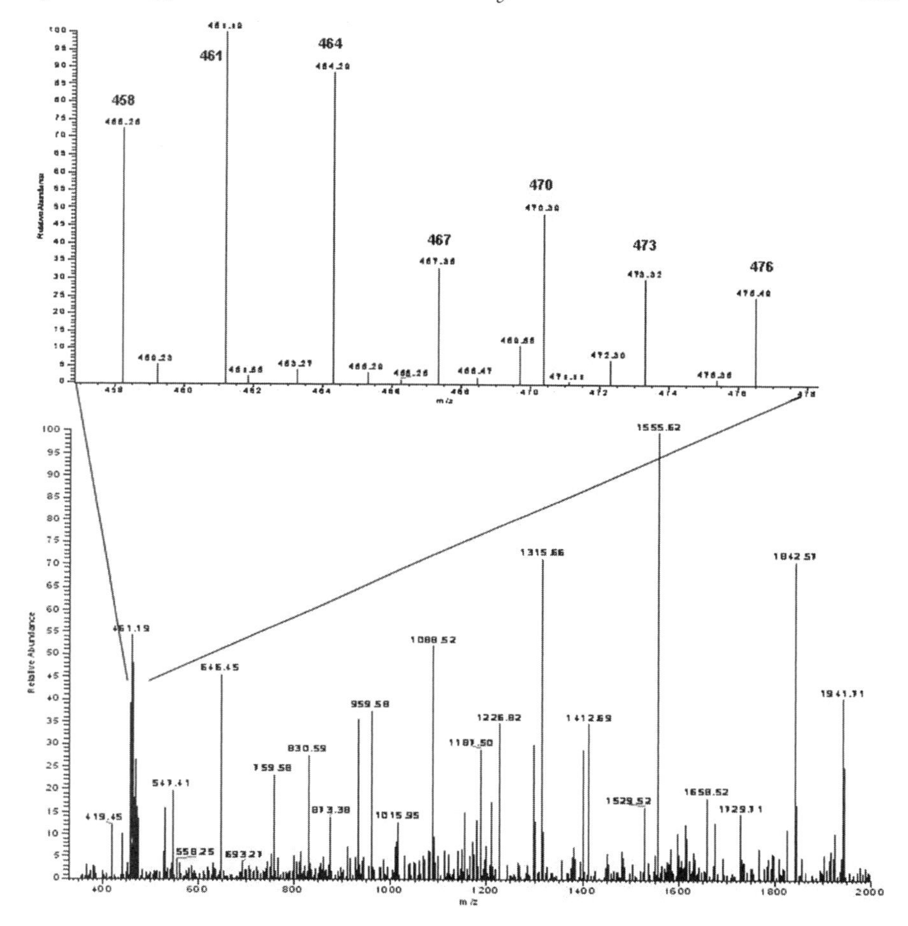

Fig. 4. MS and MS/MS spectra of isobaric mass tag–labeled cysteine-containing peptides. CPQVEEAIVQSGQK from Isoform 2 of Guanine nucleotide-binding protein-like 3 protein is identified and subsequently subjected to quantitation.

(*see* **Fig. 6** and Color Plate 7, following p. 230). The steady-state levels of many nucleolar proteins decreased to varying extents after actinomycin D treatment, similar to the observation made by Andersen et al. using the SILAC approach ***(20)***.

It is well-known that actinomycin D inhibits RNA polymerase I activity ***(22)***. Some proteins are diminished from the nucleoli fraction after actinomycin D treatment due to the inhibition of new ribosomal RNA synthesis and the export of assembled ribosomal subunits. These proteins include ribosomal proteins, RNA processing factors, exosome components, and

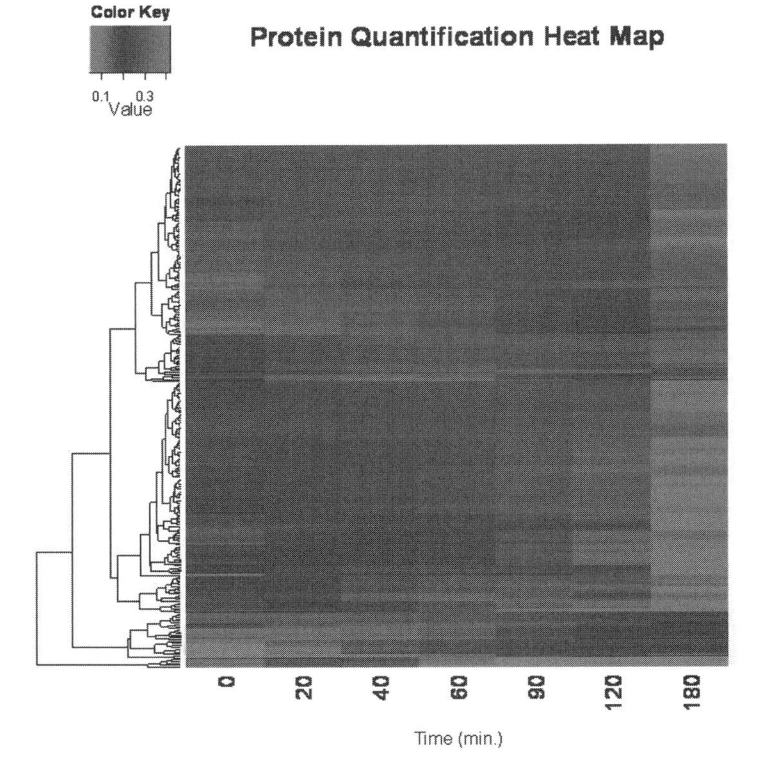

Fig. 5. Hierarchical clustering of 440 proteins. The protein list was clustered by dynamic changes in expression levels (the most downregulated in dark-green and the most upregulated in dark-red). (*see* Color Plate 6, following p. 230.)

RNA polymerase I. Our results showed that compared with the untreated sample, the levels of most quantified ribosomal proteins decrease to different extents, whereas the levels of S2, L6, L12, L17, and L21 increase **(Fig. 6A)**, suggesting that the large and small subunits of ribosomal proteins leave the nucleoli at markedly different rates.

Similar to other Cys-based protein quantification technologies, the described isobaric mass tagging technology has limitations because of the existence of proteins that have very few Cys moieties *(23)*. For example, there are 80 ribosomal proteins. Of these, only 28 could be quantified by thiol-reactive isobaric mass tagging technology. More than half of ribosomal proteins cannot be quantified because of their low Cys content (0 or 1), indicating that an amine-reactive labeling method (e.g., the recently

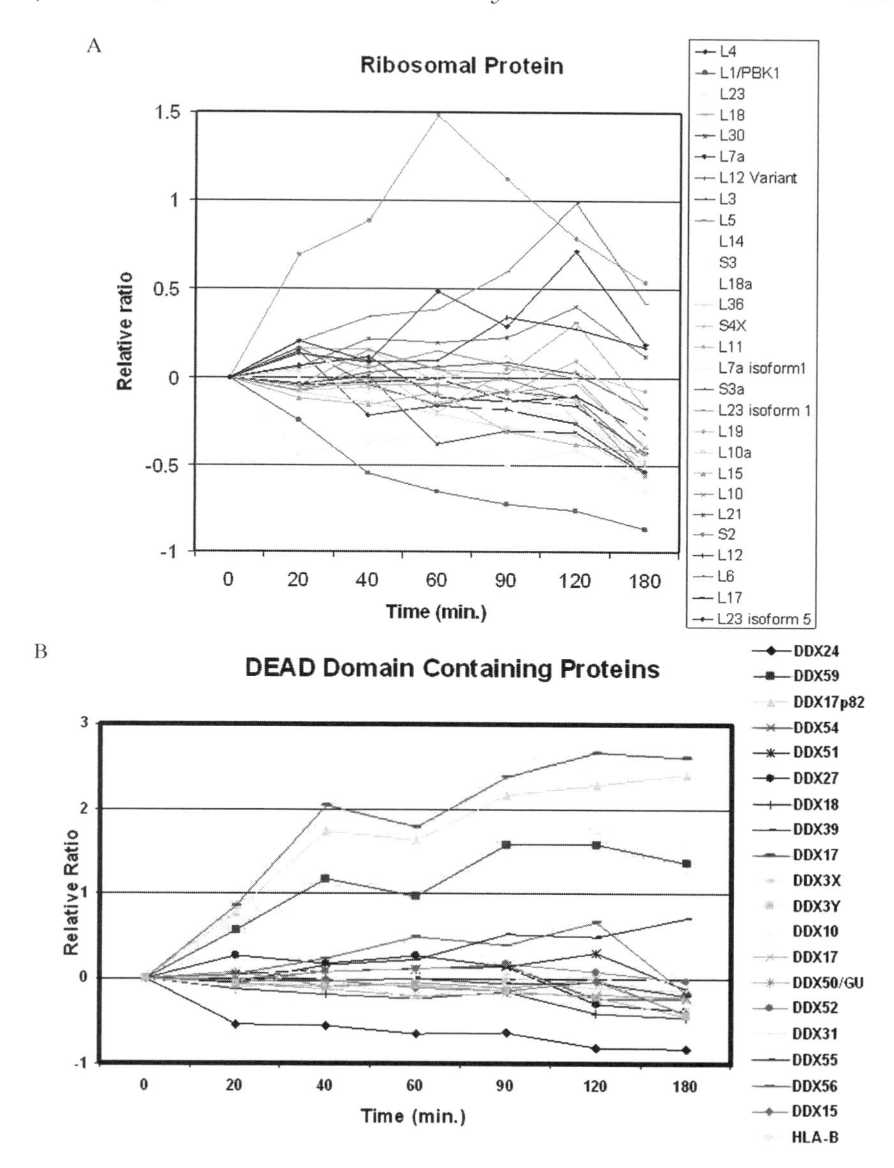

Fig. 6. Nucleolar protein profiling by the described isobaric mass tagging technology. **(A)** Changes in the abundance of various ribosomal protein subunits after actinomycin D treatment. Relative ratios of quantified ribosomal proteins are shown. **(B)** Changes in the abundances of DEAD box domain containing proteins after actinomycin D treatment. Relative fold changes of DEAD/H domain containing proteins are shown. HLA-B, HLA-B associated transcript. (*see* Color Plate 7, following p. 230.)

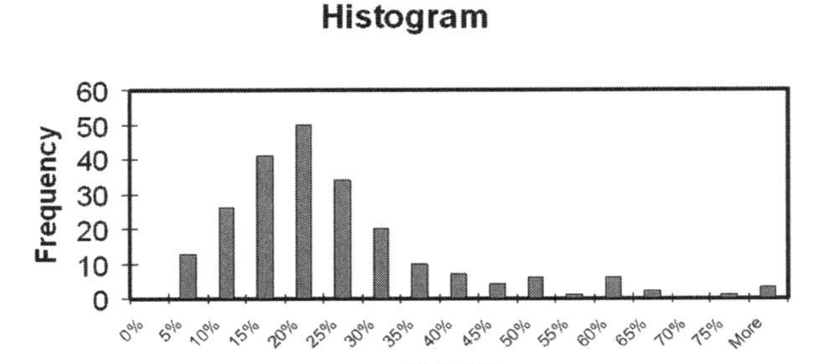

Fig. 7. Histogram of CV distribution for ExacTag 7-plex labeling in the experiment.

introduced amino-reactive ExacTag labeling reagents) should be considered to increase the quantification coverage.

Our data set clearly demonstrates that most nucleolar proteins decrease after treatment with actinomycin D, and individual proteins and specific protein classes show major differences in their rate of change after transcription inhibition. For example, nucleolar members of the DEAD box helicase family, which are likely to have distinct and non-overlapping functions in the nucleolus, respond to transcription inhibition with different kinetics (**Fig. 6B**).

The standard derivation and CV can be calculated using each of the time points from multiple scans of the same peptide (Scan CV) and from each of the scans of the multiple cysteine-containing peptides for each of the proteins (Peptide CV). The histogram indicates that one can expect to get a CV of about 20% to 25% for the 7-plex labeling, which is much lower than the common 30% to 35% of CV based on breadth of Gaussian curve fitting (**Fig. 7**).

5. CONCLUSION

A key task in functional proteomics is to quantitatively compare changes in protein expression content among samples caused by perturbations in biological processes. Biomarker discovery efforts typically require the comparison of multiple disease samples to normal control specimens. In many cases, changes in a biological program are dependent on the presence or concentration of a signaling molecule or have a strong temporal dependence on the signaling molecule. The effects of drugs and drug candidates on cells can also have strong concentration and temporal effects on cells. The described isobaric mass tagging technology was developed to monitor changes in protein expression levels and content for these important applications.

The described isobaric mass tagging set consists of multiple tags with a common amino acid composition and mass but differs in the distribution of the isotopic composition within the sequence. The aspartic acid/proline (DP) linkage is cleaved during the MS/MS analysis, fragmenting the labels in a manner whereby they are readily distinguishable on the basis of 3 Da mass differences. By quantifying the relative proportions of each of the tags, quantitative differences in protein content among samples can be assessed ***(17,18,24–27)***. The reported isobaric mass tagging labels overcome a number of problems associated with previously described protein-reactive isotopic labeling techniques. Conventional deuterium labeling methods have difficulties when used for the accurate quantification and identification of peptide pairs because labeled and unlabeled samples often do not exactly co-elute during chromatographic separations. In contrast, the mass tag labels described here all have identical compositions and hydrophobicity. The mass tag–labeled proteins and peptides obtained using the described approach co-elute during LC-based separation. The collision-induced dissociation (CID) analysis of isobaric mass tags produces dependable high and low mass fragment ions that reflect the relative abundances of the peptides from which they are derived. The described isobaric mass tagging technology shows consistent properties over a wide dynamic range, and labeled peptides can be detected with a sensitivity that is comparable with other labeling procedures (data not shown).

This study of nucleolar proteome dynamics demonstrates that the isobaric mass tagging technology can increase throughput when parallel identification and quantification of complex protein samples are required. The sevenfold or greater multiplexing capability allows for simultaneous measurements that can reduce sample preparation variation compared with other labeling technologies. Compared with previous publications using other labeling approaches, the data sets generated by the new isobaric mass tagging technology resulted in a comparable dynamic flux of the nucleolar proteome. It will be important to study the dynamic nature of endogenous proteins in other subcellular structures, compartments, and organelles whose protein compositions can also vary extensively under different growth and/or metabolic conditions.

This novel MS/MS-based isobaric mass tagging detection described produces quantified results for multiplex samples with good sensitivity, excellent signal-to-noise ratios, and a broad dynamic range. These are all features essential to the development of truly robust proteome analysis techniques. Additional capabilities should become available when additional reagents with new chemical derivatization specificities and expanded multiplexing range are developed.

ACKNOWLEDGMENT

We gratefully thank Dr. Brian Chait for innovative ideas on this technology and tremendous help on label design. We also thank Dr. Eric Kershnar for the whole technology design and development, Dr. Rixin Wang for the software development, and Paul Taylor and Christopher Smith for the MS data acquisition. Finally, we sincerely thank Dr. Paul Lizardi for the suggestions throughout the development of this technology. The described labeling technology is covered under United States Patent No. 6,824,981.

REFERENCES

1. Hanash S. Disease proteomics. Nature 2003;422:226–232.
2. Hanash SM, Strahler JR, Neel JV, et al. Highly resolving two-dimensional gels for protein sequencing. Proc Natl Acad Sci U S A 1991;88:5709–5713.
3. Unlu M, Morgan ME, Minden JS. Difference gel electrophoresis: a single gel method for detecting changes in protein extracts. Electrophoresis 1997;18: 2071–2077.
4. Appella E, Arnott D, Sakaguchi K, Wirth PJ. Proteome mapping by two-dimensional polyacrylamide gel electrophoresis in combination with mass spectrometric protein sequence analysis. Exs 2000;88:1–27.
5. Gygi SP, Rist B, Gerber SA, Turecek F, Gelb MH, Aebersold R. Quantitative analysis of complex protein mixtures using isotope-coded affinity tags. Nat Biotechnol 1999;17:994–999.
6. Shi Y, Xiang R., Crawford JK, Colangelo CM, Horvath C, Wilkins JA. A simple solid phase mass tagging approach for quantitative proteomics. J Proteome Res 2004;3:104–111.
7. Chakraborty A, Regnier FE. Global internal standard technology for comparative proteomics. J Chromatogr A 2002;949:173–184.
8. Goodlett DR, Keller A, Watts JD, et al. Differential stable isotope labeling of peptides for quantitation and de novo sequence derivation. Rapid Commun Mass Spectrom 2001;15:1214–1221.
9. Conrads TP, Alving K, Veenstra TD, et al. Quantitative analysis of bacterial and mammalian proteomes using a combination of cysteine affinity tags and 15N-metabolic labeling. Anal Chem 2001;73:2132–2139.
10. Goshe MB, Smith RD. Stable isotope-coded proteomic mass spectrometry. Current opinion in biotechnology 2003;14:101–109.
11. Hamdan M, Righetti PG. Modern strategies for protein quantification in proteome analysis: advantages and limitations. Mass Spectrom Rev 2002;21:287–302.

12. Ong SE, Blagoev B, Kratchmarova I, et al. Stable isotope labeling by amino acids in cell culture, SILAC, as a simple and accurate approach to expression proteomics. Mol Cell Proteomics 2002;1:376–386.
13. Ross PL, Huang YN, Marchese JN, et al. Multiplexed protein quantitation in Saccharomyces cerevisiae using amine-reactive isobaric tagging reagents. Mol Cell Proteomics 2004;3:1154–1169.
14. Schneider LV, Hall MP. Stable isotope methods for high-precision proteomics. Drug Discov Today 2005;10:353–363.
15. Pan S, Aebersold R. Quantitative proteomics by stable isotope labeling and mass spectrometry. Methods Mol Biol 2006;367:209–218.
16. Breci LA, Tabb DL, Yates JR 3rd, Wysocki VH. Cleavage N-terminal to proline: analysis of a database of peptide tandem mass spectra. Anal Chem 2003;75:1963–1971.
17. Skribanek Z, Mezo G, Mak M, Hudecz F. Mass spectrometric and chemical stability of the Asp-Pro bond in herpes simplex virus epitope peptides compared with X-Pro bonds of related sequences. J Pept Sci 2002;8:398–406.
18. Mak M, Mezo G, Skribanek Z, Hudecz F. Stability of Asp-Pro bond under high and low energy collision induced dissociation conditions in the immunodominant epitope region of herpes simplex virion glycoprotein D. Rapid Commun Mass Spectrom 1998;12:837–842.
19. Yu W, Vath JE, Huberty MC, Martin SA. Identification of the facile gas-phase cleavage of the Asp-Pro and Asp-Xxx peptide bonds in matrix-assisted laser desorption time-of-flight mass spectrometry. Anal Chem 1993;65:3015–3023.
20. Andersen JS, Lam YW, Leung AK, et al. Nucleolar proteome dynamics. Nature 2005;433:77–83.
21. Bradford MM. A rapid and sensitive method for the quantitation of microgram quantities of protein utilizing the principle of protein-dye binding. Anal Biochem 1976;72:248–254.
22. Zechel K. Initiation of DNA synthesis by RNA. Current topics in microbiology and immunology 1978;82:71–112.
23. Patton WF, Schulenberg B, Steinberg TH. Two-dimensional gel electrophoresis; better than a poke in the ICAT? Current Opin Biotechnol 2002;13:321–328.
24. McCormack AL, Somogyi A, Dongre AR, Wysocki VH. Fragmentation of protonated peptides: surface-induced dissociation in conjunction with a quantum mechanical approach. Anal Chem 1993;65:2859–2872.
25. Qin J, Chait BT. Matrix-assisted laser desorption ion trap mass spectrometry: efficient trapping and ejection of ions. Anal Chem 1996;68:2102–2107.
26. Qin J, Chait BT. Matrix-assisted laser desorption ion trap mass spectrometry: efficient isolation and effective fragmentation of peptide ions. Anal Chem 1996;68:2108–2112.
27. van Dongen WD, Ruijters HF, Luinge HJ, Heerma W, Haverkamp J. Statistical analysis of mass spectral data obtained from singly protonated peptides under high-energy collision-induced dissociation conditions. J Mass Spectrom 1996;31:1156–1162.

12

SELDI Technology for Identification of Protein Biomarkers

Prasad Devarajan and Gary F. Ross

Summary

The ProteinChip SELDI System and chip arrays provide a sensitive, high-throughput methodology for protein biomarker discovery. The combination of retentate chromatography and time-of-flight mass spectrometry allows for the detection of increased peak numbers even when processing microliter sample volumes. The multiple chip chemistries allow the flexibility for user-defined protocols that can facilitate protein purification strategies. In addition, ProteinChip arrays can be designed to characterize antibody-antigen and other protein-protein interactions. This review will update the reader on the current SELDI methods and protocols that are especially pertinent to protein biomarker discovery.

Key Words: biomarker; ProteinChip; proteomics; spectrometry; surface-enhanced laser desorption ionization.

1. INTRODUCTION

Expression proteomics involves the characterization of cell- and tissue-specific proteins. In addition to synthesis of primary polypeptide sequences, a variety of posttranslational modifications (e.g., glycosylation, phosphorylation, acetylation, sulfation) contribute to both the charge and mass heterogeneity for a given protein. Furthermore, proteolytic processing of specific proteins can also contribute to the complexity of a biological sample. The protein components found in a given sample are, therefore, the result of multiple synthetic, reductive-oxidative, enzymatic, transport, secretory, and degradative processes. Modification to the rates of any such process can be hypothesized to result in a diseased or abnormal physiologic state.

A critical need in clinical proteomics involves the discovery of validated protein biomarkers that may be useful in the diagnosis of specific diseased conditions. In addition to aiding in the early diagnosis and prediction,

From: *Methods in Pharmacology and Toxicology: Biomarker Methods in Drug Discovery and Development*
Edited by: F. Wang © Humana Press, Totowa, NJ

biomarkers are also needed for discerning disease subtypes, identifying the etiology, predicting the severity (risk stratification for prognostication and to guide therapy), and monitoring the response to interventions. Furthermore, biomarkers may play a critical role in expediting the drug development process. The Critical Path Initiative issued by the Food and Drug Administration (FDA) in 2004 stated that "Additional biomarkers (quantitative measures of biologic effects that provide informative links between mechanism of action and clinical effectiveness) and additional surrogate markers (quantitative measures that can predict effectiveness) are needed to guide product development". The concept of developing a new toolbox for earlier diagnosis of disease states is also prominently featured in the National Institutes of Health (NIH) Road Map for biomedical research ***(1)***.

Table 1
SELDI-TOF MS Instrumentation and Software Programs

Ciphergen Biosystems (1997–2006)	Bio-Rad Laboratories (2006–present)
Instruments	
ProteinChip System 4000 (PCS4000)	ProteinChip SELDI System
Personal Edition[a]	Personal Edition
Enterprise Edition[b]	Enterprise Edition
Autobiomarker Edition[c]	
Protein Biology Systems (PBS)	
PBS I (1997)	
PBS II (2000)	
PBS IIc (2002)	
Software	
Ciphergen Express Data Manager[d]	ProteinChip Data Manager
ProteinChip Software[e]	
Biomarker Wizard[f]	
Biomarker Pattern Software[g]	ProteinChip Pattern Analysis Software

[a]One chip array loaded manually into sample post.

[b]Autoloader allows processing of 8 × 12 cassettes, up to 14 cassettes can be read at once.

[c]Enterprise version is supplemented with Biomek 3000 (Beckman) for robotic chip processing.

[d]Instrument operation and relational database analysis for ProteinChip Systems and PBSIIc.

[e]Instrument operation and relational database analysis for Protein Biology Systems.

[f]Peak cluster analysis program, incorporated into ProteinChip Data Manager.

[g]Supervised classification software, incorporated into ProteinChip Pattern Analysis Software.

Biomarker discovery efforts require technologies enabling screening mechanisms that are sensitive, reproducible, and have high-throughput capacity. These properties can allow for reliable assessment of large sample numbers, which may be available only in limited volumes. One technology that possesses each of these features is surface-enhanced laser desorption ionization mass spectrometry (SELDI-MS). SELDI, introduced by Hutchens and Yip ***(2)***, combines selective retentate chromatography with time-of-flight mass spectrometry. SELDI techniques have been the subject of a rapidly growing literature, featuring biomarker studies in a variety fields including oncology ***(3–10)***, neurology ***(11–13)***, toxicology ***(14–17)***, infectious disease ***(18–21)***, and nephrology ***(22,23)***.

SELDI instrumentation and products were initially brought to market in 1997 by Ciphergen Biosystems, Inc (Fremont, CA). In November 2006, Bio-Rad Laboratories (Hercules, CA) purchased the rights to produce and market SELDI instrumentation, products, and services. **Table 1** reviews the nomenclature of ProteinChip systems and software currently marketed by Bio-Rad and versions previously marketed by Ciphergen.

Unique to the SELDI process are specific chip surfaces that can bind a subset of proteins present in a given sample. The ProteinChip surfaces contain either complexed chromatographic groups or preactivated chemistries to which user-defined "bait" molecules can be bound. The various ProteinChip arrays as well as their applications in biomarker discovery profiling and protein characterization are discussed below.

2. CHROMATOGRAPHIC ARRAYS

2.1. Normal Phase (ProteinChip NP20 Arrays)

The NP20 chips contain silicon oxide (SiO_2) surface groups capable of electrostatic and dipole-dipole interactions. Hydrophilic and charged amino acid residues on the protein surface can bind, and therefore the NP20 array is relatively nonselective. Buffers with a wide pH range are tolerated. Contaminating buffer salts and detergents can easily be removed by washing with distilled water. These properties allow NP20 arrays to be used in collecting mass calibration spectra using defined peptide or protein standards. In addition, the NP20 array is useful when monitoring larger-scale chromatographic column eluant fractions.

2.2. Anion Exchange (ProteinChip Q10 Arrays)

The Q10 chips contain quarternary ammonium groups, providing a cationic surface that can take part in electrostatic interactions with negatively charged aspartic acid and glutamic acid residues. Binding is highly dependent on buffer pH and ionic strength. In the profiling phase of biomarker discovery, permissive binding conditions (i.e., 50 mM Tris-HCl, pH 8 to 9) are recommended. However, more stringent binding buffers (i.e., 50 mM sodium acetate, pH 5) can also be useful in selecting for strongly acidic proteins. If the isoelectric point (pI) of a target protein is known, a buffer pH at least one unit greater than that pI is recommended.

2.3. Cation Exchange (ProteinChip CM10 Arrays)

The CM10 chips are derivatized with carboxylate groups, thus providing an anionic surface. Electrostatic interactions with positively charged arginine, lysine, and histidine residues can take place. As with the Q10 arrays, binding is dependent on buffer pH and ionic strength. Permissive binding conditions (i.e., 100 mM sodium acetate, pH 4.0) are again useful in initial biomarker discovery studies. More stringent binding buffers (i.e., 50 mM sodium Hepes, pH 7 to 8) will select for strongly basic proteins. If the pI of a target protein is known, a buffer pH at least one unit lower than that pI is recommended.

2.4. Immobilized Metal Affinity Capture (ProteinChip IMAC30 Arrays)

The IMAC30 chips incorporate nitrilotriacetic acid (NTA) groups that form stable octahedral complexes with polyvalent metal ions (i.e., Cu^{+2}, Ni^{+2}, Zn^{+2}, Fe^{+3}, Ga^{+3}). Once "charged" with a selected metal ion, two free sites on the formed octahedral complex can interact with specific amino acids (e.g., His, Cys, Trp) or phosphorylated amino acids (when using Fe^{+3} or Ga^{+3} as metal). Binding buffers (pH 6 to 8) with high ionic strength (0.5 M NaCl) are commonly used, promoting higher selectivity. Specific proteins can be eluted with increasing concentrations of competitors (such as imidazole) that can displace the protein from the coordinated metal.

2.5. Reverse Phase (ProteinChip H4 and H50 Arrays)

These surfaces contain methylene group chains of varying length: H4 (C_{16}), and H50 (C_6–C_{12}). Proteins bind via partitioning of surface hydrophobic residues (Ala, Val, Ile, Leu, Phe, Trp, Tyr) into the lipophilic chip surface. Two methods of binding can be used with the H4 and H50 arrays. In the classic reverse-phase mode, aqueous buffer with low organic content (i.e., 10% acetonitrile) is used for binding. Proteins or peptides can be selectively eluted by raising the organic solvent content of subsequent washes. These surfaces can also be used in a hydrophobic interaction chromatography (HIC) format. Here, high ionic strength (2 M ammonium sulfate) buffers are used for binding. Surface polar groups are occupied by the high salt. Surface nonpolar groups again interact with the lipophilic chip solid phase. By then reducing the buffer ionic strength, proteins can be selectively rehydrated and eluted from the chip surface.

2.6. Preactivated Surfaces (RS100 and PS20 ProteinChip Arrays)

ProteinChip RS100 Arrays are preactivated with carbonyldiimidazole groups, and ProteinChip PS20 Arrays are preactivated with epoxide groups. These two reactive SELDI chip surfaces allow specific, user-defined proteins to be covalently bound via free primary amine groups. The PS20 epoxide group can also link via free sulfhydryl (–SH) groups. Those specific "bait" molecules can then serve as affinity surfaces for selective adsorption of complimentary proteins. Typical applications include immunoassays, receptor-ligand binding studies, and transcription factor analysis. Both surfaces are susceptible to nucleophilic attack, principally by protein amine groups. Therefore, the coupling step must be carried out in amine-free buffers. In addition, free sulfhydryl compounds (i.e., 2-mercaptoethanol, dithiothreitol) should not be present.

2.7. Specialty Arrays (ProteinChip Gold, SEND-ID, and PG20 Arrays)

In the ProteinChip Gold Array, the gold chip surface is nonselective in binding characteristics. When samples are allowed to dry on the gold surface followed by matrix application, the SELDI system is comparable with a conventional matrix assisted laser desorption/Ionization time of flight (MALDI-TOF) instrument. In the ProteinChip SEND-ID Array, the SEND-ID chip surface has α-cyano-4-hydroxy cinnamic acid (CHCA) matrix incorporated as well as C_{18} chemistry. This feature greatly reduces the background noise to desorption of matrix and thereby increases the sensitivity in the low mass region. The SEND-ID chip is commonly used to characterize specific proteolytic enzyme digests. In the ProteinChip PG20 Array, protein G is coupled to the PS20 surface. The PG20 array can be used in the development of specific immunoassays.

3. SAMPLE PREPARATION

The SELDI retentive chromatography process has been shown to accommodate a wide variety of sample types including serum, plasma, urine, cerebral spinal fluid, cell and tissue lysates, subcellular fractions, conditioned cell culture media, and laser capture microdissected tissue. Because specific proteins are retained on the chip surface, the sample does not need to be desalted prior to application thus facilitating use of a wide variety of sample extraction and lysis buffers (**Note 1**).

Because biological samples often contain a complex assortment of proteins with distinct chemical properties, it can be advantageous to prefractionate samples prior to application to the SELDI chip surface. This approach is especially useful when profiling serum or plasma samples, as highly abundant proteins (albumin, immunoglobulins) can preclude binding of other proteins to the chip surface *(24)*. Strong anion exchange fractionation of serum using Bio-Rad's Serum fractionation kit and buffers is outlined in **Fig. 1**. The separate fractions are then profiled on different chip chemistries, excluding the Q10 anion exchange chip.

An alternative prefractionation technique, which also allows enrichment of lower-abundance proteins, involves Bio-Rad Proteo Miner Beads. The beads possess synthetic peptides produced in a combinatorial manner resulting in a wide variety of affinity surfaces. Proteo Miner Bead eluates have shown to increase the number of detectable SELDI peaks and sodium dodecyl sulfate polyacrylamide gel electrophoresis (SDS-PAGE) spots in plasma and urine studies *(25,26)*.

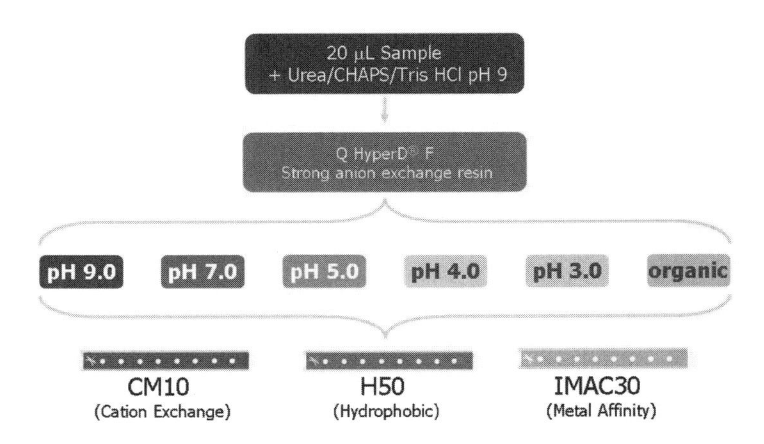

Fig. 1. Serum prefractionation scheme. Twenty microliters serum is preincubated with 30 μL 50 mM Tris-Cl, pH 9, 9 M urea, 2% CHAPS. This optional step favors dissociation or protein complexes prior to addition to Q HyperD F anion exchange resin available in filtration plate or spin column formats. Successive elution with buffers of decreasing pH and a final organic solvent wash provide six fractions (Q1 to Q6) that can be profiled on different ProteinChip arrays.

4. BASIC SELDI PROCESS

A major attribute inherent to the retentate chromatography feature of SELDI is the flexibility of experimental design allowed by the system. This point can be appreciated by an overview of SELDI chip processing (**Fig. 2**, **steps 3 to 5** are illustrated).

4.1. Selection of Chip Surface

As described above, a variety of chemical and user-designed chip surfaces can be used. The chips are sold in an 8 × 12 format. Small sample volumes (5 to 8 µL) can be applied directly onto the chip surfaces. Alternatively, the chips can be assembled within a Bioprocessor (Bio-Rad Laboratories, Hercules, CA) assembly, which includes a fitted reservoir for application of larger sample volumes (up to 400 µL).

4.2. Preequilibration Step

For a given selected chip type, a variety of binding buffers can be used. Buffer pH, ionic strength, presence of non-ionic detergents (i.e., Triton

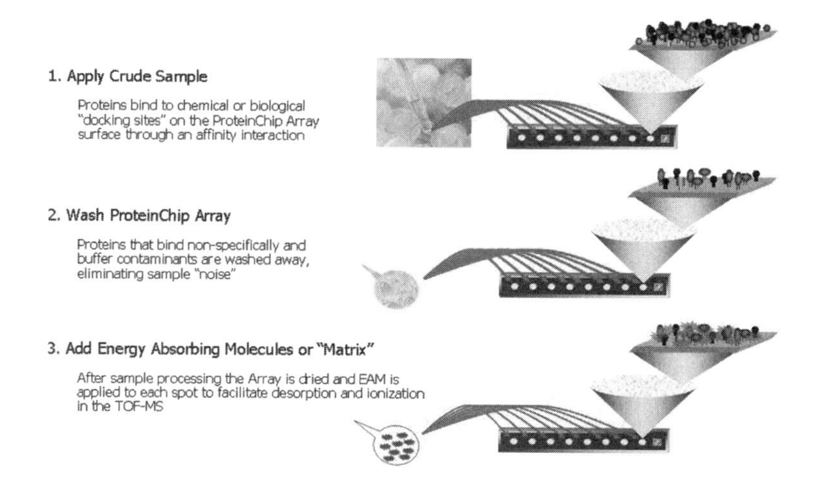

Fig. 2. SELDI chip processing. A selected ProteinChip array (**step 1**) is first preequilibrated in a suitable binding buffer (**step 2**). In **step 3**, the sample is applied, typically as a 1:10 or 1:20 dilution into binding buffer and incubated for 30 to 60 min. Based on selective affinities for the chip surface, a subset of proteins are retained and concentrated on the chip surface. In **step 4**, loosely bound proteins are removed by washing (3 × 5 min) with buffer. In **step 5**, after a brief low ionic strength wash, EAMs are added and allowed to cocrystallize with the retained proteins on the chip surface. (Steps 1 to 3 only shown in illustration.)

X-100, CHAPS) or denaturants (urea, GuHCl) can all be varied for experimental optimization. Typically, the chip arrays are preincubated in a selected binding buffer for two 5-min periods prior to addition of sample. Depending on the selected chip surface, additional preequilibration steps may be required, for example charging of the IMAC30 surface with a specific polyvalent metal ion (**Notes 2 to 4**).

4.3. Binding Step

Sample can then be diluted into the same binding buffer and incubated on the chip surface. The length of time for the binding step can be varied. Thirty- to sixty-minute binding steps are common for biomarker discovery profiling studies (**Note 5**).

4.4. Washing Step

The chip surfaces can then be washed with buffer of the same or increased stringency relative to the binding buffer. Three 5-min washes with the same binding buffer is common. However, variable wash buffers can provide valuable information regarding protein-protein interactions as well as in optimization of protein purification strategies (discussed later). Finally, a low ionic strength wash (distilled water or 5 mM sodium HEPES, pH 7) is recommended to remove high concentrations of salt ions. This will help reduce the presence of sodium adducts that may contribute to the mass heterogeneity of a given peptide/protein.

4.5. Application of Energy-Absorbing Molecules

Energy-absorbing molecules (EAMs) is a generic name for molecules that assist in desorption and ionization of the analyte. These “matrix” molecules are applied in organic solvent, solubilizing many proteins on the array surface. A standard EAMs solvent contains 50% acetonitrile, 0.5% TFA. As the EAMs solvent evaporates, proteins cocrystallize with the matrix. Commonly used EAMs include CHCA and sinipinic acid (SPA) (**Note 6**). The processed chip array is then ready to be loaded into a ProteinChip SELDI TOF-MS for spectra acquisition.

5. DATA ACQUISITION

The chip array, containing its complement of bound peptides or proteins, is transferred into the sample post of the ProteinChip SELDI-TOF Reader (**Note 7**). The sample is then equilibrated with the evacuated flight tube **(Fig. 3)**. The chip surface can then be impacted by a focused nitrogen laser (337 nm). The energy absorbed facilitates proton transfer from the

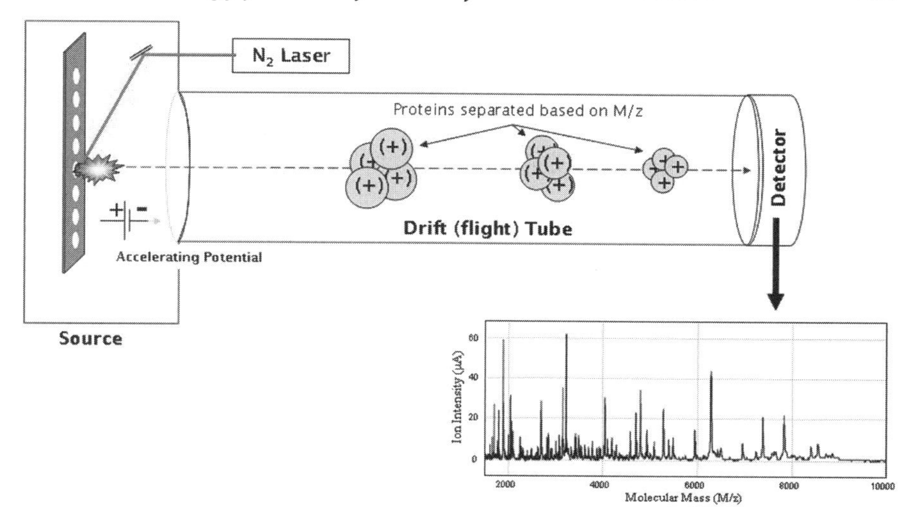

Fig. 3. SELDI-TOF schematic. The processed ProteinChip array is positioned in line with a 337-nm nitrogen laser. High-energy laser pulses induce proton transfer from EAMs, desorption of protein from the chip surface, and entry into the flight tube. The potential difference across the evacuated flight tube promotes the m/z dependent drift of H^+ ions to the detector. Impact of ions is converted to ion current and displayed chromatographically as a function of the time of flight, calibrated to molecular mass.

cocrystallized matrix compound and desorption of the protein from the chip surface. The potential difference across the flight tube draws the positively charged ions to the detector. Impact of protein ions on the detector is converted into an ion current displayed in the form of a chromatographic tracing.

In designing acquisition protocols, several variables can be adjusted to achieve well-resolved peaks with optimal recovery. The laser intensities required to desorb a given protein efficiently does increase with the mass of the peptide/protein. However, measures must be taken to control for detector saturation, signal decay, and to minimize spectral noise particularly evident in the low mass (<5 kDa) region.

5.1. Laser Intensity

The SELDI-MS has variable laser settings that must be optimized for a given target mass region. The ProteinChip SELDI System has the advanced feature that its laser settings are defined in nanojoules (nJ), such that doubling the intensity actually doubles the energy imparted on the chip target. Laser intensities from 500 to 15,000 nJ can be used.

5.2. Mass Range

The ProteinChip SELDI-MS has detection capabilities extending above 300 kDa. However, spectra with such a wide mass range will have an increasingly large accompanying data file. The mass range should fit the experimental purpose (i.e., peptide mapping can be done from 0 to 5 kDa).

5.3. Focus Mass

Peak resolution within a given mass range can be optimized by the "Focus Mass" setting. This can be especially critical in the case of peptide digest spectra.

5.4. Mass Attenuation

To guard against detector saturation due to desorbed matrix, the ProteinChip SELDI System has an adjustable mass attenuation feature that keeps the detector in the "off" position until a proscribed time (set by the corresponding mass value) has elapsed (**Note 8**). This feature is used to diminish the corresponding noise as a consequence of high laser intensity settings.

5.5. Spot Partition

The method of EAMs application allows uniform dispersion of protein/matrix cocrystals. Laser pulses can sample multiple spot positions without the need for the operator to find the best signal position or "hot spot." The SELDI System raster laser design has 215 "pixels" to which the laser can be focused and allows for complete spot coverage (**Note 9**). The uniform sample distribution also allows multiple protocols to be run on the same spot. For example, both low laser and high laser protocols can be run sequentially, each directed at different spot partitions.

5.6. Pulse Number

The laser repeatedly pulses at an individual spot position for a proscribed number of times before moving to the next position. Typically, 10 to 15 such pulses can allow for optimal peak detection before significant signal decay might ensue.

By varying the factors listed above, optimized protocols can be designed to then profile large sample numbers. Typically, reference or pooled samples are used in the protocol design step. Use of automated, optimized protocols is the essence of the high-throughput capability of the SELDI process. The ProteinChip SELDI Data Manager software can facilitate the collection of hundreds of spectra without further operator input.

6. DATA ANALYSIS

The raw data obtained in collected spectra is a data average of the multiple pulse transients specified in the acquisition protocol. The data includes time-of-flight increment plotted against ion current, displayed as intensity. Each individual spectrum can include 10^5 to 10^6 data points. Fung et al. ***(27)*** provide a detailed review of the biostatistical strategies that can be applied to groups of spectra in search of unique protein biomarkers. A few key points will be emphasized here. When considering a set of SELDI spectra to be screened for potential biomarkers, they all must contain the following characteristics:

1. Sample type (e.g., plasma), including identical steps for sample processing and preparation, including fractionation. Clarke et al. ***(9)*** discuss avoiding preanalytical biases.
2. Chip selection.
3. Matrix selection.
4. Instrument settings.

Spectra, so collected, then must be processed in an identical fashion, including baseline fit and subtraction, noise adjustment, and mass calibration (**Note 10**). Each spectrum is then normalized relative to the average total ion current across the group ***(28)***. The normalization step provides a correction for possible pipetting errors during processing of the chip arrays and also allows for identification of spectral outliers that may need to be reprofiled or eliminated from the data set.

The Bio-Rad ProteinChip Data Manager software facilitates the discovery of distinct protein biomarkers that may be differentially expressed in a subgroup of spectra. Multiple subgroups can be compared simultaneously. Spectra are subjected to Expression Difference Mapping (EDM) analysis, which creates peak clusters within a proscribed mass window. A minimum signal/noise (S/N) threshold must be met within a proscribed percentage of all spectra to assign a peak cluster at a particular m/z value. Spectra are rescreened using a second, less-stringent S/N threshold. Finally, peak intensity values are estimated using the average m/z value of the existing members of the cluster.

ProteinChip Data Manager then uses the Mann-Whitney test (nonparametric equivalent of Student's *t*-test) to calculate *p* values revealing the comparative levels of each peak across two defined groups (i.e., control vs. diseased). This univariate analysis is an initial step toward discovery of protein biomarkers that potentially could differentiate normal versus abnormal phenotypes. ProteinChip Data Manager also provides box-and-whisker plot and scatterplot **(Fig. 4)** visualization of peak intensity

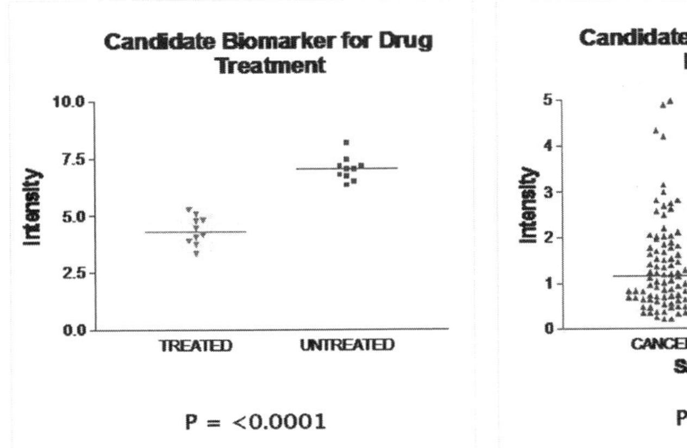

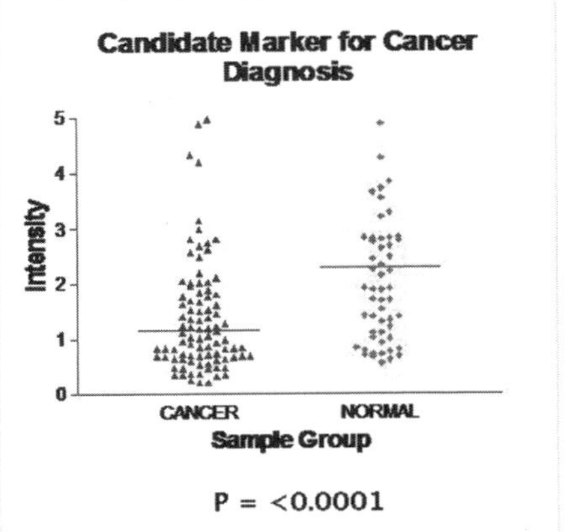

Fig. 4. Representative scatterplots using ProteinChip Data Manager. Multiple spectra obtained from defined subgroups are subjected to Expression Difference Mapping to identify peak clusters. Mann-Whitney analysis provides *p*-value calculations on each peak cluster, which may highlight biomarkers differentially expressed between two subgroups.

distributions. Such scatterplots reveal group mean differences as well as the wide distribution often characteristic of clinical samples. Even though large sample numbers may yield significant *p* values, the ability for single markers to discriminate the two groups may still be poor.

The wide variation in clinical populations points to the need for multivariate analysis of spectra groups. ProteinChip Data Manager also contains two multivariate unsupervised learning analysis methods. Principle component analysis and hierarchical clustering assess the distribution of the spectral data. These tools allow rapid visualization of spectral outliers, can indicate the presence of subgroups within the data set, and can point to key peak clusters that contribute to the variance across the total sample group.

Bio-Rad Pattern Analysis Software is a supervised learning technique, based on a classification and regression tree method *(29)*. Peak clusters are assessed for the ability to distinguish defined subgroups. A decision tree is constructed using the most discriminating peak clusters as primary splitting factors. Assignment of unknown samples (spectra) to a specific group is based on comparison with multiple threshold peak intensity values. In the development of potential diagnostic proteomic assays, use of multiple biomarkers has resulted in increased sensitivity and specificity values *(6)*.

The high-throughput capacity of SELDI data acquisition allows a large number of independent samples to be used in the construction and cross-validation of classification trees using Pattern Analysis Software. By testing a robust data set in the biomarker discovery phase, whether using univariate or multivariate analysis programs, markers with a higher degree of validation can then be the focus of identification and assay design.

7. SELDI-ASSISTED PURIFICATION

The selective binding chemistries present on the ProteinChip array surfaces are identical to chemistries present on solid-phase bead (cellulose, dextran, zirconia) supports. By varying the binding and wash conditions on a specific chip array, subsequent mass spectra can give a rapid assay for the retention of a specific biomarker. In addition, the differential binding and elution of neighboring contaminant protein peaks can be monitored. Indeed, monitoring of a specific marker to several chemical surfaces can be done in parallel. This allows for the rapid design of potential purification schemes ***(30–32)***. The basic strategy for enrichment of a protein biomarker based on SELDI-MS data will implement the following features:

1. The particular chip array to which the biomarker of interest does bind will be further characterized.
2. Alternate binding conditions (i.e., changes in pH) should be tested to optimize biomarker retention on the chip surface.
3. Various wash procedures are then screened to establish conditions for elution of the biomarker from the chip array.
4. The conditions established for selective binding and elution are then implemented at a larger scale on a compatible solid phase. For example, Bio-Rad Q HyperD zirconia resins are compatible with the ProteinChip Q10 surface chemistry. Fractions collected from the solid phase resin can be monitored using the nonselective ProteinChip NP20 array.
5. When the best enrichment step is established, a control sample, in which the same biomarker is known to be deficient (as in the discovery phase of the study), should be processed in parallel.
6. A second enrichment step can then be carried out using an alternate ProteinChip array.

Ultimately, the specific biomarker needs to be identified. Often, the selective enrichment steps are adequate to allow subsequent SDS-PAGE isolation of the proteins.

8. PEPTIDE MAPPING PROTOCOLS

After SDS-PAGE, standard protocols for Coomassie Blue staining can be used. Subsequent protease digestion of selected bands can be achieved using the following approaches: in-gel digestion; passive elution of protein from the gel followed by protease digestion in solution; passive elution of protein followed by protease digestion on-chip.

8.1. In-Gel Digestion

8.1a. To carry out in-gel protease digestion, SDS must first be removed from the gel. The specific Coomassie-stained gel band (destained in solution containing 50% methanol, 10% acetic acid) should be excised, cut into 1-mm sections, and transferred to an 0.5-mL microcentrifuge tube. The corresponding molecular weight (MW) region of a control lane (containing the processed "deficient" sample) should also be excised. Alternatively, a protein-free sample lane can be processed in parallel.

8.1b. Add 0.4 mL 50% methanol, 10% acetic acid and incubate with agitation for 1 h. Replace the solution with fresh 50% methanol, 10% acetic acid; incubate with agitation for an additional 30 min. Remove solution.

8.1c. Incubate in 0.4 mL 50% acetonitrile, 0.1 M ammonium bicarbonate, with agitation for 1 h. Remove solution.

8.1d. Add 50 μL 100% acetonitrile. Incubate with agitation for 15 min. Remove solution.

8.1e. Dry the gel pieces in a SpeedVac for 15 min or until completely dry. Prepare trypsin solution. Sequencing grade trypsin is dissolved in 10 mM HCl (0.2 μg/μL) = 10X stock stored at –20°C or –80°C. A 1X trypsin solution is made by diluting with 9 volumes of 25 mM ammonium bicarbonate, pH 8.0.

8.1f. Add 10 μL 1X trypsin solution to the dried gel piece(s) in an 0.5-mL microcentrifuge tube. Incubate at room temperature for 15 min to rehydrate the gel. Process the "no protein" control tube in the same manner.

8.1g. If necessary, add up to 40 μL 25 mM ammonium bicarbonate, pH 8.0, to cover the gel. Only add as much as is necessary to cover the pieces so that the resultant digest solution is as concentrated as possible. Seal the tube tightly and incubate at 37 °C for up to 16 h (**Note 11**).

8.2. Passive Elution and Protease Digestion in Solution

8.2a. After destaining, transfer the gel piece to a fresh microcentrifuge tube and crush the gel piece at the bottom of the tube with a dissection needle.

8.2b. Passive elution of protein from polyacrylamide gels can be accomplished using either of two extraction solutions (**Note 12**). Add 40 μL elution solution for an entire protein band or 20 μL for half of a band. The solution should completely cover the gel pieces.

8.2c. Sonicate for 30 min at room temperature. Vortex for 4 h at room temperature. The samples can be stored overnight at 4°C.

8.2d. Collect the eluant in a fresh microcentrifuge tube; rinse the gel pieces with an equal volume of fresh elution solution. Combine the solutions.

8.2e. Evaporate the eluant in a SpeedVac. Redissolve the eluted protein in 5 to 10 μL water (for low-molecular-weight proteins) or 0.5% *n*-octyl glucopyranoside in water (for high-molecular-weight proteins). Vortex for 15 min at room temperature.

8.2f. The samples are ready for analysis. If you are not able to recover the protein after the SpeedVac drying, remove the liquid, wash the vial with a small volume of 20% acetonitrile, and examine. If a specific ProteinChip array was used to detect the protein of interest, the same type of array should be used to analyze the retrieved protein. Binding and washing conditions should be similar to those used originally. Use 1 or 2 μL of recovered protein solution to confirm that the protein of interest was eluted from the gel.

8.2g. The eluted protein can then be incubated with protease (i.e., add an equal volume of 50 mM ammonium bicarbonate, containing 2X trypsin) at 37°C.

8.2h. The proteolytic digests resulting from these two protocols can then be profiled on several ProteinChip arrays. For NP20: Apply 1 to 2 μL of the digested sample. Let dry. Apply 2 × 0.5 μL aliquots CHCA (20% saturation) allowing 5 min for matrix to dry between each application. For H4 or H50: Apply 1 to 2 μL sample as above. The hydrophobic chains allow for a brief wash step with 2 μL water to reduce salt contaminants. Apply 2 × 0.5 μL (20%) CHCA. For SEND-ID: Dilute the digested sample with an equal volume 50% acetonitrile, 0.2% TFA. Apply 1 to 2 μL onto the SEND array and let dry. As CHCA is already incorporated onto the surface, the sample is ready to be inserted into the ProteinChip reader.

8.3. Protease Digestion On-Chip

8.3a. Protein that has been passively eluted from a gel band or purified by alternate methods can be digested directly on the chip surface. From the initial profiling study, a specific ProteinChip array as well as the best binding-washing conditions can be adopted.

8.3b. Add the eluted protein to the array surface. Rinse with water for 1 min. Repeat two times for a total of three washes.

8.3c. Prepare a spot with elution solution alone and no protein as a control to identify background signal from protease autolysis. Dry the spots for 5 min at 60°C (heat denaturation of protein before digestion).

8.3d. Add 4 to 4.5 μL of buffer for digestion (i.e., 25 mM ammonium bicarbonate). Add 0.5 to 1 μL of protease (1 to 10 ng/μL). (**Note 13**).

8.3e. Incubate for 2 h at optimal temperature. Dry the spot for 2 min at 60°C.

8.3f. Add 1 μL of saturated CHCA solution in 50% acetonitrile/0.05% TFA. Allow the array to dry.

8.3g. Read the array using the ProteinChip Reader (**Note 14**).

8.3h. The unique proteolytic fragment m/z values are can then searched (using Mascot or ProFound) against protein databases (SwissProt, NCBI) for the identification of the protein biomarker ***(33,34)***.

9. DIRECT SEQUENCING APPLICATIONS

Sequence analysis of peptides bound to Bio-Rad ProteinChip arrays can be obtained using a ProteinChip Tandem MS Interface ***(35,36)*** as a front-end SELDI ion source for the Applied Biosystems/MDS Sciex QStar Hybrid LC/MS/MS System (Model QStar XL). The ProteinChip Interface uses a 337-nm nitrogen laser with a lensed fiber optic, delivering 150 mJ of energy per pulse at 30 pulses/s. Peptides up to 5 kDa can be selected in Q1 and transmitted to Q2 for collisional induced dissociation, using an applied collisional energy of 50 eV/kDa. Sample spot scanning is controlled via ProteinChip Interface Control Software (Bio-Rad Laboratories), and data analysis and acquisition are carried out using the QStar System's Analyst software (Applied Biosystems, Inc., Foster City, CA). This system is capable of high attomole–low femtomole MS and tandem mass spectrometry (MS/MS) sensitivities. The resulting sequence obtained from spectral data can be submitted to the database mining tools Mascot or ProFound for identification.

10. PROTEIN-PROTEIN INTERACTION ASSAYS

As mentioned in the discussion of the ProteinChip RS100 and PS20 arrays, user-defined ligands can be coupled to the chip creating an affinity surface for specific proteins or protein complexes (**Fig. 5**). This strategy can be used via antibody capture to probe for specific antigens *(37–39)*. Similarly, affinity surfaces have been used for the detection of autoantibodies *(40,41)*, DNA transcription factors *(42,43)*, and a variety of other protein-ligand interactions *(44–46)*. Several key points must be considered in designing optimal affinity chip surfaces and assays:

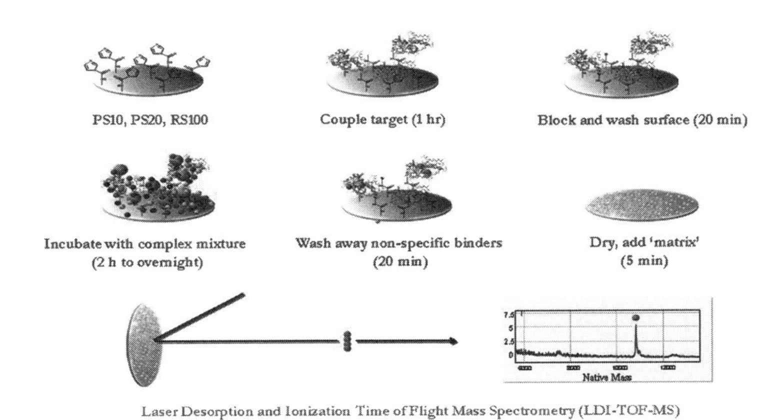

Fig. 5. Scheme for use of preactivated ProteinChip arrays for protein-protein interaction. User-selected proteins (i.e., antibodies or ligands) can be covalently coupled to the ProteinChip PS20 or RS100 surfaces. A blocking step is then necessary to occupy residual reactive sites. Subsequent addition of sample, wash steps, and EAMs addition is similar to steps outlined in **Fig. 2**.

1. Retention of binding capacity for the "bait" molecule:

   A. When coupling antibodies to a PS20 or RS100 chip surface, it is useful to test several pH conditions. Although coupling with lysine amino groups can occur at basic pH (0.1 M sodium bicarbonate, pH 9), preferential involvement of the alpha-amino group can be favored in lower pH solutions. Coupling in PBS, pH 7.0, or 50 mM sodium acetate, pH 5.5, should be tested to assess the most productive binding of antigen.
   B. Avoid stearic hindrance of ligand adjacent to the chip surface (**Notes 15, 16**).

2. Blocking of unreacted epoxy or carbonyldiimidazole groups can also be achieved using nonspecific (BSA) protein solutions (0.1 mg/mL in Tris buffer).
3. Binding conditions (time and buffer components) must be optimized.
4. Nonspecific binding may be reduced by inclusion of high salt and 1 M urea washes.
5. Relevant controls (i.e., nonspecific or preimmune antibodies or scrambled DNA sequences) should be incorporated to help define the specifically bound m/z factors.

11. NOTES

1. As the SELDI MS system involves positive ion flight, it is recommended that SDS be removed from cell lysis buffers. Non-ionic detergents (i.e. Triton X-100, CHAPS) can be substituted.
2. IMAC chips are charged with selected metals (e.g., 50 μL 0.1 M copper sulfate) for 5 min. When using copper metal, an additional neutralization wash with 0.1 M sodium acetate, pH 4, is included prior to preincubation with binding buffer.
3. The hydrophobic carbon chains on the H4 and H50 chips are activated by incubation with 50% acetonitrile (or methanol) for 5 min prior to preincubation with binding buffer.
4. When using the preactivated RS100 and PS20 arrays to which specific "bait" molecules have been covalently bound, the coupling step is followed by a wash step to remove unbound ligand and then incubation in amine buffer (50 mM Tris-HCl, pH 8, or 0.5 M ethanolamine). This serves to block unreacted sites on the chip surface. Finally, the chips can be preequilibrated in the selected binding buffer.
5. Appropriate binding times will be expected to vary when using ligand-coupled RS100 or PS20 arrays. For example, binding of specific antigens by antibody-coupled arrays may require overnight incubation at 4°C.
6. CHCA is useful in peptide profiling as matrix ions are generally <600 Da. However, production of multiply protonated protein ions can be seen. The matrix ions for SPA can be >1000 Da, with a lesser tendency to form multiply protonated ions. SPA is more commonly used over a wide mass range (2 to 200 kDa) biomarker discovery study.

7. Bio-Rad Laboratories markets the PCS4000 edition ProteinChip Reader. An earlier version, PBSIIc, accounts for a larger number of SELDI instruments in the field and is also supported by Bio-Rad.
8. The PBSIIc instrument uses laterally positioned (–) charged deflector plates that can be activated, transiently, suppressing ion entry into the drift tube. This strategy diminishes the impact of matrix ions onto the detector, also limiting detector saturation.
9. The PBSIIc laser makes a focused lateral impact on the spot. The lateral spot positions are on a 0 to 100 scale. Positions between 15 and 85 are commonly used for quantitative analysis.
10. Spectra can be externally calibrated by application of known m/z peptide or protein standards on an NP20 chip. Standards should be chosen to span the mass range of interest. The standard spectrum should be collected using matrix and instrument settings matching the corresponding sample set. A second polynomial equation relating TOF to the standard m/z values can then be saved and applied to the spectral group being analyzed.
11. Shorter incubation times may be adopted. Remove 1-μL aliquots for analysis throughout the digestion period to determine the optimum/minimum digestion time.
12. FAPH solution (50% formic acid, 25% acetonitrile, 15% isopropanol, 10% water, v/v/v/v) is recommended for proteins greater than 10 kDa. ACN/TFA solution (70% acetonitrile, 30% 0.2% trifluoroacetic acid, v/v) is effective for proteins less than 10 kDa.
13. Protein to protease molar ratio should be at least 50:1. Significant excess of protease over the protein will produce ion suppression and nonspecific cleavage.
14. Accurate m/z estimates can be achieved using peptide standards, most effectively when mixed with the sample for internal calibration.
15. Binding of biotin-DNA molecules to a streptavidin-coated chip surface can benefit from the use of biotin adducts possessing long-chain linkers and the use of concatamers of the specific DNA sequences.
16. Use of the ProteinChip PG20 array, containing bound protein G, can provide the correct orientation for adsorbed immunoglobulins used to screen for specific antigens.

REFERENCES

1. Zerhouni E. The NIH Roadmap. Science 2003;302:63–65.
2. Hutchens TW, Yip TT. New desorption strategies for the mass spectrometric analysis of macromolecules. Rapid Commun. Mass Spectrom 1993;7:576–580.
3. Yip TT, Lomas L. SELDI ProteinChip Array in oncoproteomic research. Technol Cancer Res Treatment 2002;1:273–280.
4. Wadsworth JT, Somers KD, Cazares LH, Malik G, Adam BL, Stack BC, et al. Serum protein profiles to identify head and neck cancers. Clin Cancer Res 2004;10:1625–1632.

5. Wilson LL, Tran L, Morton DL, Hoon DS. Detection of differentially expressed proteins in early-stage melanoma patients using SELDI-TOF mass spectrometry. Ann N Y Acad Sci 2004;1022:317–322.
6. Zhang Z, Bast RC Jr, Yu Y, Li J, Sokoll LJ, Rai AJ, et al. Three biomarkers identified from serum proteomic analysis for the detection of early stage ovarian cancer. Cancer Res 2004;64:5882–5890.
7. Semmes OJ, Feng Z, Adam BL, Banez LL, Bigbee WL, Campos D, et al. Evaluation of serum protein profiling by surface-enhanced laser desorption/ionization time-of-flight mass spectrometry for the detection of prostate caner: I. Assessment of platform reproducibility. Clin Chem 2005;51:102–112.
8. Fung ET, Yip TT, Lomas L, Wang Z, Yip C, Meng XY, et al. Classification of cancer types by measuring variants of host response proteins using SELDI serum assays. Int J Cancer 2005;115:783–789.
9. Clarke CH, Buckley JA, Fung E. SELDI-TOF-MS proteomics of breast cancer. Clin Chem Lab Med 2005;43:1314–1320.
10. Li J, Orland R, White CN, Rosenzweig J, Zhao J, Seregni E, et al. Independent validation of candidate breast cancer serum biomarkers identified by mass spectrometry. Clin Chem 2005;51:2229–2235.
11. Beher D, Wrigley J, Owens AP, Shearman MS. Generation of C-terminally truncated amyloid-beta peptides is dependent on gamma-secretase activity. J Neurochem 2002;82:563–575.
12. Carrette O, Demalte I, Scherl A, Yalkinoglu O, Corthals G, Burkhard P, et al. A panel of cerebrospinal fluid potential biomarkers for the diagnosis of Alzheimer's disease. Proteomics 2003;3:1486–1494.
13. Furuta M, Shiraishi T, Okamoto H, Mineta T, Tabuchi K, Shiwa M. Identification of pleiotrophin in conditioned medium secreted from neural stem cells. Dev Brain Res 2004;152:189–197.
14. Chu R, Zhang W, Lim H, Yeldandi AV, Herring C, Brumfield L, et al. Profiling of acyl-CoA oxidase-deficient and peroxisome proliferator Wy14, 643-treated mouse liver. Gene Expression 2002;10:165–177.
15. Dare TO, Davies HA, Turton JA, Lomas L, Williams TC, York MJ. SELDI detection and identification of urinary parvalbumin-alpha. Electrophoresis 2002;23:3241–3251.
16. Yuan M, Carmichael WW. Detection and analysis of the cyanobacterial peptide hepatotoxins microcystin and nodularin. Toxicon 2004;44:561–570.
17. Vermeulen R, Lan Q, Zhang L, Gunn L, McCarthy D, Woodbury RL, et al. Decreased levels of CXC-chemokines in serum of benzene-exposed workers identified by array-based proteomics. Proc Natl Acad Sci USA 2005;102: 17041–17046.
18. Papadopoulos MC, Abel PM, Agranoff D, Stich A, Tarelli E, Bell BA, et al. A novel and accurate diagnostic test for human African trypanosomiasis. Lancet 2004;363:1358–1363.
19. Gravett MG, Novy MJ, Rosenfeld RG, Reddy AP, Jacob T, Turner , et al. Diagnosis of intra-amniotic infection by proteomic profiling and identification of novel biomarkers. JAMA 2004;292:462–469.

20. Kang X, Xu Y, Wu X, Liang Y, Wang C, Guo J, et al. Proteomic fingerprints for potential application to early diagnosis of severe acute respiratory syndrome. Clin Chem 2005;51:56–64.
21. Buhimschi IA, Christner R, Buhimschi CS. Proteomic biomarker analysis of amniotic fluid for identification of intra-amniotic inflammation. Br J Obstet Gynaecol 2005;112:173–181.
22. Schaub S, Wilkins J, Weiler T. Urine protein profiling with surface-enhanced laser-desorption/ionization time-of-flight mass spectrometry. Kidney Int 2004;65:323–332.
23. Nguyen M, Ross G, Dent C, Devarajan P. Early prediction of acute renal injury using urinary proteomics. Am J Nephrol 2005;25:318–326.
24. Koopman J, Zhang Z, White N, Rosenzweig J, Fedarko N, Jagannath S, et al. Serum diagnosis of pancreatic adenocarcinoma using surface-enhanced laser desorption and ionization mass spectrometry. Clin Cancer Res 2004;10: 860–868.
25. Thulasiraman V, Lin S, Gheorghiu L, Lathrop J, Lomas L, Hammond D, Boschetti E. Reduction of the concentration difference of proteins in biological liquids using a library of combinatorial ligands. Electrophoresis 2005;26: 3561–3571.
26. Castagna A, Cecconi D, Sennels L, Rappsilber J, Guerrier L, Fortis F, et al. Exploring the hidden human urinary proteome via ligand library beads. J Proteome Res 2006;4:1917–1930.
27. Fung ET, Weinberger SR, Gavin E, Zhang F. Bioinformatics approaches in clinical proteomics. Expert Rev Proteomics 2005;2:847–862.
28. Fung E, Enderwick C. ProteinChip clinical proteomics: computational challenges and solutions. Computat Proteomics 2002;32:S34–S41.
29. Breiman L, Friedman J, Olshen R, Stone C. Classification and Regression Trees. Pacific Grove, CA: Wadsworth, 1984.
30. Weinberger SR, Boschetti E, Santambien P, Brenac V. SELDI—a new method for rapid development of process chromatography conditions. J Chromatogr B 2002;782:307–316.
31. Shiloach J, Santambien P, Trinh L, Schapman A, Boschetti E. Endostatin capture from Pichia pastoris culture in a fluidized bed From on-chip process optimization to application. J Chromatogr B 2003;790:327–336.
32. Prahalad AK, Hickey RJ, Huang J, Hoelz J, Dobrolecki L, Murthy S, et al. Serum proteome profiles identifies parathyroid hormone physiologic response. Proteomics 2006;6:3482–3493.
33. Thulasiraman V, McCutchen-Maloney SL, Motin VL, Garcia E. Detection and identification of virulence factors in Yersinia pestis using SELDI ProteinChip System. BioTechniques 2001;30:428–432.
34. Wang S, Diamond DL, Hass GM, Sokoloff R, Vessella RL. Identification of prostate-specific membrane antigen (PMSA) as the target of monoclonal antibody 107–1A4. Int J Cancer 2001;92:871–876.

35. Merchant M, Weinberger SR. Recent advancements in surface-enhanced laser desorption-ionization-time of flight-mass spectrometry. Electrophoresis 2000;21:1164–1177.
36. Reid G, Gan BS, She YM, Ens W, Weinberger S, Howard JC. Rapid identification of probiotic lactobacillus biosurfactant proteins. Appl Environ Microbiol 2002;68:977–980.
37. Davies H, Lomas L, Austen B. Profiling of amyloid beta peptide variants using SELDI ProteinChip Arrays. BioTechniques 1999;27:1258–1262.
38. Landuyt B, Jansen J, Wildiers H, Goethals L, De Boeck G, Highley M, et al. Immuno affinity purification combined with mass spectrometry detection for the monitoring of VEGF isoforms in patient tumor. J Separation Sci 2003;26: 619–623.
39. Wang Z, Yip C, Ying Y, Wang J, Meng XY, Lomas L, et al. Mass spectrometric analysis of protein markers for ovarian cancer. Clin Chem 2004;50:1939–1942.
40. Winer S, Tsui H, Lau A, Song A, Li X, Cheung RK, et al. Autoimmune islet destruction in spontaneous type 1 diabetes is not beta-cell exclusive. Nat Med 2003;9:198–205.
41. Ueki A, Isozaki Y, Tomokuni A, Hatayama T, Ueki H, Kusaka M, et al. Intramolecular epitope spreading among anti-caspase-8 autoantibodies. Clini Exp Immunol 2002;129:556–561.
42. Forde CE, McCutchen-Maloney SL. Characterization of transcription factors by mass spectrometry and the role of SELDI-MS. Mass Spectrom Rev 2002;21:419–439.
43. Bane TK, LeBlanc JF, Lee TD, Riggs AD. DNA affinity capture and protein profiling by SELDI-TOF mass spectrometry—effect of DNA methylation. Nucleic Acids Res 2002;30:e69.
44. Zhu Y, Valdes R, Simmons CQ, Linder MW, Pugia MJ, Jortani SA. Analysis of ligand binding by bioaffinity mass spectrometry. Clin Chim Acta 2006;371: 71–78.
45. Hinshelwood J, Spencer DIR, Edwards YJK, Perkins SJ. Identification of the C3b binding site in a recombinant. J Mol Biol 1999;294:587–599.
46. Howard JC, Heinemann C, Thatcher BJ, Martin B, Gan BS, Reid G. Identification of collagen-binding proteins in Lactobacillus spp. with SELDI time of flight Protein Chip technology. Appl Environ Microbiol 2000;66:4396–4400.

13

Sandwich ELISA Microarrays: *Generating Reliable and Reproducible Assays for High-Throughput Screens*

Rachel M. Gonzalez, Susan M. Varnum, and Richard C. Zangar

Summary

The sandwich ELISA microarray is a powerful screening tool in biomarker discovery and validation because of its ability to simultaneously probe for multiple proteins in a miniaturized assay. The technical challenges of generating and processing the arrays are numerous. However, careful attention to possible pitfalls in the development of one's antibody microarray assay can overcome these challenges. In this chapter, we describe in detail the steps that are involved in generating a reliable and reproducible sandwich ELISA microarray assay.

Key Words: antibody microarrays; biomarker discovery; high-throughput protein assays; proteomics; sandwich ELISA

1. INTRODUCTION

The availability of complete genome sequences and comprehensive DNA microarray chips has amplified the quantity and complexity of data available to model and predict disease states *(1–3)*. There has also been a major shift in biomarker research to use proteomic technologies that can screen for proteins that predict the presence, subtype, or therapeutic response of various diseases *(4–7)*. Antibody-based microarrays have the potential for creating snapshots of disease states at a particular point in time and can be used to validate disease biomarkers identified by discovery technologies such as proteomics or DNA microarrays *(8–11)*.

Antibody microarrays hold great promise for the quantitative high-throughput screening of multiple proteins in parallel ***(6,8,12–14)***. One significant advantage of multiplexing antibodies in a microarray format is the

From: *Methods in Pharmacology and Toxicology: Biomarker Methods in Drug Discovery and Development*
Edited by: F. Wang © Humana Press, Totowa, NJ

ability to screen biological samples after dilution in volumes as small as 15 µL ***(10,15–19)***. In contrast with gene arrays, where synthesizing the components can be done relatively inexpensively, reagents (antibodies and recombinant proteins) for antibody microarrays must be individually prepared and evaluated ***(20–22)***. Therefore, many research groups are developing new strategies to create affinity reagents in high-throughput, less-costly systems. However, success in this area has been slow, and diverse issues such as licensing restrictions and standardized quality control have encumbered reagent development.

Antibody arrays are based on immobilizing antibodies to a planar surface. The immobilized antibody is called the *capture antibody* because it binds and concentrates the targeted analyte to the planar surface. The planar surface can be a filter, plate, or a glass slide that is chemically treated or coated with an organic film to enhance protein binding. Examples of organic films that achieve good passive adsorption are polyacrylamide hydrogels ***(23)***, nitrocellulose ***(24)***, and poly-L-lysine ***(25)***. Stronger attachment can be achieved by creating a reactive surface, which covalently cross-links the antibodies to the slide ***(26–29)***. Chemicals such as aldehyde silane, epoxy silane, activated aminosilane, and hydrogel aldehyde are commonly used examples of reactive surfaces that bind amino groups on the antibody.

There are two main types of antibody microarray. The first is the *single antibody* or *two-color* microarray where potentially thousands of antibodies can be arrayed on a single chip. Two protein samples are labeled with different fluorescent dyes prior to combining and incubating on a chip ***(12, 17,29,30)***. Although this approach offers great potential for high numbers of assays per chip, success with this assay system has been mixed due to inherit problems in nonspecific protein-antibody interactions ***(17,30)***. The second type of antibody array, which is described in this chapter, is the sandwich enzyme-linked immunosorbent assay (ELISA) antibody microarray ***(11,18, 31–33)***. The sandwich microarray requires immobilized capture antibodies, protein samples, and labeled detection antibodies done in an assay similar to the standard 96-well microplate ELISA. The important steps in sandwich microarrays are shown in **Fig. 1**. The primary advantages of the sandwich ELISA microarrays is the increased specificity and sensitivity that results from the use of two different antibodies for each analyte. Additionally, these assays use purified proteins to generate standard curves and accurately calculate levels of free analyte. The major disadvantage to the sandwich ELISA is that every assay needs a purified antigen and two antibodies that bind to different sites on the antigen. Generating two antibodies and the purified antigen is challenging if the goal is to validate a large number of novel candidate biomarkers.

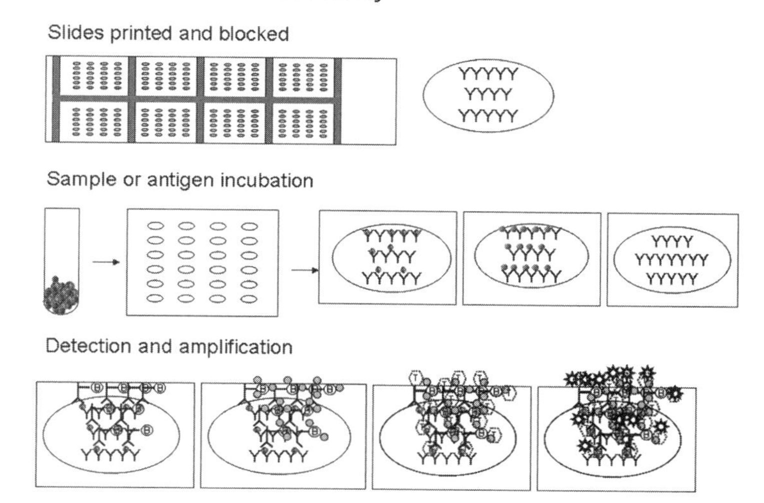

Fig. 1. The sandwich ELISA microarray assay allows the high-throughput screening of multiple analytes at various concentrations in a sample. This process can be divided into three major steps: the creation of the chip **(top panel)**, incubation of the sample **(middle panel)**, amplification of detection signal **(bottom panel)**. Key: single assay, oval spot; capture antibody, Y; detection antibody, Y-B; balls, analyte; streptavidin HRP, solid spots; biotins added by TSA amplification, T in hexagon; signal (fluorescence tags), 8-pointed stars.

2. MATERIALS

2.1. Instruments

1. Contact printer MicroGrid II printer (Apogent Discoveries; Hudson, NY)
2. Noncontact printer NanoPlotter NP 2.0 (GeSiM, Großerkmannsdorf, Germany)
3. IEC Centra MP4R Plate centrifuge (ThermoFisher Scientific, Waltham, MA)
4. MicroMax RF/refrigerated micro centrifuge (ThermoFisher Scientific, Waltham, MA)
5. Orbital Shaker (Belly Dancer, Stovall Life Science; Greensboro, NC)
6. ScanArray ExpressHT (PerkinElmer, Boston, MA)
7. Vacuum Sealer (Vac1075; www.foodsaver.com)

2.2. General Supplies

1. Aminosilane coated glass slides (Erie Scientific, Portsmouth, NH)
2. Antibodies (multiple sources)
3. Antigens (multiple sources)

4. 1% Casein (block solution) in phosphate-buffered saline (PBS; BioRad, Hercules, CA)
5. Coplin Jar (VWR Scientific, Brisbane, CA)
6. Deionized water (18 MΩ)
7. Desiccant packages (Humidity Sponges; VWR)
8. Forceps
9. 100% HPLC-grade Methanol (Fisher, Pittsburgh, PA)
10. Micro Cleaning Solution (ArrayIt; TeleChem International, Inc., Sunnyvale, CA)
11. PAP Pen (Beckman Coulter, Fullerton, CA)
12. Phosphate buffered saline (PBS), pH 7 to 7.4 (NERL Diagnostics, East Providence, RI)
13. Phosphate buffered saline containing 0.05% Tween 20 (PBS-T), pH 7 to 7.4
14. Potassium hydroxide (KOH) (Sigma Chemicals, St Louis, MO)
15. Reagent reservoir (Costar, Fisher, cat. no. 13-681-101)
16. Standard 384-well microplate
17. Tissue Tek H4465 24-slide holder (VWR cat. no. 25608-868)
18. Tissue Tek H4455 Staining dish (VWR cat no. 25608-906)
19. Tween 20 (Sigma Chemicals)
20. Storage containers with sealable lids
21. Tyramide amplification system (TSA Biotin System, PerkinElmer)

3. GENERATING THE CHIP METHODOLOGY

The first step in the ELISA microarray process is generating the chip. Every facet of chip construction from the printer to the targets, slide type, and block step influences the fidelity of the chip. This section will emphasize strategies we employ to improve quality and reproducibility of chips.

3.1. *Printer and Setup*

Microarray printers can be divided into contact and noncontact printers. Contact printers commonly use stainless steel pins to pick up sample through capillary action and then print by quickly touching the pin tip to the slide. Noncontact printers commonly use piezoelectric micropipette tips to dispense uniform droplets on the slide. The main advantage of the noncontact printer for protein microarrays is improved spot uniformity.

Our lab uses a contact MicroGrid II printer (Apogent Discoveries, Hudson, NY) and a noncontact Nano-Plotter NP2 printer (GeSiM, Quantum Analytics, Foster City, CA). For both types of instruments, it is important to keep the pins on the printer very clean and to remove particulates from all buffers and water involved in the printing process with an 0.2-μm filter. To prevent evaporative concentration of the capture antibody solutions during printing, the printer should have a humidity control device and a plate chiller. Typically, the humidity is set at 50% to 70% and the plate chiller set between 10°C to 12°C for both instruments. Precise settings need to be corrected for the dew point. Ideally, the microplate cooler should be set at ~1°C to 3°C above the dew point, which is defined by the air temperature and humidity inside the instrument.

When programming the print parameters, consideration should be made to optimize the slide layout in order to streamline the processing steps for large sample numbers. On standard 25 × 75 mm slides, printing chips in two columns, eight rows, and 9-mm spacing allows the use of an eight-channel pipettor to transfer samples or reagents from 96-well plates or reagent reservoirs. Arraying replicate spots in a row is commonly the fastest way to print. However, printing replicate spots within a chip in a distal pattern (e.g., four replicates with one in each quadrant) minimizes effects of localized background.

3.2. *Pin Cleaning*

We have found that protein buildup is a common problem and that all pins, contact or noncontact, need occasional vigorous cleaning to maintain consistent printing. We have also found that new pins will not function properly unless vigorously cleaned prior to use. Cleaning steel pins requires 50 mM KOH, Micro Cleaning Solution, distilled water, and a sonicating bath. Always handle pins with gloves. Wash steps are done in a 250-mL Pyrex beaker containing a plastic sheet with holes that suspend the pin such that the pin tips are in the wash solution. Routine cleaning of contact pins should take less than an hour, but washes for problem pins can take 3 h.

1. Add 150 mL 50 mM KOH to the 250-mL beaker and carefully place the pins so that the tips are submerged in the KOH solution. Place in sonicating bath for 10 min or 1 h. Use 10 min if the pins are printing well and 1 h if the pins are printing poorly. (*Note*: Pins should never touch the sides or bottom of the beaker when sonicating because this will damage the pin.)
2. Poor off the KOH and rinse the pins twice with distilled water. (*Note*: metal pins perform poorly if they are rinsed with de-ionized water.)
3. Add 150 mL 2% Micro Cleaning Solution and carefully replace the pins in the cleaning solution. Sonicate for ten minutes or 1 h.
4. Remove from the cleaning solution and rinse three times with distilled water.
5. Add 150 mL distilled water and carefully replace the pins. Sonicate for 10 min or 1 h.
6. Repeat **step 5**.
7. Dry the pins and examine them under the microscope to verify that no particulate matter remains on the tip.

For noncontact pins, we typically follow the instructions provided by GeSiM for the NanoPlotter pins. In brief, we aspirate a 30% KOH (w/v in water) into the tip, manually dip the tip in water to remove KOH off the outside of the tip, and let it sit for 10 min prior to washing.

3.3. Capture Antibody

High-quality antibodies are crucial to generating reliable chips. The capture antibodies must have the following features: high sensitivity and specificity over a large dynamic range for their targeted analyte, low cross-reactivity with other analytes in the sample, and work in an ELISA assay. Additionally, primary amine buffers (e.g., Tris or glycine) and carrier protein in the print solution should be avoided. Primary amines in these reagents will compete with the capture antibody for the *N*-hydroxysuccinimide (NHS) esters that bind antibodies to the slides. A second antibody is required for each analyte that targets a different antigenic region, has high sensitivity, good specificity, and is labeled. Antibodies meeting these criteria can be met either by commercial or in-house antibody sources.

Preparation of capture antibodies can be done in advance and stored for at least 1 week at 4°C and for a month or more at –20°C, provided the plate is sealed to prevent evaporation. Requirements for printing include antibodies, a refrigerated centrifuge, 0.2-μm filtered dilution buffer, low-protein-binding standard 384-well plate, 3 M Aluminum Microplate Sealing Tape, plate centrifuge, glycerol, and aerosol filter pipette tips. Finally, filter tips for the pipettor are strongly recommended to prevent the addition of particulates and aerosolized contamination to the samples.

Preparation for Standard 384-Well Plate

1. Filter the dilution buffer using a 0.2-μm filter. Antibodies are brought up in PBS for the noncontact printer and PBS with 10% glycerol for the contact printer.
2. Resuspend antibodies to a concentration of 0.5 to 1.0 mg/mL. Allow the antibody to equilibrate 15 to 30 min at room temperature. Remix antibody solution again and transfer to a sterile 1.5 mL microcentrifuge tube.
3. Elimination of particulates and bubbles that could interfere with printing are removed by spinning the antibodies at 14,000 rpm for 10 min at 4°C. After spinning, carefully remove the tubes and transfer the upper 90% of the antibody solution to a fresh 1.5 mL microcentrifuge tube. Most antibodies are stable for at least 6 to 12 months after reconstituting if kept free of bacterial contamination.
4. Transfer 12 μL of the antibody to low-protein-binding standard 384-well plate. (*Note*: Positive and negative controls should be printed in every chip. Our positive control is an IgG label with the same fluorescent tag used for detection in the assay. The negative control is PBS, which ensures that there is no carryover between spots.)
5. For extended storage, cover the 384-well plate with Aluminum Microplate Sealing Tape and store at 4°C until needed.

6. Carefully remove sealing tape and centrifuge plates on a microplate centrifuge for 1 min at 1500 × *g*. Check wells for bubbles; if present, repeat spin step until all the bubbles have disappeared. Bubbles can prevent the efficient uptake and/or spotting by the pins resulting in missing assays on your chip.

3.4. Slide Preparation

Aminosilane coated slides are purchased from Erie Scientific and come stamped with a hydrophobic barrier that creates 16 individual sample wells per slide. We commonly enhance this hydrophobic barrier by tracing over it with a PAP pen. Aminosilane slides require activation with 0.5 mM (bis)sulfosuccinimidyl suberate (BS^3) to create NHS ester sites. The NHS ester reacts with primary amines, of the lysine side chains or the N-terminus, to form stable amide bonds that cross-link to the slide.

3.4.1. Slide Activation for 10 Slides or Less

The BS^3 or disuccinimidyl suberate (DSS) (see below) cross-linker should be prepared immediately before use, as the NHS ester hydrolyzes readily and quickly becomes nonreactive.

1. Dissolve 0.3 mg BS^3 per mL of PBS to make a 0.5 mM BS^3 solution.
2. Arrange the slides on a flat surface, using an eight-channel pipettor to add 100 μL of 5 mM BS^3 solution to each well, and incubate at room temperature for 5 min.
3. Tap off the BS^3 solution and rinse the slides with 100% methanol either by spraying slides with constant stream or dipping slides in Coplin jar.
4. Rapidly dry the slides using a high-pressure stream of argon or nitrogen gas.
5. The slides are ready for printing and should be used as soon as possible.

3.4.2. Slide Activation for 11 Slides or More

For large numbers of slides, we use DSS, which is a less expensive, water-insoluble analog of BS^3, to generate reactive NHS esters on the slide. To uniformly process a large number of slides, we use a 24-slide rack, staining dish, ScanArray ExpressHT, PBS, and 100% ethanol.

1. Dissolve 0.054 g disuccinimidyl suberate (DSS) in 600 μL of DMSO.
2. Add the 600 μL of DSS to 270 mL of 100% HPLC-grade methanol in a Tissue Tek staining dish to make a 0.5 mM DSS solution.
3. Arrange the slides in a Tissue Tek 24 slide rack and submerge the slides in DSS solution and incubate at room temperature for 5 min.
4. Tap off the DSS solution and rinse the slides twice in Tissue Tek staining dish containing 270 mL 100% HPLC-grade methanol.
5. Rapidly dry the slides using a high-pressure stream of argon, nitrogen, or other inert gas. Alternatively, dry the slide rack in a vacuum chamber for 5 min.
6. Immediately place the slides on a printer and begin printing.

7. When the print job is complete, the slides are left in the printer with humidity set at 70% or placed in a humid chamber for 1 h to allow the reaction between the NHS ester and antibody to go to completion.
8. Carefully remove the slides, place on a tray (spot side up), and allow the slides to completely air dry.
9. Once the slides are dry, a hydrophobic barrier should be drawn around each chip using a PAP pen. For hundreds of chips, slides such as those produced by Erie Scientific that have a stamped hydrophobic barrier make this process easier. (*Note*: It is important to generate the hydrophobic barrier before the block step because you will not be able to visualize the spots after the block step.)
10. As a quality control step, scan the chips with ScanArray ExpressHT using the Red Reflect setting and the 633-nm laser to catch any errors in the printing process, as illustrated in **Fig. 2**. This setting allows the imaging of salt crystals and provides a rapid method to identify missing spots. However, this step does not work well for buffers that contain glycerol, as these spots do not completely dry and have few salt crystals. Glycerol-containing spots are relatively easy to visualize by the naked eye.

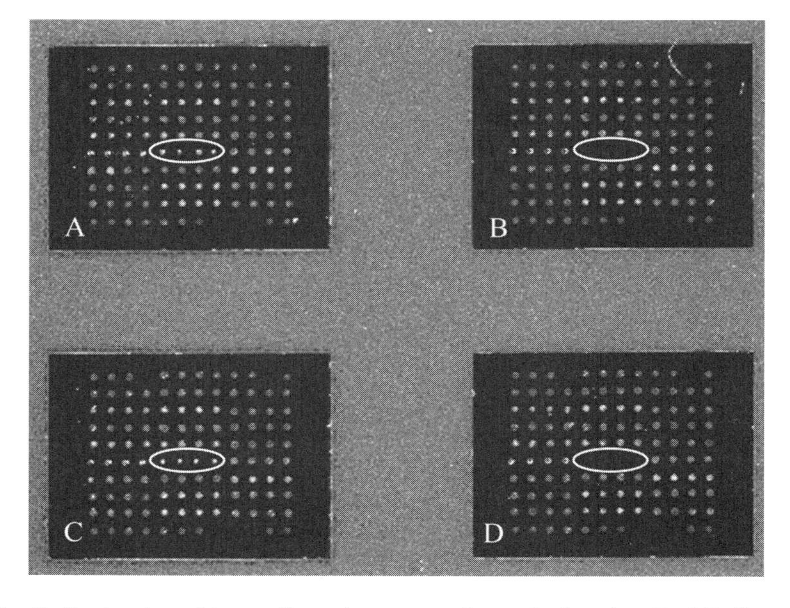

Fig. 2. Evaluating chip quality prior to sample analysis using Red Reflect. This image shows four chips on a single slide where one assay (enclosed by oval) failed to print in chips (**B**) and (**D**) but was fine in chips (**A**) and (**C**). The image was scanned before the block step to verify print success. The image was generated using the ScanArray ExpressHT using 633-nm wavelength, laser power 75, and PMT gain 75 at a 50-μm resolution.

3.5. Blocking and Storage of Slides

Blocking is done to inactivate unused cross-linker and prevent nonspecific binding of proteins during the assay. One of the problems that exist is the "comets" that form by the addition of the block solution (**Fig. 3**). We have found that slow pipetting of the block solution, onto the edge of the chip where there are no spots, reduces comet formation. In contrast, when dipping slides, rapid submersion greatly reduces the formation of comets. The following are required for blocking: 1% casein in PBS, 24-slide holder, staining dish, plastic container with sealable lid, PBS, PBS-T, paper towels, deionized water, Coplin jar, forceps, food vacuum sealer, food vacuum seal bags, and desiccant packages.

3.5.1. Block Steps for a Few Slides

Direct Application of Block Solution to Chips

1. Lay the slides in a humid chamber. This is created by placing a wet paper towel in the plastic container, with a rigid plastic sheet over the towel to keep the slides from coming in direct contact with water.

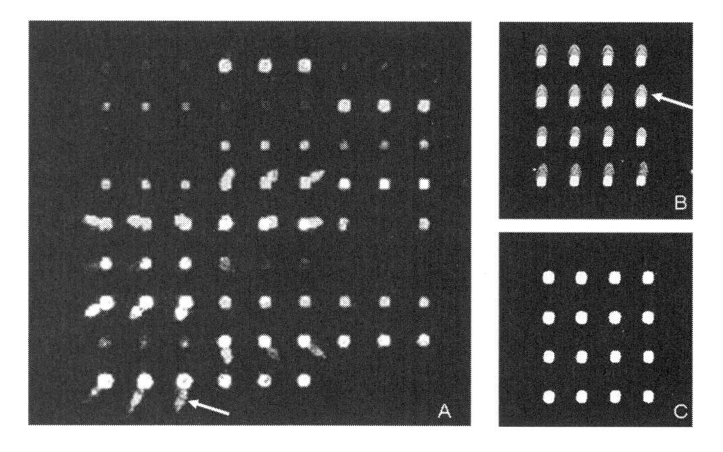

Fig. 3. Comet formation can result from the blocking procedures. The arrows in images (**A**) and (**B**) point to an example of a comet formed after the addition of block. Image (**A**) illustrates comet pattern if block solution is rapidly pipetted directly on the center of the chip. The chip was printed with a series of capture antibodies at 0.8 mg/mL and then used to screen a serum sample. Images (**B**) and (**C**) are chips of our control Cy3 labeled IgG printed at 1X (rows 1 and 2) and 0.5X (rows 3 and 4) that were printed and processed at the same time. Image (**B**) illustrates comets induced by submerging the slide too slowly in block. Image (**C**) illustrates the lack of comets when slides are rapidly submerged.

2. Add 50 to 100 μL of 1% casein in PBS to each well using eight-channel pipettor. As noted above, applying the block very slowly decreases comet formation. Block for 1 h at room temperature or overnight at 4°C.

3.5.2. Block Steps for a Large Number of Slides

1. Load up to 24 slides in the Tissue Tek slide holder.
2. Fill the Tissue Tek staining dish with 270 mL of 1% casein in PBS.
3. Add the slides quickly without disturbing the buffer. Block for 1 h at room temperature or overnight at 4°C.
4. Wash the slides twice in PBS-T for 2 min each. The slides can be stored using **steps 5 to 7** or used immediately.
5. For storage, quickly dip the slides twice in deionized water.
6. Air-dry the slides for 20 min. For faster drying, spin the rack in a plate centrifuge for 1 min at 500 × *g*. Flip the rack and spin again for 1 min.
7. Place the slides in storage boxes; put the storage boxes in vacuum seal bags with desiccant packages and vacuum seal them. Store slides until they are needed at 4°C or –20°C.

4. STANDARDS, SAMPLES, AND CHIP INCUBATION PROTOCOL

As with the single-protein sandwich ELISA, the multiplex ELISA microarray assay requires purified antigens of known concentrations, antibody chips, detection antibodies, and the biological samples.

Antibody microarrays are miniaturized assays that require amplification for signal detection. We use the tyramide amplification system (PerkinElmer) based on horseradish peroxidase (HRP) enzyme and detection antibodies labeled with biotin. The advantage of incorporating tyramide in the assay is increased assay sensitivity without loss of resolution or an increase in background signal. Fluorescence (e.g., cyanine 3; Cy3) is used for signal detection because of its dynamic range and availability of instruments that can scan fluorescence in a microarray format.

4.1. Standards

Similar to a single-analyte ELISA, standard curves for multiplex ELISAs are generated using a mixture of analytes with known concentrations. Ideally, the range of concentrations of each antigen standard will bracket the range of the antigen in the samples at a particular dilution. In order to bracket the concentrations of all of the antigens assayed on a chip, it is usually necessary to analyze several dilutions of each sample.

1. A standard mixture is prepared that contains all the targeted analytes diluted in 0.1% casein/PBS at their highest concentration used in the standard curve. This mixture can be prepared in advance and stored as frozen aliquots at –20°C to –80°C. Some antigens may not retain antigenicity when diluted or frozen, and these will have to be added to the antigen mixture immediately before use.
2. Using 0.1% casein/PBS, prepare a threefold dilution series containing at least seven dilutions, and a blank.

4.2. Samples

Sample preparation depends on the type of biological fluid used. A general protocol is presented here for blood plasma. Plasma samples are stored at –80°C with minimal freeze-thaw cycles prior to use.

1. Immediately before use, samples are thawed at room temperature, mixed gently, and 50 μL is transferred to a new 0.5-mL microcentrifuge tube.
2. Samples are centrifuged for 15 min at 14,000 rpm and 4°C to remove particulates.
3. Without disturbing the sample, transfer the top 45 μL to an 0.5-mL tube containing 5 μL of 1% casein/PBS. Mix gently.
4. Make 1:4, 1:49, and 1:499 dilutions of each sample using 0.1% casein. (*Note*: Multiple dilutions are needed to ensure that all antigen concentrations are well characterized by the standard curve.)

4.3. Chip Incubation Protocol

All incubation steps are done at room temperature.

1. Remove the vacuum sealed slides from the –20°C and allow slides to come to room temperature before opening. Warming prevents condensation from forming on slides that may be used at a later date.
2. Arrange samples and standards in 96-well plate in the same format as desired for the slides. Place extra volume in each well to ensure transfer of the desired sample volume. It is worthwhile to make a slide map as a guide and to record any changes in the sample delivery plan.
3. Slide are removed from the vacuum-sealed container, placed in a Tissue Tek 24-slide holder and are washed in PBS-T for 2 min.
4. Place the slide in the humid chamber, aspirate excess liquid, and cover chamber with lid. (*Note*: It is important that the slides do not dry out at any step because drying may increase the amount of nonspecific background signal.)
5. Open one container at a time and transfer samples to slides in the container using an eight-channel pipettor. This process usually takes 1 to 2 min per six slides. Seal the container and move to the next group of slides.
6. Place slides on an orbital shaker and incubate for 1 h to overnight at room temperature.
7. Make the detection antibody mix at the beginning of day two (or in advance) and place in reservoir. Typically, the detection antibodies are diluted to 25 μg/mL in

0.1% casein/PBS buffer. (*Note*: The optimal concentration for each antibody is determined experimentally beforehand. We find that most detection antibodies work best in a multiplex assay between 20 and100 ng/mL. These concentrations are less than typically used for single-analyte ELISAs but are necessary to minimize the increasing background signal that results with multiplexing.) The optimal detection antibody dilution is determined by running the complete mixture of antigens at all dilutions with twofold serial dilutions of the detection antibody mixture. The dilution for each antibody that gives the best dynamic range on the standard curve and has a high signal-to-noise ratio is used.

8. Open one humid chamber at a time. Aspirate all samples from all the slides, quickly dip each slide in PBS-T, and place slide horizontally in holder. (*Note*: The horizontal position keeps the slides wet while processing the slides from the next humid chamber.)
9. Do three washes with PBS-T buffer for 2 min each.
10. Incubate each chip with 20 μL of detection antibody diluted in 0.1% casein/PBS buffer on an orbital rocker for 2 h at room temperature in humidified chamber.
11. Do three washes with PBS-T buffer for 2 min each.
12. Place the slide in the humid chamber, aspirate excess liquid, and cover chamber with lid until you are ready to add sample.
13. Incubate each array with 20 μL streptavidin-HRP conjugate (PerkinElmer Tyramide Kit) solution diluted 1:100 in PBS-T on an orbital rocker for 30 min at room temperature.
14. Open one humid chamber at a time. Fast dip each slide in PBS-T, and place slide in slide holder.
15. Do three washes with PBS-T buffer for 2 min each.

Fig. 4. An image of a slide that was used to generate standard curves for 22 assays using a fourfold dilution series of the antigen mix. Capture antibodies were printed at a protein concentration of 0.5 mg/mL or 0.8 mg/mL using the GeSiM NanoPlotter piezoelectric printer. Each antibody was printed four times (once in each quadrant of the chip) on each of the 16 chips. The chips were printed on a BS^3-activated, aminosilane-coated glass slide that contained an enhanced hydrophobic barrier. Each row of two chips were incubated at room temperature overnight with 15 μL/chip of a diluted antigen mixture. Relative dilutions of the standard mixture were (row and dilution): **(A)** 0.25X, **(B)** and **(H)** 0.0625X, **(C)** 0.0156X, **(D)** 0.0039X, **(E)** 0.0097X, **(F)** 0.0X, and **(G)** 1.0X. The slide was subsequently washed and incubated with a mixture of 22 detection antibodies; the signal was amplified with the biotinyltyramide system (PerkinElmer) and detected with Cy3-conjugated streptavidin. In order that this slide could be readily visualized, it was scanned using ScanArray Express HT and the 543-nm laser, 10 μm/pixel resolution, laser power setting of 90, and photo multiplier tube (PMT) gain of 70. However, to generate data for the actual standard curves, we would use a lower laser power and PMT gain to prevent saturation of the signal for the most concentrated of the standard mixtures.

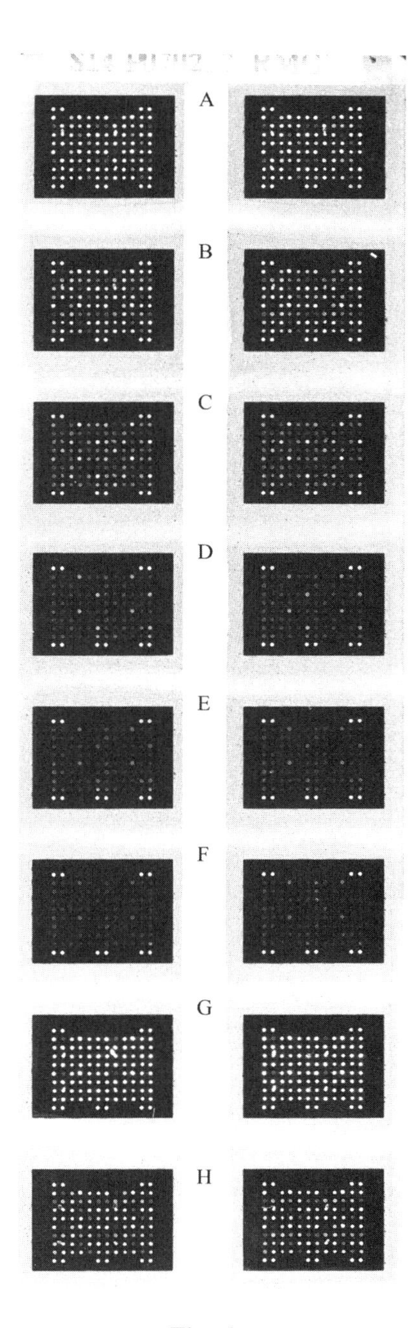

Fig. 4.

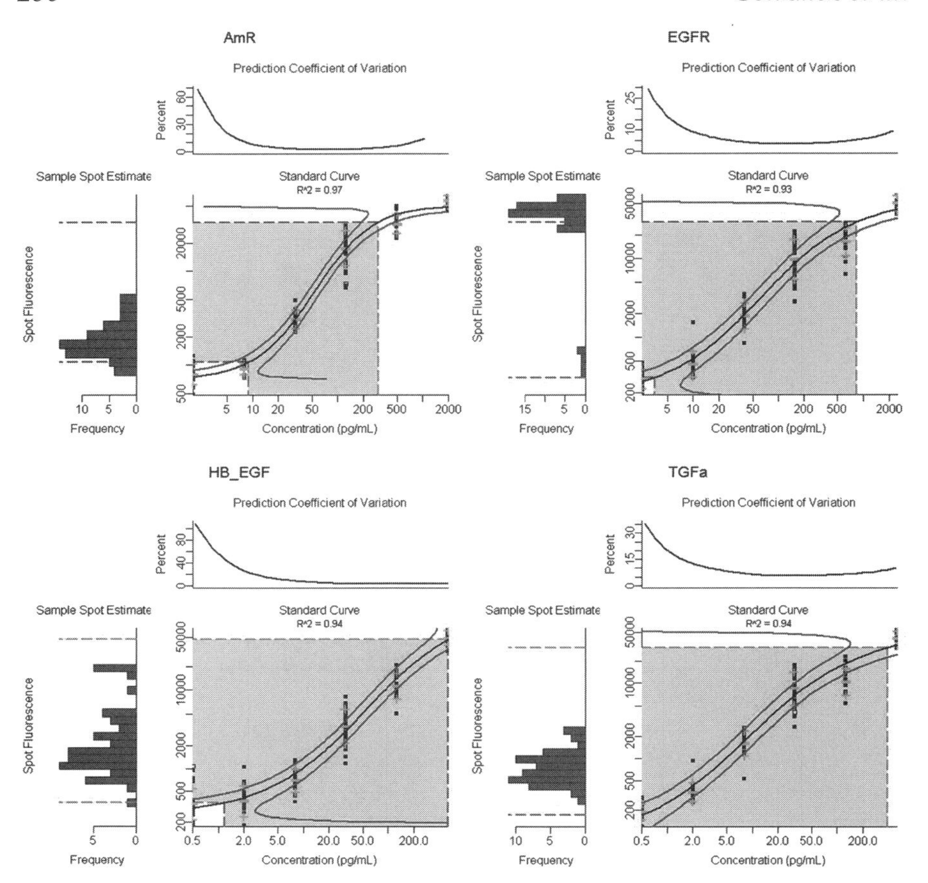

Fig. 5. Representative diagnostic images generated by ProMAT for four assays from a multiplex ELISA experiment. For each image, the standard curve is shown on the bottom right, a histogram of the sample fluorescent values (average of four spot measurements for each chip) is shown on the bottom left, and a graphical representation of the accuracy of the sample concentration measurements over the range of the standard curve is shown on the top. As can be readily seen, the standard curves are a reasonable fit with the standard data. Similarly, the range of the sample signal intensities generally falls within the optimal range of the standard curves for amphiregulin (AmR), heparin binding epidermal growth factor (HB_EGF), and transforming growth factor alpha (TGFa). However, for epidermal growth factor receptor (EGFR), the samples were too concentrated and need to be diluted to obtain quantitative estimates of concentration.

16. Repeat **step 4**. Place the slide in the humid chamber, aspirate excess liquid, and cover chamber with lid until you are ready to add the solutions for the next step.
17. Add 20 μL biotinyltyramide (Tyramide Kit; PerkinElmer) solution diluted 1:100 in 100 mM borate pH 8.5, 0.001% H_2O_2, and incubate each array on an orbital rocker for 10 minutes. Carefully time this step for each slide, incubations longer than 10 minutes will increase background.
18. Open one humid chamber at a time. Fast dip each slide sequentially in two containers of PBS-T, and place slide in slide holder.
19. Do three washes with PBS-T buffer for 2 min each.
20. Place the slide in the humid chamber, aspirate excess liquid, and cover chamber with lid until you are ready to add sample.
21. Incubate each array with 20 μL of streptavidin-Cy3 at 1 μg/mL in PBS-T buffer on an orbital rocker for 0.5 to 1 h in the dark.
22. Open one humid chamber at a time. Aspirate sample from each chip, fast dip each slide in PBS-T, and place slide in slide holder.
23. Do two washes with PBS-T buffer for 2 min each.
24. Rinse twice in deionized water and air dry.
25. The slides are scanned with ScanArray ExpressHT using the Cy3 emission filter and the 543 laser. **Figure 4** shows an example of a slide that has serial dilutions of the antigen mix that are placed on different chips to generate the standard curves.
26. Spot fluorescent is evaluated using ScanArray Express Quantitative software.
27. Standard curves are generated and sample concentrations are calculated using ProMAT, a free software program developed at our institution specifically for this purpose *(34)* (http://www.pnl.gov/statistics/ProMAT/). **Figure 5** shows an example of ProMAT diagnostic images for four assays. Each image includes the standard curve, a histogram of the sample fluorescence intensities on the same scale as the standard curve, and a curve that indicates the accuracy of the predicted concentration values.

For large numbers of slides, several of the processing steps, including the detection antibody, streptavidin-HRP conjugate, and streptavidin-Cy3 conjugate steps, can be done by submersing the slides in the reagent mixtures. An example of how we do this is described here for the detection antibody step. After the wash at **step 10**, place the slides in the rack in Tissue Tek staining dish containing 270 mL of label antibody diluted in 0.1% casein/PBS. Incubate for 2 h at room temperature. Do not use a stir bar or put the bucket on an orbital shaker because the current from the motion will generate side-to-side variation of the slide.

5. CONCLUSION

The ultimate goal with high-throughput ELISA microarray screens is to produce reliable and reproducible results. Reliability of the sandwich ELISA microarray depends on the uniform creation of the chip, the specificity of the assays on the chip, and the dynamic range of the targeted analytes in the screen. Reproducibility occurs when all samples and chips are uniformly processed. Therefore, considerable thought should be given to developing protocols for high-throughput screens. In this chapter, we described our basic protocols for the sandwich ELISA microarray when the goal is to analyze a small number of samples or to do high-throughput screens of larger sample numbers.

REFERENCES

1. Copland JA, Davies PJ, Shipley GL, Wood CG, Luxon BA, Urban RJ. The use of DNA microarrays to assess clinical samples: the transition from bedside to bench to bedside. Recent Prog Horm Res 2003;58:25–53.
2. Frolov AE, Godwin AK, Favorova OO. [Differential gene expression analysis by DNA microarrays technology and its application in molecular oncology.] Mol Biol (Mosk) 2003;37(4):573–584.
3. Sevenet N, Cussenot O. DNA microarrays in clinical practice: past, present, and future. Clin Exp Med 2003;3(1):1–3.
4. Finnskog D, Jaras K, Ressine A, et al. High-speed biomarker identification utilizing porous silicon nanovial arrays and MALDI-TOF mass spectrometry. Electrophoresis 2006;27(5–6):1093–1103.
5. Ho L, Sharma N, Blackman L, Festa E, Reddy G, Pasinetti GM. From proteomics to biomarker discovery in Alzheimer's disease. Brain Res Brain Res Rev 2005;48(2):360–369.
6. Hudelist G, Pacher-Zavisin M, Singer CF, et al. Use of high-throughput protein array for profiling of differentially expressed proteins in normal and malignant breast tissue. Breast Cancer Res Treat 2004;86(3):281–291.
7. Zangar RC, Varnum SM, Covington CY, Smith RD. A rational approach for discovering and validating cancer markers in very small samples using mass spectrometry and ELISA microarrays. Dis Markers 2004;20(3):135–148.
8. Angenendt P, Glokler J, Sobek J, Lehrach H, Cahill DJ. Next generation of protein microarray support materials: evaluation for protein and antibody microarray applications. J Chromatogr A 2003;1009(1–2):97–104.
9. Haab BB, Geierstanger BH, Michailidis G, et al. Immunoassay and antibody microarray analysis of the HUPO Plasma Proteome Project reference specimens: systematic variation between sample types and calibration of mass spectrometry data. Proteomics 2005;5(13):3278–3291.
10. Miller JC, Zhou H, Kwekel J, et al. Antibody microarray profiling of human prostate cancer sera: antibody screening and identification of potential biomarkers. Proteomics 2003;3(1):56–63.
11. Woodbury RL, Varnum SM, Zangar RC. Elevated HGF levels in sera from breast cancer patients detected using a protein microarray ELISA. J Proteome Res 2002;1(3):233–237.

12. Haab BB, Zhou H. Multiplexed protein analysis using spotted antibody microarrays. Methods Mol Biol 2004;264:33–45.
13. Wingren C, Borrebaeck CA. High-throughput proteomics using antibody microarrays. Expert Rev Proteomics 2004;1(3):355–364.
14. Rucker VC, Havenstrite KL, Herr AE. Antibody microarrays for native toxin detection. Anal Biochem 2005;339(2):262–270.
15. Gehring AG, Albin DM, Bhunia AK, Reed SA, Tu SI, Uknalis J. Antibody microarray detection of Escherichia coli O157:H7: quantification, assay limitations, and capture efficiency. Anal Chem 2006;78(18):6601–6607.
16. Hamelinck D, Zhou H, Li L, et al. Optimized normalization for antibody microarrays and application to serum-protein profiling. Mol Cell Proteomics 2005;4(6):773–784.
17. Poetz O, Ostendorp R, Brocks B, et al. Protein microarrays for antibody profiling: specificity and affinity determination on a chip. Proteomics 2005;5(9):2402–2411.
18. Varnum SM, Woodbury RL, Zangar RC. A protein microarray ELISA for screening biological fluids. Methods Mol Biol 2004;264:161–172.
19. Knecht BG, Strasser A, Dietrich R, Martlbauer E, Niessner R, Weller MG. Automated microarray system for the simultaneous detection of antibiotics in milk. Anal Chem 2004;76(3):646–654.
20. Pavlickova P, Schneider EM, Hug H. Advances in recombinant antibody microarrays. Clin Chim Acta 2004;343(1–2):17–35.
21. Kusnezow W, Hoheisel JD. Antibody microarrays: promises and problems. Biotechniques 2002;Suppl:14–23.
22. Steinhauer C, Wingren C, Khan F, He M, Taussig MJ, Borrebaeck CA. Improved affinity coupling for antibody microarrays: engineering of double-(His)6-tagged single framework recombinant antibody fragments. Proteomics 2006;6(15):4227–4234.
23. Brueggemeier SB, Kron SJ, Palecek SP. Use of protein-acrylamide copolymer hydrogels for measuring protein concentration and activity. Anal Biochem 2004;329(2):180–189.
24. Kusnezow W, Hoheisel JD. Solid supports for microarray immunoassays. J Mol Recognit 2003;16(4):165–176.
25. Stillman BA, Tonkinson JL. FAST slides: a novel surface for microarrays. Biotechniques 2000;29(3):630–635.
26. Frederix F, Bonroy K, Reekmans G, et al. Reduced nonspecific adsorption on covalently immobilized protein surfaces using poly(ethylene oxide) containing blocking agents. J Biochem Biophys Methods 2004;58(1):67–74.
27. Danczyk R, Krieder B, North A, Webster T, HogenEsch H, Rundell A. Comparison of antibody functionality using different immobilization methods. Biotechnol Bioeng 2003;84(2):215–223.
28. Lee Y, Lee EK, Cho YW, et al. ProteoChip: a highly sensitive protein microarray prepared by a novel method of protein immobilization for application of protein-protein interaction studies. Proteomics 2003;3(12):2289–2304.

29. Haab BB. Methods and applications of antibody microarrays in cancer research. Proteomics 2003;3(11):2116–2122.
30. Olle EW, Sreekumar A, Warner RL, et al. Development of an internally controlled antibody microarray. Mol Cell Proteomics 2005;4(11):1664–1672.
31. Nielsen UB, Geierstanger BH. Multiplexed sandwich assays in microarray format. J Immunol Methods 2004;290(1–2):107–120.
32. Geierstanger BH, Saviranta P, Brinker A. Antibody microarrays using resonance light-scattering particles for detection. Methods Mol Biol 2006;328:31–50.
33. Saviranta P, Okon R, Brinker A, Warashina M, Eppinger J, Geierstanger BH. Evaluating sandwich immunoassays in microarray format in terms of the ambient analyte regime. Clin Chem 2004;50(10):1907–1920.
34. White AM, Daly DS, Varnum SM, Anderson KK, Bollinger N, Zangar RC. ProMAT: protein microarray analysis tool. Bioinformatics 2006;22(10): 1278–1279.

14

LC-MS Metabonomics Methodology in Biomarker Discovery

Xin Lu and Guowang Xu

Summary

Metabonomics as an important component of functional genomics and system biology has gained increasing interest in drug development and biomarker discovery. Comprehensive metabonomics investigations are primarily a challenge for analytical chemistry. The main advantages of liquid chromatography–mass spectrometry (LC-MS) technique include high sensitivity, high resolution, and wide dynamic range. LC-MS–based technology platforms combined with multivariate statistical methods are used to explore the complex metabolite patterns of biofluids. In this chapter, the methodology and application of high-performance liquid chromatography (HPLC)-MS–based metabonomics in biomarker discovery are reviewed.

Key Words: biomarker; HPLC-MS; metabolic profiling; metabonomics

1. INTRODUCTION

The increased number of new biomarker candidates is directly linked to recent developments in discovery, informatics, and assay technologies. Biomarker discovery tools in genomics, proteomics, and metabonomics have advanced rapidly over the past decade. Metabonomics has shown the ability to quickly assess changes in the abundance of a large number of endogenous metabolites representing multiple compound classes *(1)*. *Metabonomics* is defined as "a quantitative measurement of multi-parametric metabolic responses of multi-cellular systems to pathophysiological stimuli or genetic signaling" *(2)*. Metabonomics has recently begun to have a more important role in efforts at biomarker identification. Metabolic biomarkers are closely identifiable with real biological end points and provide a global systems interpretation of biological effects, including the interactions between multiple genomes such as humans and their gut microflora *(3)*.

From: *Methods in Pharmacology and Toxicology: Biomarker Methods in Drug Discovery and Development*
Edited by: F. Wang © Humana Press, Totowa, NJ

Metabonomics is typically performed on biofluids, such as serum, urine, saliva, and cerebrospinal fluid. It is useful for physiologic evaluation, drug safety assessment, diagnosis of human disease, drug therapy monitoring, and characterization of genetically modified animal models of disease ***(4,5)***. In combination with genomics, transcriptomics, and proteomics, metabonomic analysis is being increasingly used in drug discovery and development. Much, therefore, depends on the ability of the analytical technique employed to detect often-subtle differences in the complex mixtures found for biofluids. Metabonomics has had a long history in toxicology ***(5)***. An extensive literature exists on the use of metabonomics to evaluate nephrotoxicants and hepatic toxicants ***(6–8)***.

Figure 1 gives the scheme for the biomarker discovery based on metabonomics. From the point of view of analytical chemistry, it contains two key steps, that is, data collection of metabolite fingerprints (metabolome) and

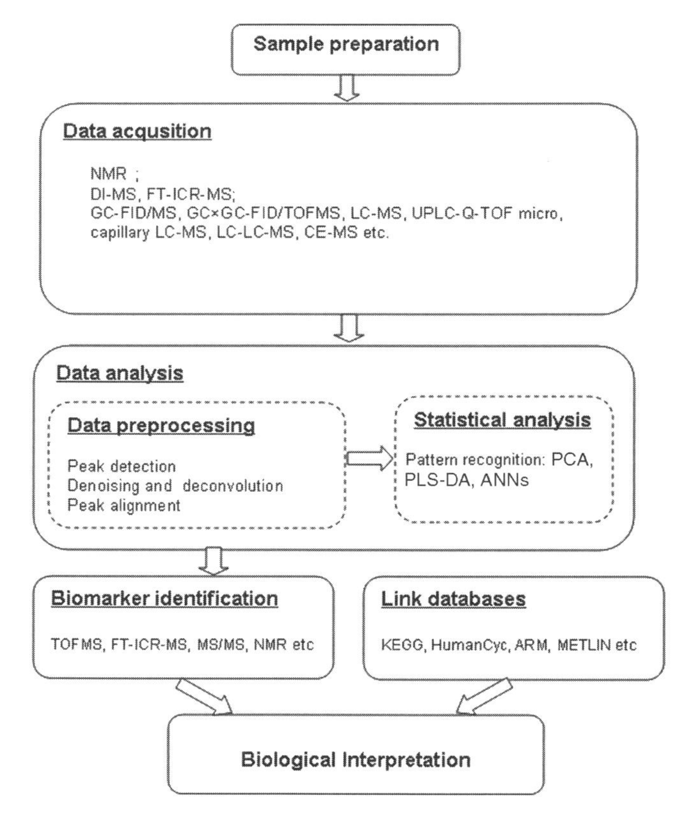

Fig. 1. Scheme for biomarker discovery based on metabonomics.

information mining. The comprehensive investigation of the metabolome is being complicated by its enormous complexity and dynamics. An ideal analytical technique can be performed directly on the samples, without the need for sample pretreatment. It should be a high-throughput screening, unbiased with the whole metabolites, robust, reproducible, sensitive, and accurate. In additional, it should have a wide dynamic range. In an ideal world, all of these desirable characteristics would also be combined with sufficiently high information content to enable the identification of the key metabolites identified via the postanalysis multivariate statistical analysis of the data. A number of different analytical strategies are employed, with the ultimate goal of measuring a large fraction or all of the metabolites present. Realistically, there is no technique currently available that can provide all of the desired properties. The major analytical techniques that are employed for metabonomics investigations are based on nuclear magnetic resonance (NMR) spectroscopy ***(2,5,9)*** and mass spectrometry (MS) ***(10–13)***. Conventionally, metabonomics investigations have been performed using high-field NMR spectroscopy ***(5,8,14–16)***. In this type of application, NMR has many advantages, such as high information content of the resulting spectra, the relative stability of NMR-chemical shifts, the ease of quantification, and the lack of any need to preselect the conditions employed for the analysis. MS, besides NMR spectroscopy, is the main analytical technique widely used for metabonomic studies. Its main features are high sensitivity, high resolution, wide dynamic range, coverage of a wide chemical diversity, robustness, and feasibility to elucidate the molecular weight (MW) and structure of unknown compounds. MS is inherently considerably more sensitive than NMR spectroscopy ***(17)***, but hyphenated gas chromatography (GC) after chemical derivatization or high-performance liquid chromatography (HPLC) enables preseparation of metabolite. The identification of differences between samples has normally involved different multivariate tools such as principal component analysis (PCA), hierarchical cluster analysis (HCA), discriminate analysis, or correlative network analysis. Ideally, the unprocessed MS files would be subjected directly to multivariate analysis, but certain problems make preprocessing of raw data before multivariate analysis essential, including noisy baseline, drifting retention time, variations in peak shape, and differences in recovery between the analyzed samples. In the meantime, in the identification step, because of lack of standard samples and database, an integrated strategy including various chromatography-MS techniques and NMR have to be used.

Recently, HPLC-MS has begun to be employed in metabonomic studies ***(13)***, either alone ***(11,18,19)*** or in combination with NMR spectroscopy ***(20–22)***. Although most investigations have been on plant metabolism and model cell system extracts, its applications to mammalian studies are recently increasing ***(10,13,23)***. The first reported rodent toxicology investigation of HPLC-MS–based metabonomics method was by Plumb and co-workers ***(11)***. In this investigation, urine samples obtained in a long-term toxicity study, at different dose levels, were analyzed using reversed-phase liquid chromatography electrospray ionization time-of-flight (LC-ESI/TOF) MS analysis and negative ion detection. Using PCA, it was possible to show clear differences between dose groups and the controls. Five unique m/z values were identified from the PCA loading plots as strong contributors to the observed clustering. Other examples of metabolic profiling using LC-ESI/MS have predominately involved toxicology. Several studies focused on the effects of nephrotoxins on the urinary metabolite profiles of rats that had been administered mercuric chloride ***(20)***, cyclosporine ***(21)***, gentamicin ***(22)***, and D-serine ***(24)***.

With the future technological, methodological, and informatics advances, LC-MS technique will increase its impact on metabonomics investigations and biomarker discovery. The improvement in the mass accuracy and resolution of mass spectrometers will enable better prediction of metabolite biomarker identities. Many TOF and Fourier transform (FT)/MS mass spectrometers already provide 3 ppm mass accuracy, and further improvement can be expected. The sensitivity of mass spectrometers will continue to increase and will further broaden its applicability to trace analyte detection in complex biological media. As mass spectrometry technologies improve, methodologies and informatics will be developed to facilitate new capabilities. Increased use of immunoaffinity ***(25)***, multidimensional chromatography, and other novel separations methods coupled with mass spectrometry will further increase the sensitivity and dynamic range of trace biomarker detection. The use of capillary LC and ultra-performance liquid chromatography (UPLC) methods will improve the coverage of the metabolome provided by HPLC-MS as a result of reduced ion suppression. The difficulty with exchanging a spectra library from LC-MS analysis is a major drawback of that technology, particularly for metabolite profiling, where the identification of a large number of compounds is desirable. Increasing our knowledge in ionization mechanisms in LC-MS may lead to the possibility to exchange spectra libraries between laboratories and research groups; that exchange will have a substantial impact on the entire field of metabolome analysis. Increased incorporation of stable isotope labeling into biomarker identification and quantification methods will greatly improve assay validation and biomarker qualification. Software and database advancements will complement advances in mass spectrometry technology. A metabolite database that allows quick mass spectrometric identification of endogenous metabolites may advance metabonomics to the user level of genomics. And bioinformatics advances that allow systems biology understanding of RNA, protein, and metabolite changes in the same biological systems will lead to faster evaluation of novel biomarker candidates.

In this section, the main LCMS technologies and data analysis methods used in metabonomics are summarized, and the workflow of LC-MS–based metabonomics for biomarker discovery is described.

2. LC-MS–BASED METABONOMICS

2.1. Sampling and Sample Preparation

Sample acquisition is primarily driven by the experimental design and the experimental type. If possible, a power analysis should be performed to ensure that a sufficient number of samples are acquired and to reduce the influence of biological variability and obtain statistically validated data. In particular, when studying human samples, the influences of diet, gender, age, and genetic factors have to be considered. Therefore, in metabolic fingerprinting analysis, a large number of samples are commonly analyzed to detect biologically relevant sample clustering. In addition, representative quality control samples, such as sample replicates, analytical replicates, and blanks including method blanks, have to be analyzed. While processing biological samples, special care must be taken to minimize the formation or degradation of metabolites after sampling due to remaining enzymatic activity or oxidation processes. Several techniques have been used to inhibit metabolism such as freezing in liquid nitrogen, freeze clamping, acid treatment, *(26)* or quenching in salt containing aqueous methanol at low temperatures *(27)*.

In metabolic target analysis or quantitative metabolic profiling, the analytes are known, and standard compounds or stable isotope-labeled standards can be used to optimize the extraction procedure and matrix removal. However, in metabonomics investigation, all small molecules are the targets and only salts and macromolecules, such as proteins or larger peptides, can be considered as matrix. Consequently, a sample preparation step should be as simple and universal as possible. Sample preparation and sample introduction methods for the analysis of biological samples include direct injection, liquid-liquid extraction (LLE), solid-phase extraction (SPE), supercritical fluid extraction, accelerated solvent extraction, microwave-assisted extraction, protein precipitation, and membrane methods such as dialysis or ultracentrifugation.

Common sample preparation techniques of serum or plasma in metabolite profiling involve protein precipitation through the addition of an organic solvent such as methanol or acetonitrile, LLE, or acid protein precipitation ***(28–30)***. Organic solvents are the most widely used protein precipitants in metabonomics investigations. Acetonitrile was found to be the superior organic plasma protein precipitant. Urine samples or diluted urine samples have been directly injected for LC-ESI/MS–based metabolic fingerprinting ***(20,31)***. LLE for biofluids is the method of choice for tissue extraction ***(32)***. Upon collection, tissues are typically flash frozen in liquid nitrogen. After they are flash frozen, tissues should either be kept frozen below −70°C or lyophilized to avoid recovery of enzymatic activity. Prior to sample extraction, different methods can be used to break up tissues and cells. The most common methods include grinding in a liquid N_2–cooled mortar and pestle ***(33)*** or homogenization by an electric tissue homogenizer directly in the extraction solvent ***(34)***. Either wet frozen or freeze-dried samples were used to metabolite extraction. Methanol/chloroform/water were recommended as optimum extraction solvents for homogenizing wet tissues ***(35)***.

2.1.1. Data Collection

In general, metabonomics investigations by HPLC-MS have been performed using solvent gradients, on reversed-phase (RP) packing materials, C18 3.0 or 4.6 mm i.d. columns, of length between 5 and 25 cm in containing 3 to 5 μm packing materials. A typical HPLC-MS total ion current chromatogram (positive electrospray ionization) of a serum sample obtained from a chronic hepatitis B patient is shown in **Fig. 2**. There are many ions present in these total ion chromatograms (TICs). Using a variety of statistical approaches (e.g., partial least-squares discriminant analysis; PLS-DA) to compare such metabolic profiles, the hepatitis group was clearly separated from the healthy control group.

Many biofluids, particularly urine, contain a vast array of highly polar molecules that are not retained well on reversed-phase chromatography. Normal phase techniques, which result in the elution of less polar molecules first and thus the retention of more polar molecules, require a different solvent system to that used by reversed-phase chromatography, typically containing no aqueous. One option for the HPLC-MS analysis of such polar compounds using the so-called HILIC (hydrophilic interaction chromatography) has been demonstrated for rat urine after SPE (on Waters Oasis HLB) ***(36)***. The unretained material in RP-HPLC was then analyzed on a ZIC-HILIC column using a solvent gradient over 15 min (followed by a further 4-min isocratic elution before returning to the starting conditions for reequilibration). HILIC approaches combined with ESI-MS techniques have already been applied to the analysis of dichloroacetic acid in rat blood and tissues ***(37)***, and with atmospheric pressure chemical ionization (APCI)

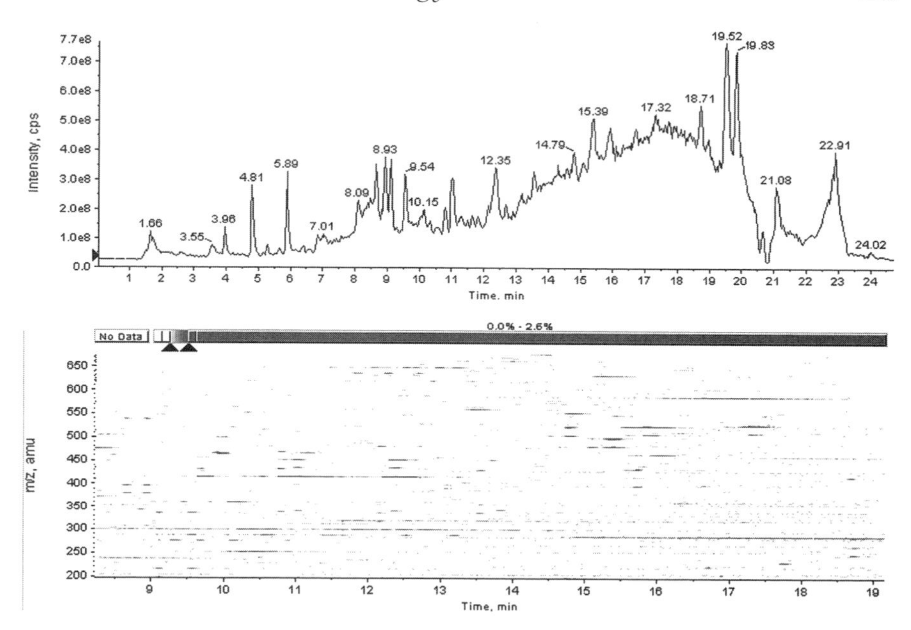

Fig. 2. A typical gradient positive-ion LC-ESI/MS of patient serum showing both the total ion current mass chromatogram **(top)** and a 2D mass chromatogram **(bottom)**.

mass spectrometry for the determination of 5-fluorouracil in plasma and tissues *(38)*.

Conventional RP-HPLC separation is often insufficient for the separation of complex biological samples, resulting in poor metabolite resolution. The recently introduced UPLC is a combination of a 1.7-μm reversed-phase packing material and a chromatographic system, operating at pressures in the 6000 to 15,000 psi range. This has enabled better chromatographic peak resolution and increased speed and sensitivity to be obtained for complex mixture separation. Reduction of the particle diameter from 5 to 1.7 μm will, in principle, result in a 3-fold increase in efficiency, a 1.7-fold increase in resolution, a 1.7-fold increase in sensitivity, and a 3-fold increase in speed. The typical peak widths generated by the UPLC system are of the order 1 to 2 s for a 10-min separation. Because of the much improved chromatographic resolution of UPLC, the problem of ion suppression from coeluting peaks is greatly reduced. Orthogonal-acceleration time-of-flight (oaTOF) MS, and a combination of quadrupole and oaTOF–MS (Q–TOF) coupled to UPLC, have proved to be a powerful tool for the identification of trace constituents of complex mixtures and/or for confirming their presence. Such instruments enable accurate mass measurement with accuracies of < 5 ppm, which dispel interpretation ambiguities.

There have been an increasing number of applications of UPLC-MS to the analysis of biological fluids in the field of metabonomics ***(12,18,39–43)***. We compared UPLC-MS to conventional HPLC-MS methods to analyze urinary nucleosides and other metabolites with cis-diol structure to distinguish between cancer patients and healthy persons. **Figure 3** show the HPLC-MS and UPLC-MS chromatograms from a caner patient urine sample. About 11,000 ions were detected with the faster gradient UPLC-TOFMS method, whereas with HPLC-TOFMS, 8000 ions were detected. The UPLC-TOFMS method needed about one third of the time required by HPLC-TOFMS and achieved a much better resolution and more information. PCA was used to

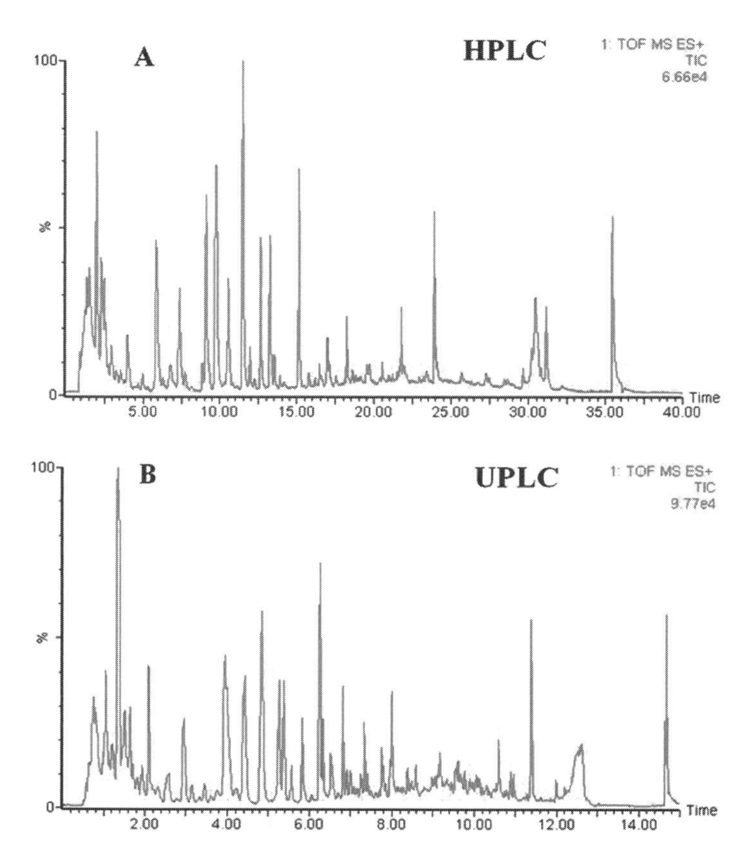

Fig. 3. Typical chromatogram of a urinary sample: **(A)** HPLC-MS and **(B)** UPLC-MS.

distinguish between cancer patients and healthy controls. With the HPLC-TOFMS data, the two groups were not sufficiently separated, four patients entered into the region of healthy persons, and three healthy persons were classified as patients. Using the UPLC-TOFMS data, the cancer patients and healthy controls were located in two different cluster regions; only one healthy person entered into the region of cancer patients, and much better separation was obtained.

For metabonomics studies, ESI is most commonly used coupled with LC-MS ***(44,45)***. In order to obtain a comprehensive profile, ionization must be performed in both positive and/or negative mode. ESI utilizes a high electric field to produce charged droplets from a liquid solution, ultimately leading to the formation of gas phase ions ***(46)***. The main advantages of the ESI ion source are soft ionization, no need for derivatization, ability to ionize compounds of a large mass range, suitability for nonvolatile and polar compounds, and excellent quantitative analysis and high sensitivity. A drawback of the ESI process is its liability for ion suppression due to competition effects in the ionization process.

The ability to form analyte ions from the electrospray process depends on the pH and mobile phase composition. To generate analyte ions from the mobile phase, a volatile organic acid such as formic acid or acetic acid is added to the mobile phase. The addition of acids favors the generation of positive ions, $[M+H]^+$, but is not the first choice for negative ions, $[M-H]^-$. Electrospray is a soft ionization technique, generating $[M+H]^+$ or $[M-H]^-$ ions even for very thermally unstable and nonvolatile molecules. The choice of mobile phase is very important in that it must be volatile and not have strong ion pairing properties. For example, HPLC mobile phases containing acetic acid, formic acid, and ammonium acetate are acceptable. However, HPLC mobile phases containing nonvolatile buffers, such as phosphate buffers, are not acceptable because their strong ion pairing properties will favor the formation of neutral products. Stronger acids such as trifluoroacetic acid (TFA), although frequently used with HPLC, are less desirable because strong acid anions pair with analyte cations, thus reducing the analyte ion abundance.

The MS parameters, such as drying and nebulizer nitrogen flow rates, drying temperature, fragmenter voltage, and so forth, are optimized to provide maximum response for the positive ions, $[M+H]^+$, or negative ions, $[M-H]^-$. Evaluation of spectral information can be obtained in full-scan mode. Because the electrospray interface performs optimally at low flow rates, usually below the 0.5 mL/min range, the typical flow rate is 0.10 to 0.20 mL/min for ESI. Using a conventional HPLC column (4.6 mm i.d. × 250 mm), the HPLC effluent flow is split such that approximately 100 μL

min^{-1} actually passes through the interface (10:1 split) while the remainder goes to a waste container. Currently, UPLC columns (e.g., 2.1 mm i.d.) and slower flow rates are commonly used to achieve the desirable flow rates. The advantage of this approach is that improved separation efficiency and faster separations are also achieved.

There are many types of mass analyzers available for interfacing with HPLC. The basic common mass analyzers include single quadrupoles, triple quadrupoles (TQ), TOF, ion traps, orbitrap, and Fourier transform ion cyclotron resonance (FT-ICR). In addition to these types of mass analyzers, an increasing number of hybrid systems exist that combine two basic types of mass spectrometers, such as Q-TOF instruments ***(47)***, quadrupole linear ion traps (Q-Trap) ***(48,49)***, or ion trap FT mass spectrometers ***(50)***. TOF instruments feature fast scanning capabilities, wide mass range, and high resolution [5000 to 20,000 full width, half maximum (FWHW)] and mass accuracy. It is extremely useful for profiling complex metabolic mixtures. In order to perform MS-MS experiments with TOF instruments, another mass analyzer has to be combined. In quadrupole-TOF instruments, the last quadrupole of the triple quad configuration is substituted by a TOF analyzer. These hybrid instruments combine the stability and robustness of the quadrupole analyzer with the fast scanning capabilities, accuracy (< 5 ppm) and high sensitivity of TOF mass analyzers. Orbitrap ***(51,52)*** and FT-MS ***(53,54)*** offer the highest resolution available and high accuracy fragment masses. The unsurpassed resolution (> 100,000 FWHW) and mass accuracy (< 2 ppm) of FT mass spectrometers leads to formula candidates and supports metabolite identification. For high-end hybrid instruments like quadrupole FT-MS and quadrupole linear ion trap FT-MS instruments, ion selection and fragmentation can be performed outside the cyclotron of the FT-MS. The performance and application depth of FT-MS is thereby significantly expanded, although quantitative aspects may be compromised due to the ion gating function of the ion trap.

Metabolic profiling approaches require a sensitive full scan mode and exact masses. Therefore, Q-TOF instruments or linear ion trap FT-MS instruments are advantageous. In contrast, for targeted analysis of selected metabolites, triple quadrupole instruments, and especially Q-Trap instruments (with their capability for multiple reaction monitoring), are frequently used. The development of the hybrid quadrupole-orthogonal time-of-flight mass spectrometer allowed the generation of exact mass information with greater accuracy and precision; these mass values can then be used to produce candidate empirical formulae that, at the 3- to 5-ppm error range, significantly reduce the number of possible structures of putative metabolites with molecular masses of a few hundred daltons.

2.2. Data Processing and Data Analysis

Data processing is an essential step to properly analyze and interpret metabonomics data. Data processing for mass-based metabonomics has been reviewed by Katajamaa *(55)*. Data handling tasks in metabonomics investigations include data processing and data analysis. The data processing stage consists of processing of raw data and combining data between measurements. Then the raw data were transformed into format that is easy to use in the subsequent data analysis steps. The data analysis stage includes tasks for analysis and interpretation of processed data. This typically includes multivariate analyses such as clustering of metabolic profiles or discovering important differences between groups of samples.

In LC-MS–based metabonomics, a set of raw data files were obtained, and each file corresponded with a single biological sample. The basic aim of data processing is to transform raw data files into representation that facilitates easy access to characteristics of each observed ion. Typical data processing workflow consists of multiple stages, including filtering, peak detection, curve resolution or deconvolution, alignment, and normalization.

Filtering methods process the raw measurement signal with aim of removing effects like measurement noise or baseline. LC-MS data is transformed into a two-dimensional matrix, with one index corresponding with the retention time scans and another to fixed m/z values, and matrix values represent the ion intensities. Random noise from the measurement signal was removed using traditional signal processing techniques such as filtering with moving average window *(56)*, median filter *(57)* in chromatographic direction, and Savitzky-Golay method *(58)*. Chemical noise was eliminated by baseline correction. Peak detection is used to detect representations of measured ions from the raw signal. Peak detection could be carried out using peak detection method such as finding peaks independently in both retention time and m/z direction or slicing the whole data set to extracted ion chromatograms and processing them independently or model fitting. Curve resolution or deconvolution methods are mainly applied for data processing that results in a multivariate profile for each sample. Deconvolution can also be used to reduce the complexity of chromatograms obtained with soft ionization techniques by filtering multiple charged species, clusters, and adducts *(59)*. Alignment methods cluster measurements across different samples. Most alignment methods work in pairwise fashion by aligning either only pairs of samples or multiple samples against a selected reference sample or a template. In general, the choice of reference sample has effect on the alignment results. A simple, much explored alignment strategy is mapping the retention time axis of one TIC to another. The correlation optimized warping (COW) method developed by Nielsen et al. is one well-known method for alignment ***(60)***. In addition to the COW method, various other types of warping algorithms for aligning chromatograms have been developed ***(61–65)***. Because total ion chromatogram represents a collapsed view of the raw data, using the full two-dimensional raw data can potentially lead to better alignment results. Several commercial and open source routines for automatic alignment, de-noising, deconvolution, and extraction of peak have been proposed ***(59,66–71)***, such as MS Resolver and ReOrder (Pattern Recognition Systems, Bergen, Norway) ***(71)***, MZmine ***(67,68)***, XCMS ***(66,69)***, and MET-IDEA ***(66)***. Many instrument manufacturers have produced their own software, such as MarkerLynx (www.waters.com), MassHunter (www.chem.agilent.com), and MarkerView (www.appliedbiosystems.com). Normalization stage is needed to remove or minimize the systematic error in data. There are two different normalization methods of metabolic profile data. One is to use statistical models to derive optimal scaling factors for each sample based on complete data set, such as normalization by unit norm ***(72)*** or median ***(58)*** of intensities. Another normalization method utilizes information from a single or multiple internal (i.e., added to sample prior to extraction) or external (i.e., added to sample after extraction) standard compounds ***(73)***.

After processing, LC-MS data is exported as a peak height or area matrix. This matrix can then be further processed with chemometric tools available for statistical analyses of multivariate data. Multivariate data analysis is used for interpretation of metabolic profiling data. The purpose of data analysis in metabonomics is versatile, including identification of similarities and difference in data, classification of samples, identification and quantification of analytes, and more. Both univariate and multivariate statistics follow the data reduction step, and both supervised and unsupervised strategies are employed for multivariate statistics and model building for prediction and classification of outcomes ***(74)***. The most common chemometric tool used in the evaluation of a metabonomics study is PCA. PCA is always recommended as a starting point for analyzing multivariate data and will rapidly provide an overview of the information hidden in the data. PCA can be used to investigate clustering tendency, to detect outliers, and to visualize data structure ***(75)***. In addition, there are many other unsupervised methods, such as nonlinear mapping and hierarchical cluster analysis. PCA gives a simplified representation of the information contained in the spectra and cannot generally use additional information about data, such as class information. Therefore, PCA is often followed by a supervised analysis technique such as partial least squares discriminant analysis (PLS-DA). PLS is one of the widely used supervised methods. It relates a data matrix containing independent variables from samples, such as spectral intensity values (an X matrix), to a matrix containing dependent variables (e.g., measurements of response, such as toxicity scores) for those samples (a Y matrix). PLS-DA is performed in order to enhance the separation between groups of observations. Orthogonal projection on latent structure discriminant analysis (O-PLS-DA), which is the most recent advanced development of PLS-DA, can improve the interpretation of models *(76)*. Methods from the field of artificial intelligence, in particular artificial neural networks (ANN), have been successfully applied to metabonomics. It is capable of learning patterns and relations from input data, making good pattern recognition engines and robust classifiers. ANNs are used for building nonlinear classification and regression models. Recently, an approach to the mining of highly complex metabolomics data is to apply evolutionary supervised learning techniques, including genetic algorithms, genetic programming, evolutionary programming, and genomic computing, which could be ideal strategies for mining such high-dimensional data as that obtained from metabolomic studies *(77)*. An overview of how the underlying philosophy of chemometrics is integrated throughout metabonomics studies has been given by Trygg et al. *(78)*.

2.3. Metabolite Identification

Once a potential feature has been identified from the metabolic profiling investigation, the identification of the potential biomarker is required. Experimental approaches for the structure elucidation by mass spectrometry are extensively described in the literature *(79–81)*. Often, a combination of different mass spectrometric techniques is required for the structural elucidation of unknowns. Valuable information can be obtained by FT-ICR-MS because of its ultrahigh resolution and mass accuracy, which allows derivation of molecular empirical formulae and its capability for high-resolution MS^n experiments. It also provides detailed information about molecular structure units. A less expensive alternative to FT-ICR-MS for multiple mass spectrometry (MS^n) experiments is a 3D ion trap mass analyzer. High-resolution Q-TOF instruments can be used for MS/MS experiments providing high-resolution data on fragment ions. Directly coupled chromatography-NMR spectroscopy methods can also be used. The most general of these "hyphenated" approaches is HPLC-NMR-MS *(82)*, in which the eluting HPLC peak is split, with parallel analysis by directly coupled NMR and MS techniques. This can be operated in on-flow, stopped-flow, and loop-storage modes and thus can provide the full array of NMR- and MS-based molecular identification tools.

In addition, metabolites can be identified by library search. A more detailed overview on currently available LC-MS and gas chromatography (GC)-MS libraries was recently published ***(83)***. The database is a comprehensive package of valuable resources for characterizing known and unknown metabolites. In contrast with the well-annotated gene and protein databases that can be searched easily, at present, no such comprehensive tools exist for metabolite researchers. However, current metabolite databases, although incomplete, offer a starting point for characterization. Biochemical databases can be used to identify unknown metabolites, for example, to identify structure from known elemental composition or to determine the biological function of the identified metabolites. Among the databases currently available, the most widely used are the NIST database, which includes mass spectral data for some known metabolites (http://www.nist.gov/srd/nist1.htm), as well as the KEGG, HumanCyc, ARM, and METLIN databases. The KEGG database is a valuable resource for metabonomics researchers (http://www.genome.jp/kegg/ligand.html). HumanCyc (http://biocyc.org) includes known metabolites as well as those predicted by algorithms that project metabolic pathways from a genomic sequence. A database constructed as part of the Atomic Reconstruction of Metabolism (ARM) project compiles metabolite structures together with MW and MS fragmentation data (http://www.metabolome.jp). The University of Alberta hosts a mini-library of full mass spectra of newer drugs, metabolites, and some breakdown products, (http://www.ualberta.ca/_gjones/mslib.htm). The METLIN database (http://metlin.scripps.edu/) catalogues metabolites, MS/MS spectra, and LC-MS profiles of human plasma and urine samples ***(84)***. The Spectral Database for Organic Compounds (SDBS) provides access to a wealth of spectra of organic compounds (NMR, MS, IR). Another metabolite database is the "tumor metabolome" database, established at the Justus-Liebig University Giessen in Germany (http://www.metabolic-database.com). More specific lipidomics databases exist, such as Lipid Maps (http://www.lipidmaps.org/data/index.html), SphinGOMAP (http://sphingomap.org/), and Lipid Bank (http://lipidbank.jp/index00.shtml), which contain structural and nomenclatural information as well as standard analytical protocols. General information about physicochemical properties of metabolites can be obtained by searching general chemical databases such as PubChem or CAS. To date, existing metabonomics databases aim primarily at the structural identification of metabolites in various biological samples. However, once a better annotation of the metabolome in various organisms is achieved, the generation of databases containing quantitative metabolite data can be expected. An example for this type of database is the human metabolite database that contains more than 1400 metabolites found in the human body. Each metabolite is described by a MetaboCard designed to contain chemical data, clinical data, and molecular biology/biochemistry data (http://www.hmdb.ca/) ***(85)***.

Despite the usefulness of mass spectra data, the lack of comprehensive mass spectral libraries often precludes identification of molecules based on this data alone. The combination of many technologies will be required to identify unknown metabolites in biofluids, including high-sensitivity capillary NMR, which can provide metabolite structure characterization down to the low microgram level, chemical modification for functional group identification, and finally chemical synthesis of potential candidates for verification.

3. APPLICATION

As one of our LC-MS–based metabonomics investigations, we studied the effect of the acute deterioration of liver function in chronic hepatitis B on the sera metabolic profiles to find potential biomarkers that may indicate the character of the disease ***(86)***.

Sera from 37 chronic hepatitis B patients hospitalized for acute deterioration of liver function were collected. The control group consisted of sera samples from 50 healthy individuals who all had normal liver biochemistry tests and had no evidence of diseases. After centrifugation at 3000 ×*g* for 10 min at 4°C, serum was stored –70°C until analysis. The sera were thawed before analysis. Then, 150 μL acetonitrile was added to 150 μL of the serum and shaken vigorously (30 s), and then the mixture was allowed to stand for 5 min and centrifuged at 12,000 rpm for 3 min. The supernatant was lyophilized. Serum material obtained after acetonitrile precipitation was dissolved in 150 μL 80% acetonitrile solutions. A 10-μL aliquot of the reconstructed solution was injected onto a 4.6 mm × 150 mm Zorbax Eclipse XDB-C8 5-μm column. The column was eluted with a linear gradient of 2% to 98% 0.1% formic acid in acetonitrile (B) over 0.1 to 18 min; the composition was held at 98% of B for 2 min then returned to 2% of B at 22 min; the composition was held at 2% of B for another 3 min at flow rate of 1 mL min^{-1}. The mass spectrometer was operated in positive ion mode on a tandem mass spectrometer equipped with an electrospray source after split with 1:5 split ratio. The "Enhanced MS" mode was employed to scan from 100 to 800 m/z with Q^0 trapping on and 50 ms linear ion trap (LIT) fill time. After data acquisition using HPLC-MS, the de-noising, baseline correction, and peak detection were followed. Metabolite ID Software (Applied Biosystems Instrument Co.) was employed to accomplish the de-noising and peak detection. The steps included the following: subtract average blank spectra from sample for time slices (0.2 min), combine masses into a unique mass list, calculate extracted ion chromatograms (XIC), find peaks above the MS threshold (200,000 cps, the noise ion level of the front baseline), remove isotope peaks, and remove peaks found in the blank based on XIC peak height ratio (2:1). The peak list was exported to a CSV file, which includes the information of the m/z, retention time, height, and area of the XIC peaks. Peak alignment was carried out using an in-house program written in MATLAB software (Mathworks, Natick, MA).

The 7347 peaks were found in the final reference peak list after being merged. PLS-DA and coefficient of correlation analysis were used for potential biomarker selection and identification. The score and loading plot of the PLS-DA are given in **Fig. 4**. The hepatitis group is obviously separated from the healthy control group. Components that played important roles in the separation were picked out according to the parameter VIP (variable importance in the projection). According to VIP value, the most important 20 variables (potential biomarkers) were first selected as the candidate of potential biomarkers. Then *t*-test was employed for these 20 variables; significant differences ($p < 0.05$) were found except for Var_6014 and Var_6461. A total of 18 variables were considered as the potential biomarkers (**Table 1**). A trend plot of Var_5229 is shown in **Fig. 5**. The value of the control group is obviously higher than that of the hepatitis patient groups.

For potential biomarker identification, Var_2266, which contributed to the differentiation, was given as an example. The mass spectrum and tandem mass spectrum of Var_2266 are shown in **Fig. 6**. The corresponding quasi-molecular ion is 450.3, and m/z 432.4 and 414.4 are ions of neutral loss. It could be presumed that two hydroxyl groups are in the structure. When subjected to MS/MS analysis, further information about fragmentation patterns was obtained. The database of Pubchem (pubchem.ncbi.nlm.nih.gov) and KEGG (www.kegg.com) were searched based on the clues that were obtained from the above process. As a result, the potential biomarker was tentatively identified as glycochenodeoxycholic acid (GCDCA) or its isomer glycodeoxycholic acid (GDCA). It was confirmed by comparison of tandem mass spectrum of authentic standard (GCDCA). The authentic sample of GDCA is unavailable, and it could not be removed from the candidate list. Using this method, five potential biomarkers were identified: Lysophosphatidyl Choline (LPC) C18:0, LPC C18:1, LPC C18:2, LPC C16:0, and GCDCA (or its isomer GDCA). Four

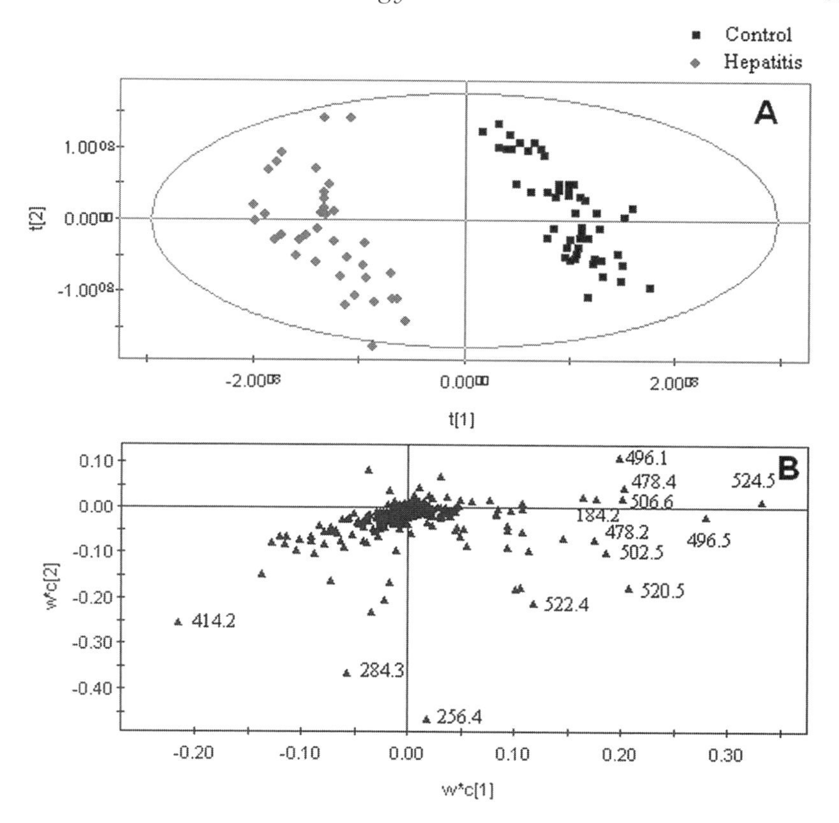

Fig. 4. **(A)** PLS-DA score plot for the first two components showing the separation between the hepatitis B group and healthy control: (♦) hepatitis patients; (■) healthy control. **(B)** PLS-DA variable loadings plot for the two first components (w*c2/w*c1) explaining the above separation between hepatitis B and healthy groups *(86)*.

of them were LPC. LPC regulates a variety of biological process including cell proliferation, tumor cell invasiveness, and inflammation. Another potential biomarker, GCDCA or GDCA, is one of the coagulated bile acids, which are helpful for the digestion of lipids. It was related to a part of the traditionally diagnostic marker CA (cholic acid)/CDCA (chenodeoxycholic acid). Compared with the case report of the samples, the same trend was found between the bile acid (traditional marker) and GCDCA. LPC levels in hepatitis patients were downregulated and GCDCA (or GDCA) levels were upregulated.

Table 1
Potential Biomarkers and Their Identification Results (*86*)

Variable No.	VIP[a]	Retention Time (min)	m/z	Identification Result
VAR_5229	41.234	17.22	524.5	LysoPC[b] C18:0
VAR_4177	33.074	15.33	496.5	LysoPC C16:0
VAR_4167	20.353	15.25	478.2	Fragment of LysoPC C16:0
VAR_5226	18.784	17.21	506.6	Fragment of LysoPC C18:0
VAR_3850	18.038	14.75	520.5	LysoPC C18:2
VAR_686	17.280	7.78	235.2	Unknown
VAR_644	16.333	7.55	235.2	Unknown
VAR_4422	16.163	15.85	522.4	LysoPC C18:1
VAR_3849	12.549	14.75	502.5	Fragment of LysoPC C18:2
VAR_2266	12.204	12.34	414.2	Fragment of GCDCA[c]
VAR_6169	11.453	19.81	282.4	Unknown
VAR_4417	10.064	15.84	504.4	Fragment of LysoPC C18:1
VAR_4022	9.7592	15.05	478.4	Fragment of LysoPC C16:0
VAR_4024	9.5824	15.05	496.1	LysoPC C16:0
VAR_4021	8.1953	15.05	184.2	Phosphatidylcholine moiety of LPC C16:0
VAR_5104	7.9564	16.89	524.4	LysoPC C18:0
VAR_4178	7.5850	15.34	479.3	Isotope compound of 478.4
VAR_741	5.628	7.99	235.3	Unknown

[a]VIP: Variable Importance in the Projection, which reflects the influence on y of every term (x_k) in the model, is the sum over all model dimensions of the contributions of variable influence.[87]

[b]LysoPC: Lysophosphatidyl Choline.

[c]GCDCA: Glycochenodeoxycholate acid.

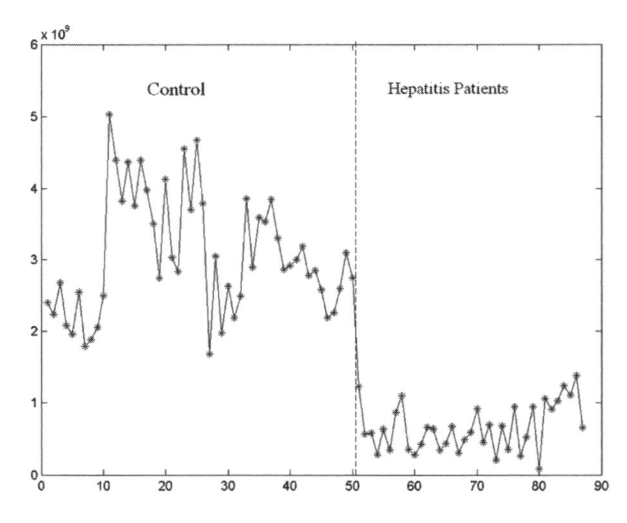

Fig. 5. Trend plot of Var_5229.

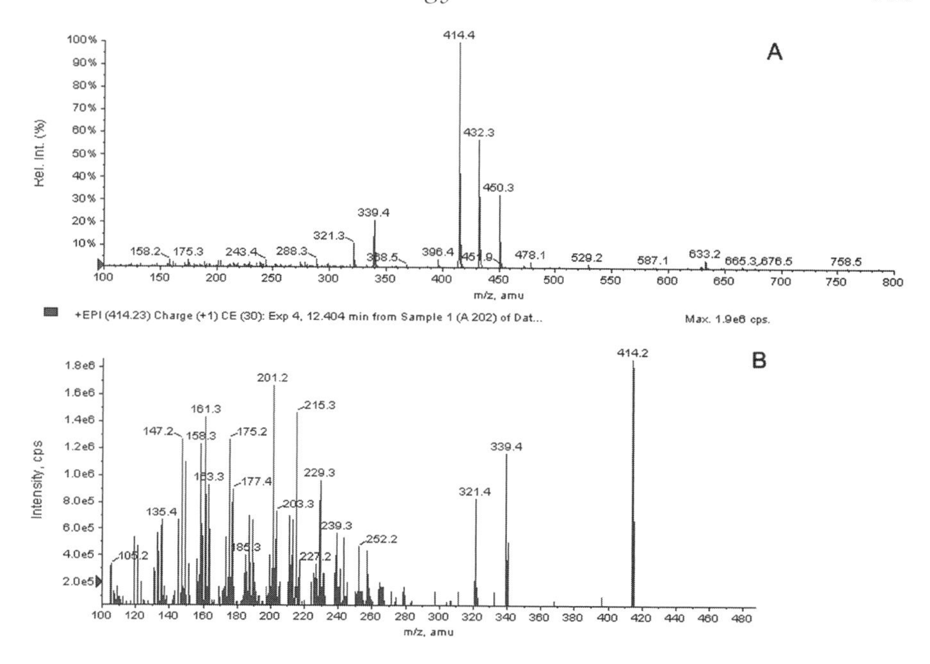

Fig. 6. IDA view of Var_2266. **(A)** Mass spectrum of Var_2266. **(B)** Tandem mass spectrum of the ion m/z 414.

4. CONCLUSION

There is no question that the HPLC-MS technique will continue as an important technology for identification and quantification of biomarkers. The application of HPLC-MS–based metabonomics analysis inevitably gave a great benefit to biomarker discovery. The studies have demonstrated considerable promise for HPLC-MS–based analytical strategies for metabonomic and biomarker research.

ACKNOWLEDGMENTS

The study has been supported by the grant (no. 20425516) for Distinguished Young Scholars and a grant (no. 20675082) from the National Natural Science Foundation of China; the Sino-German Center for Research Promotion (DFG and NSFC, GZ 364); the National Basic Research Program (2006CB503902) of the State Ministry of Science and Technology of China; and the Knowledge Innovation Program of the Chinese Academy of Sciences (KSCX1-YW-02).

REFERENCES

1. Brown SC, Kruppa G, Dasseux JL. Metabolomics applications of FT-ICR mass spectrometry. Mass Spectrom Rev 2005;24(2):223–231.
2. Nicholson JK, Lindon JC, Holmes E. 'Metabonomics': understanding the metabolic responses of living systems to pathophysiological stimuli via multivariate statistical analysis of biological NMR spectroscopic data. Xenobiotica 1999;29(11):1181–1189.
3. Lindon JC, Holmes E, Nicholson JK. Metabonomics techniques and applications to pharmaceutical research & development. Pharm Res 2006;23(6):1075–1088.
4. Lindon JC, Holmes E, Bollard ME, Stanley EG, Nicholson JK. Metabonomics technologies and their applications in physiological monitoring, drug safety assessment and disease diagnosis. Biomarkers 2004;9(1):1–31.
5. Nicholson JK, Connelly J, Lindon JC, Holmes E. Metabonomics: a platform for studying drug toxicity and gene function. Nat Rev Drug Discov 2002;1(2): 153–161.
6. Anthony ML, Sweatman BC, Beddell CR, Lindon JC, Nicholson JK. Pattern-recognition classification of the site of nephrotoxicity based on metabolic data derived from proton nuclear-magnetic-resonance spectra of urine. Mol Pharmacol 1994;46(1):199–211.
7. Anthony ML, Rose VS, Nicholson JK, Lindon JC. Classification of toxin-induced changes in H-1-NMR spectra of urine using an artificial neural-network. J Pharm Biomed Anal 1995;13(3):205–211.
8. Robertson DG, Reily MD, Sigler RE, Wells DF, Paterson DA, Braden TK. Metabonomics: evaluation of nuclear magnetic resonance (NMR) and pattern recognition technology for rapid in vivo screening of liver and kidney toxicants. Toxicol Sci 2000;57(2):326–337.
9. Brindle JT, Antti H, Holmes E, et al. Rapid and noninvasive diagnosis of the presence and severity of coronary heart disease using H-1-NMR-based metabonomics. Nat Med 2002;8(12):1439–1444.
10. Dettmer K, Aronov PA, Hammock BD. Mass spectrometry-based metabolomics. Mass Spectrom Rev 2007;26(1):51–78.
11. Plumb RS, Stumpf CL, Gorenstein MV, et al. Metabonomics: the use of electrospray mass spectrometry coupled to reversed-phase liquid chromatography shows potential for the screening of rat urine in drug development. Rapid Commun Mass Spectrom 2002;16(20):1991–1996.
12. Wilson ID, Nicholson JK, Castro-Perez J, et al. High resolution "ultra performance" liquid chromatography coupled to oa-TOF mass spectrometry as a tool for differential metabolic pathway profiling in functional genomic studies. J Proteome Res 2005;4(2):591–598.
13. Wilson ID, Plumb R, Granger J, Major H, Williams R, Lenz EA. HPLC-MS-based methods for the study of metabonomics. J Chromatogr B 2005;817(1): 67–76.
14. Griffin JL. Metabonomics: NMR spectroscopy and pattern recognition analysis of body fluids and tissues for characterisation of xenobiotic toxicity and disease diagnosis. Curr Opin Chem Biol 2003;7(5):648–654.

15. Lenz EM, Bright J, Wilson ID, Morgan SR, Nash AFP. A H-1 NMR-based metabonomic study of urine and plasma samples obtained from healthy human subjects. J Pharm Biomed Anal 2003;33(5):1103–1115.
16. Pelczer I. High-resolution NMR for metabomics. Curr Opin Drug Discov Dev 2005;8(1):127–133.
17. Dunn WB, Ellis DI. Metabolomics: current analytical platforms and methodologies. Trac-Trends Anal Chem 2005;24(4):285–294.
18. Plumb RS, Granger JH, Stumpf CL, et al. A rapid screening approach to metabonomics using UPLC and oa-TOF mass spectrometry: application to age, gender and diurnal variation in normal/Zucker obese rats and black, white and nude mice. Analyst 2005;130(6):844–849.
19. Wagner S, Scholz K, Donegan M, Burton L, Wingate J, Volkel W. Metabonomics and biomarker discovery: LC-MS metabolic profiling and constant neutral loss scanning combined with multivariate data analysis for mercapturic acid analysis. Anal Chem 2006;78(4):1296–1305.
20. Lenz EM, Bright J, Knight R, Wilson ID, Major H. A metabonomic investigation of the biochemical effects of mercuric chloride in the rat using H-1 NMR and HPLC-TOF/MS: time dependant changes in the urinary profile of endogenous metabolites as a result of nephrotoxicity. Analyst 2004;129(6):535–541.
21. Lenz EM, Bright J, Knight R, Wilson ID, Major H. Cyclosporin A-induced changes in endogenous meta-bolites in rat urine: a metabonomic investigation using high field H-1 NMR spectroscopy, HPLC-TOF/MS and chemometrics. J Pharm Biomed Anal 2004;35(3):599–608.
22. Lenz EM, Bright J, Knight R, et al. Metabonomics with H-1-NMR spectroscopy and liquid chromatography-mass spectrometry applied to the investigation of metabolic changes caused by gentamicin-induced nephrotoxicity in the rat. Biomarkers 2005;10(2–3):173–187.
23. Robertson DG. Metabonomics in toxicology: a review. Toxicol Sci 2005;85(2):809–822.
24. Williams RE, Major H, Lock EA, Lenz EM, Wilson ID. D-Serine-induced nephrotoxicity: a HPLC-TOF/MS-based metabonomics approach. Toxicology 2005;207(2):179–190.
25. Sen JW, Bergen HR, Heegaard NHH. On-line immunoaffinity-liquid chromatography-mass spectrometry for identification of amyloid disease markers in biological fluids. Anal Chem 2003;75(5):1196–1202.
26. Theobald U, Mailinger W, Reuss M, Rizzi M. In-vivo analysis of glucose-induced fast changes in yeast adenine-nucleotide pool applying a rapid sampling technique. Anal Biochem 1993;214(1):31–37.
27. Maharjan RP, Ferenci T. Global metabolite analysis: the influence of extraction methodology on metabolome profiles of Escherichia coli. Anal Biochem 2003;313(1):145–154.
28. Polson C, Sarkar P, Incledon B, Raguvaran V, Grant R. Optimization of protein precipitation based upon effectiveness of protein removal and ionization effect in liquid chromatography-tandem mass spectrometry. J Chromatogr B 2003;785(2):263–275.

29. Want EJ, Smith CA, Qin CA, VanHorne KC, Siuzdak G. Phospholipid capture combined with non-linear chromatographic correction for improved serum metabolite profiling. Metabolomics 2006;2(3):145–154.
30. Daykin CA, Foxall PJD, Connor SC, Lindon JC, Nicholson JK. The comparison of plasma deproteinization methods for the detection of low-molecular-weight metabolites by H-1 nuclear magnetic resonance spectroscopy. Anal Biochem 2002;304(2):220–230.
31. Plumb RS, Stumpf CL, Granger JH, Castro-Perez J, Haselden JN, Dear GJ. Use of liquid chromatography/time-of-flight mass spectrometry and multivariate statistical analysis shows promise for the detection of drug metabolites in biological fluids. Rapid Commun Mass Spectrom 2003;17(23):2632–2638.
32. Mullen W, Graf BA, Caldwell ST, et al. Determination of flavonol metabolites in plasma and tissues of rats by HPLC-radiocounting and tandem mass spectrometry following oral ingestion of [2-C-14]quercetin-4' glucoside. J Agric Food Chem 2002;50(23):6902–6909.
33. Weckwerth W, Wenzel K, Fiehn O. Process for the integrated extraction identification, and quantification of metabolites, proteins and RNA to reveal their co-regulation in biochemical networks. Proteomics 2004;4(1):78–83.
34. Pears MR, Cooper JD, Mitchison HM, Mortishire-Smith RJ, Pearce DA, Griffin JL. High resolution H-1 NMR-based metabolomics indicates a neurotransmitter cycling deficit in cerebral tissue from a mouse model of Batten disease. J Biol Chem 2005;280(52):42508–42514.
35. Lin CY, Wu HF, Tjeerdema RS, Viant MR. Evaluation of metabolite extraction strategies from tissue samples using NMR metabolomics. Metabolomics 2007;3(1):55–67.
36. Idborg H, Zamani L, Edlund PO, Schuppe-Koistinen I, Jacobsson SP. Metabolic fingerprinting of rat urine by LC/MS Part 1. Analysis by hydrophilic interaction liquid chromatography-electrospray ionization mass spectrometry. J Chromatogr B 2005;828(1–2):9–13.
37. Delinsky AD, Delinsky DC, Muralidhara S, Fisher JW, Bruckner JV, Bartlett MG. Analysis of dichloroacetic acid in rat blood and tissues by hydrophilic interaction liquid chromatography with tandem mass spectrometry. Rapid Commun Mass Spectrom 2005;19(8):1075–1083.
38. Pisano R, Breda M, Grassi S, James CA. Hydrophilic interaction liquid chromatography-APCI-mass spectrometry determination of 5-fluorouracil in plasma and tissues. J Pharm Biomed Anal 2005;38(4):738–745.
39. Plumb RS, Wilson ID. High throughput and high sensitivity LC/MS-OA-TOF and UPLC/TOF-MS for the identification of biomarkers of toxicity and disease using a metabonomics approach. Abstracts of Papers of the American Chemical Society 2004;228:U164-U.
40. Plumb RS, Johnson KA, Rainville P, et al. The detection of phenotypic differences in the metabolic plasma profile of three strains of Zucker rats at 20 weeks of age using ultra-performance liquid chromatography/orthogonal acceleration time-of-flight mass spectrometry. Rapid Commun Mass Spectrom 2006;20(19):2800–2806.

41. Plumb RS, Johnson KA, Rainville P, et al. UPLC/MSE; a new approach for generating molecular fragment information for biomarker structure elucidation. Rapid Commun Mass Spectrom 2006;20(13):1989–1994.
42. Yin PY, Zhao XJ, Li QR, Wang JS, Li JS, Xu GW. Metabonomics study of intestinal fistulas based on ultraperformance liquid chromatography coupled with Q-TOF mass spectrometry (UPLC/Q-TOF MS). J Proteome Res 2006;5(9):2135–2143.
43. Zhao XJ, Wang WZ, Wang JS, Yang J, Xu GW. Urinary profiling investigation of metabollites with cis-diol structure from cancer patients based on UPLC-MS and HPLC-MS as well as multivariate statistical analysis. J Separation Sci 2006;29(16):2444–2451.
44. Gaskell SJ. Electrospray: principles and practice. J Mass Spectrom 1997;32(7):677–688.
45. Huhman DV, Sumner LW. Metabolic profiling of saponins in Medicago sativa and Medicago truncatula using HPLC coupled to an electrospray ion-trap mass spectrometer. Phytochemistry 2002;59(3):347–360.
46. Enke CG. A predictive model for matrix and analyte effects in electrospray ionization of singly-charged ionic analytes. Anal Chem 1997;69(23):4885–4893.
47. Morris HR, Paxton T, Dell A, et al. High sensitivity collisionally-activated decomposition tandem mass spectrometry on a novel quadrupole/orthogonal-acceleration time-of-flight mass spectrometer. Rapid Commun Mass Spectrom 1996;10(8):889–896.
48. Hopfgartner G, Husser C, Zell M. Rapid screening and characterization of drug metabolites using a new quadrupole-linear ion trap mass spectrometer. J Mass Spectrom 2003;38(2):138–150.
49. Le Blanc JCY, Hager JW, Ilisiu AMP, Hunter C, Zhong F, Chu I. Unique scanning capabilities of a new hybrid linear ion trap mass spectrometer (Q TRAP) used for high sensitivity proteomics applications. Proteomics 2003;3(6):859–869.
50. Syka JEP, Marto JA, Bai DL, et al. Novel linear quadrupole ion trap/FT mass spectrometer: Performance characterization and use in the comparative analysis of histone H3 post-translational modifications. J Proteome Res 2004;3(3): 621–626.
51. Makarov A, Denisov E, Lange O, Horning S. Dynamic range of mass accuracy in LTQ Orbitrap hybrid mass spectrometer. J Am Soc Mass Spectrom 2006;17(7):977–982.
52. Makarov A, Denisov E, Kholomeev A, et al. Performance evaluation of a hybrid linear ion trap/orbitrap mass spectrometer. Anal Chem 2006;78(7):2113–2120.
53. Marshall AG, Hendrickson CL, Jackson GS. Fourier transform ion cyclotron resonance mass spectrometry: a primer. Mass Spectrom Rev 1998;17(1):1–35.
54. Amster IJ. Fourier transform mass spectrometry. J Mass Spectrom 1996;31(12):1325–1337.
55. Katajamaa M, Oresic M. Data processing for mass spectrometry-based metabolomics. J Chromatogr A 2007;1158:318–328.

56. Radulovic D, Jelveh S, Ryu S, et al. Informatics platform for global proteomic profiling and biomarker discovery using liquid chromatography-tandem mass spectrometry. Mol Cell Proteomics 2004;3(10):984–997.
57. Hastings CA, Norton SM, Roy S. New algorithms for processing and peak detection in liquid chromatography/mass spectrometry data. Rapid Commun Mass Spectrom 2002;16(5):462–467.
58. Wang WX, Zhou HH, Lin H, et al. Quantification of proteins and metabolites by mass spectrometry without isotopic labeling or spiked standards. Anal Chem 2003;75(18):4818–4826.
59. Jonsson P, Bruce SJ, Moritz T, et al. Extraction, interpretation and validation of information for comparing samples in metabolic LC/MS data sets. Analyst 2005;130(5):701–707.
60. Nielsen NPV, Carstensen JM, Smedsgaard J. Aligning of single and multiple wavelength chromatographic profiles for chemometric data analysis using correlation optimised warping. J Chromatogr A 1998;805(1–2):17–35.
61. Jaitly N, Monroe ME, Petyuk VA, Clauss TRW, Adkins JN, Smith RD. Robust algorithm for alignment of liquid chromatography-mass spectrometry analyses in an accurate mass and time tag data analysis pipeline. Anal Chem 2006;78(21):7397–7409.
62. Sadygov RG, Maroto FM, Huhmer AFR. ChromAlign: a two-step algorithmic procedure for time alignment of three-dimensional LC-MS chromatographic surfaces. Anal Chem 2006;78(24):8207–8217.
63. Prince JT, Marcotte EM. Chromatographic alignment of ESI-LC-MS proteomics data sets by ordered bijective interpolated warping. Anal Chem 2006;78(17):6140–6152.
64. Nordstrom A, O'Maille G, Qin C, Siuzdak G. Nonlinear data alignment for UPLC-MS and HPLC-MS based metabolomics: quantitative analysis of endogenous and exogenous metabolites in human serum. Anal Chem 2006;78(10):3289–3295.
65. America AHP, Cordewener JHG, van Geffen MHA, et al. Alignment and statistical difference analysis of complex peptide data sets generated by multidimensional LC-MS. Proteomics 2006;6(2):641–653.
66. Broeckling CD, Reddy IR, Duran AL, Zhao XC, Sumner LW. MET-IDEA: Data extraction tool for mass spectrometry-based metabolomics. Anal Chem 2006;78(13):4334–4341.
67. Katajamaa M, Miettinen J, Oresic M. MZmine: toolbox for processing and visualization of mass spectrometry based molecular profile data. Bioinformatics 2006;22(5):634–636.
68. Katajamaa M, Oresic M. Processing methods for differential analysis of LC/MS profile data. BMC Bioinformatics 2005;6:12.
69. Smith CA, Want EJ, O'Maille G, Abagyan R, Siuzdak G. XCMS: processing mass spectrometry data for metabolite profiling using nonlinear peak alignment, matching, and identification. Anal Chem 2006;78(3):779–787.

70. Yang J, Xu GW, Zheng WF, et al. Strategy for metabonomics research based on high-performance liquid chromatography and liquid chromatography coupled with tandem mass spectrometry. J Chromatogr A 2005;1084(1–2):214–221.
71. Idborg H, Zamani L, Edlund PO, Schuppe-Koistinen I, Jacobsson SP. Metabolic fingerprinting of rat urine by LC/MS Part 2. Data pretreatment methods for handling of complex data. J Chromatogr B 2005;828(1–2):14–20.
72. Scholz M, Gatzek S, Sterling A, Fiehn O, Selbig J. Metabolite fingerprinting: detecting biological features by independent component analysis. Bioinformatics 2004;20(15):2447–2454.
73. Bijlsma S, Bobeldijk L, Verheij ER, et al. Large-scale human metabolomics studies: a strategy for data (pre-) processing and validation. Anal Chem 2006;78(2):567–574.
74. Schlotterbeck G, Ross A, Dieterle F, Senn H. Metabolic profiling technologies for biomarker discovery in biomedicine and drug development. Pharmacogenomics 2006;7(7):1055–1075.
75. Martens H, Naes T, eds. Multivariate Calibration. New York: John Wiley & Sons Inc., 1989.
76. Trygg J, Wold S. Orthogonal projections to latent structures (O-PLS). J Chemometr 2002;16(3):119–128.
77. Goodacre R. Making sense of the metabolome using evolutionary computation: seeing the wood with the trees. J Exp Bot 2005;56(410):245–254.
78. Trygg J, Holmes E, Lundstedt T. Chemometrics in metabonomics. J Proteome Res 2007;6(2):469–479.
79. Prakash C, Shaffer CL, Nedderman A. Analytical strategies for identifying drug metabolites. Mass Spectrom Rev 2007;26(3):340–369.
80. Ma SG, Chowdhury SK, Alton KB. Application of mass spectrometry for metabolite identification. Curr Drug Metab 2006;7(5):503–523.
81. Nassar AEF, Talaat RE. Strategies for dealing with metabolite elucidation in drug discovery and development. Drug Discov Today 2004;9(7):317–327.
82. Williams RE, Lenz EM, Lowden JS, Rantalainen M, Wilson ID. The metabonomics of aging and development in the rat: an investigation into the effect of age on the profile of endogenous metabolites in the urine of male rats using H-1 NMR and HPLC-TOF MS. Mol Biosyst 2005;1(2):166–175.
83. Halket JM, Waterman D, Przyborowska AM, Patel RKP, Fraser PD, Bramley PM. Chemical derivatization and mass spectral libraries in metabolic profiling by GC/MS and LC/MS/MS. J Exp Bot 2005;56(410):219–243.
84. Smith CA, O'Maille G, Want EJ, et al. METLIN—a metabolite mass spectral database. Ther Drug Monit 2005;27(6):747–751.
85. Wishart DS, Tzur D, Knox C, et al. HMDB: the human metabolome database. Nucleic Acids Res 2007;35:D521–D6.
86. Yang J, Zhao XJ, Liu XL, et al. High performance liquid chromatography-mass spectrometry for metabonomics: potential biomarkers for acute deterioration of liver function in chronic hepatitis B. J Proteome Res 2006;5(3):554–561.
87. User's Guide to SIMCA-P, SIMCA-P+version 11.0. Umetrics AB: Umeå, 2005;397.

15

GC-MS-Based Metabolomics

Sally-Ann Fancy and Klaus Rumpel

Summary

Gas chromatography–mass spectrometry (GC/MS)-based metabolomics profiling methods have been developed and used for plant metabolite profiling since the 1980s. Only during the past few years has the technology been more widely used for metabolomics studies in animals and humans with the aim of toxicology and biomarker discovery, disease diagnosis and classification.

GC-MS is ideally suited and has traditionally been used for analysis of nonpolar analytes like synthetic organic compounds and hydrophobic natural products. Nonpolar compounds have to be derivatized to make them amenable to analysis by GC-MS. This means that care has to be taken to ensure that the variability inevitably introduced by this preprocessing step is kept to an absolute minimum. Over the past few years, robust sample preparation and analysis methods have been developed, which will be described in this chapter. In addition, a brief introduction to gas chromatography and mass spectrometry will be given for readers who are less familiar with these subjects.

GC-MS is distinctly different from the two other more established analytical methods that are used in metabolomics: nuclear magnetic resonance (NMR) and liquid chromatography (LC)-MS, in that it covers a unique range of analyte polarities. It has been found that the overlap between GC-MS and LC-MS metabolomic profiling data is limited because of the differences in separation and ionization mechanisms.

GC-MS has very high sensitivity and can therefore be used for the analysis of less commonly encountered types of samples that might only be available in minute amounts. Examples for this application area will also be given in this chapter.

Key Words: biomarker; derivatization; GC-MS; metabolomics

1. INTRODUCTION

Metabolite levels can be regarded as the amplified output of biological systems in response to genetic and environmental changes *(1)*. Their concentrations and composition are influenced by both internal factors (e.g., genetic makeup, state of health, age) and external factors (e.g., environment, drug treatment, diet). Metabolite analysis is an excellent starting point for

From: *Methods in Pharmacology and Toxicology: Biomarker Methods in Drug Discovery and Development*
Edited by: F. Wang © Humana Press, Totowa, NJ

non–hypothesis-based investigations into disease mechanism and diagnosis for several reasons. Firstly, compared with the fields of transcriptomics and particularly proteomics, certain experimental challenges in metabolomics seem to be less demanding: Compared with the number of mRNA and protein species that are present in an organism, the number of metabolites is significantly smaller. The number of known metabolites in many organisms has been estimated to be 10- to 100-fold lower than the number of genes or proteins *(2)*. Secondly, the integrated nature of the metabolite pool can be used to great advantage: Minor and complex changes at the transcript and protein level, which can be very difficult to detect, can still lead to major effects at the metabolite level, which can be much more amenable to detection and quantification.

However, the diverse chemical properties of metabolites—from small, highly charged to large and extremely hydrophobic—and the large dynamic range of metabolite concentrations still presents major analytical challenges in metabolite profiling. Metabolites in the cell span a concentration range of 10^9 and a log polarity range of $\sim$ –6 to 14 *(3)*. The ultimate aim of metabolomic profiling studies is to get a complete "snapshot" of the concentration of every single metabolite in an organism at a given point in time. Given that many different metabolomes exist in parallel in higher organisms (e.g., serum, urine, tissues) and that at the present day only a fraction of all metabolites have been identified, such an analysis is far beyond what is currently achievable. Another issue is throughput: For biomarker research, the value of a study comes from being able to compare larger numbers of data sets (tens to thousands) obtained from individuals under different conditions at different points in time. The above points clearly define the analytical prerequisites for metabolite profiling: An ideal method must (i) have sufficient sensitivity and an appropriate dynamic range, (ii) have adequate separation power for the complexity of the mixtures applied, (iii) have reasonable throughput, and (iv) be analytically robust (experimental variability small compared with analyte variability). Ideally, it will also allow for identification of metabolites of interest.

Gas chromatography (GC) is a separation technique that has been used for more than 50 years. Analytes are bound to the surface of a column; individual components are then sequentially eluted according to their volatility using a temperature gradient and give a response in a detector. A commonly used detection method is flame ionization detection (FID). It gives a response that is proportional to the carbon content of an analyte and is very useful for analyte quantitation. However, GC can also be used in combination with mass spectrometric detection. This has the significant advantage that in addition to the detection of the analytes as peaks in an ion chromatogram, further information can be obtained: Depending on the mass spectrometric ionization method used, this can be the mass of the analyte or a characteristic fragmentation pattern, which can be used for identification purposes.

GC is well suited for analysis of relatively hydrophobic molecules, typically synthesized by organic chemists or extracted from natural sources using organic solvents. Polar, water-soluble analytes are more problematic because they tend to bind irreversibly to the GC column material. However, they can be derivatized prior to analysis to make them more hydrophobic so that they can be eluted off the column. Given the diverse chemical nature of metabolites, it is obvious that for good coverage of the metabolome by GC, derivatization will be necessary prior to analysis.

Since the 1970s, the potential of gas chromatography–mass spectrometry (GC/MS) for analysis of metabolites in plasma and urine has been recognized, and targeted quantitative analysis for disease diagnosis was successfully established in clinical settings. More than 40 known inborn disorders of metabolism could be diagnosed by Jellum et al. as early as 1973 ***(4)***. In 1991, a landmark paper was published by Shoemaker and Elliott ***(5)*** on the automated screening of urine samples by GC-MS that describes the quantitation of well over 100 carbohydrates and organic acids in clinical samples in a single GC-MS run. More recently, the technology has been used to study a wide range of mutations of inborn errors of metabolism ***(6)***.

These studies used GC-MS in a targeted, hypothesis-driven way, focusing on a number of specific metabolites. The task in metabolomics is significantly more complex; the goal is to analyze and quantify as many metabolites as possible in a non-hypothesis driven way to generate data sets that can be used for biomarker discovery.

Metabolomics (i.e., the profiling and quantitation of metabolites in body fluids) is the most recent addition to the field of "omics" studies. NMR-based profiling has been successfully used for many years; liquid chromatography (LC)-MS–based methods are now also being widely used. More and more examples for the application of GC-MS methods for metabolite profiling have been published over the past few years, and the field is developing and evolving rapidly. Fast progress is also due to the fact that GC-MS–based plant metabolomics was established in the 1980s and 1990s, and many of the methods and learnings are now being transferred to metabolomic profiling of samples from other biological sources *(7)*.

2. GC-MS TECHNOLOGY

Gas chromatography separates molecules according to their volatility. Its first use was described in 1952 *(8)*. Separation is achieved by initially adsorbing the analytes to the surface of a GC column at slightly elevated temperature. The GC column is situated in an oven, which allows for rapid heating and cooling of the column. Once the analytes are bound, the temperature is ramped up, which drives them off the column surface in order of decreasing volatility. A carrier gas (*mobile phase*, usually helium) is used to transport the analytes along the column surface toward the detector once they have been thermally desorbed. A detailed review on the technical side of GC with relevance to its application in metabolomic profiling has been published by Kopka ***(9)***.

A brief description of the components of a GC system will be given in the following. More details can be found in Refs. ***(10)*** and ***(11)***.

2.1. Sample Introduction

The sample is vaporized in the injection system before it enters the column. There are several injection systems that can be used to introduce the sample onto the column. The heated split/splitless interface is most commonly used .

2.1.1. Split/Splitless Interface

2.1.1.1. Split Mode

To control the amount of sample reaching the column and prevent overloading, a sample splitter, controlled by the operating system software, is used to deliver only a small part of the sample injected onto the column, the remainder going to waste. High-concentration samples can easily overload the GC column, resulting in all active sites on the column becoming occupied and leading to additional analyte not being retained and as a consequence to poor chromatographic resolution.

2.1.1.2. Splitless Mode

For trace analysis, which may be more useful for targeted measurements, the injector can be used in splitless mode, which injects the entire volume of sample vaporized in the injector onto the column. For trace components close to the limit of detection of the instrument, this injection mode will increase the amount of sample introduced. As mentioned above, care has to be taken not to overload the column in this mode when analyzing more abundant components.

2.1.2. Programmed Temperature Vaporizer

An alternative to the split/splitless interface is the programmed temperature vaporizer (PTV). Samples are injected onto a cool (40ºC to 60ºC) PTV where they can be trapped and preconcentrated before the inlet is rapidly heated to ~300ºC to desorb the sample onto the column. Different liners for the PTV are available with a range of packings with different selective retention properties that can be used to trap compounds depending on the analyte of interest. Very hydrophobic analytes can bind very strongly to the liner material, which can lead to carryover between samples. A PTV can be particularly useful as a thermal desorber for volatiles or semivolatiles that may be present in biological samples only at low concentrations. By rapidly desorbing the sample onto a cool column, condensation and focusing of the analytes is improved, which reduces smearing and excessive tailing during subsequent chromatography. The risk of introducing artifacts caused by oxidation and thermal degradation of sensitive analytes is much reduced if one uses a PTV.

Direct thermodesorption (DTD) is a variation of this injection method in which the sample is introduced at 40°C and then injected onto the column by a shallow thermal gradient (4°C/min). In combination with automated liner exchange, DTD has been shown

to avoid cross-contamination and to enable nondestructive analysis of thermolabile compounds such as lipids ***(1)***.

2.2. Columns

A range of different selectivities and sizes of columns has been used for GC-MS–based metabolomic analysis. The most commonly used phase is 5% phenyl, 95% methyl siloxane. For metabolomic applications where analytes with a wide range of volatilities have to be separated, a column with the most generic selectivity should be chosen.

Column lengths can be varied depending upon the particular application with longer columns having generally a higher number of theoretical plates, (i.e., better resolution). Columns 25 to 30 m will provide the highest resolution and are available in most phases. Shorter columns, 15 to 20 m, can provide faster analysis times but will have a reduced resolving capability. For targeted measurements, higher throughput methods on short columns (5 to 10 m) can be developed to focus on one or a few components.

An important point for all capillary GC-MS work is the need to condition the column prior to running valuable samples. Sangster et al. have recommended that several quality control samples be run at the beginning of a sample batch to condition the column ***(12)***. It is absolutely essential to minimize the variability introduced through the hardware to ensure that even small differences between samples can be reproducibly detected.

2.3. Detectors

2.3.1. Flame Ionization Detection

For decades, flame ionization detection (FID) has been the detector of choice in gas chromatography. It is relatively inexpensive, has high sensitivity, a wide dynamic range, and gives a response based on the total carbon content of the analyte.

2.3.2. *Mass Spectrometry–Based Detectors*

MS-based detectors are used to great advantage in metabolomic analysis; there are many different types of mass analyzers, and the most relevant ones for coupling to GC systems will be described in the following. A good overview over all aspects of mass spectrometry can be found in Ref. ***(13)***.

2.3.2.1. Single Quadrupole

A widely used and relatively inexpensive mass analyzer is the single quadrupole. The ions move along the axis of four parallel rods to which a DC and an AC voltage are applied. These voltages affect the trajectory of ions traveling down the flight path between the rods in a way that only ions of a given m/z are transmitted at a given point in time. Compared with other instruments types, scan speeds are rather low on quadrupole instruments. Considering the very high separation power of GC with peak widths of only a few seconds, this means that it will be difficult to acquire several spectra across the width of a typical peak on a single quadrupole instrument.

2.3.2.2. Time-of-Flight

Time-of-flight (TOF) instruments are the most commonly used mass analyzers in GC-MS–based metabolomics. In this mass analyzer, the ions are accelerated in an electric field. All ions with the same charge will have the same kinetic energy but their velocity depends on their m/z ratio. The ions then enter a field-free region (*flight tube*) where they separate according to their m/z; the time it takes them to reach the detector (situated at the end of the flight tube) is proportional to their m/z.

TOF instruments have the fastest scan rate of all mass analyzers. This means that a significant number of spectra can be acquired across each peak, leading to higher sensitivity and better spectral quality.

The application of TOF instruments in GC-MS with particular emphasis on quantitation has been recently reviewed by Williamson and Bartlett ***(14)***.

2.3.2.3. Fourier Transform Ion Cyclotron Resonance Mass Spectrometry (FT-ICR-MS or FT-MS)

Fourier transform ion cyclotron resonance mass spectrometers have the highest resolving capability of all mass analyzers; the high mass accuracy data can be used to derive molecular formulae that, combined with isotope abundance information, can be a big advantage for structure elucidation of unknown metabolites ***(15)***. The use of FT-MS in direct infusion electrospray mode for metabolomics has been reported ***(16)*** but not in electron impact ionization (EI) or chemical ionization (CI) mode as yet.

2.3.2.4. Orbitrap

A more recent development is the Orbitrap instrument; it is an electrostatic ion trap using fast Fourier transform to obtain mass spectra. It provides high mass accuracy and resolution (100,000). This means that is has similar strengths to an FT-MS but at lower cost.

2.4. Ionization Modes for GC-MS Experiments

2.4.1. Electron Impact Ionization

In EI, analytes that elute from the GC column are vaporized into the ion source and collide with an electron beam. Ions produced through the loss of an electron from the molecule are then accelerated, separated by mass to charge ratio and focused onto an electron multiplier for detection. As a result of the energy which the electrons impart to the vaporized molecules, fragmentation can occur at several positions along the molecule, providing useful structural information. EI spectra are highly reproducible from instrument to instrument, and spectral libraries are available that can be used to search for the identities of unknown compounds based on m/z and intensity ratios of the observed fragment ions. A disadvantage of EI is that fragmentation is usually so efficient that the intensity of the molecular ion—giving the mass of the analyte—can be extremely low or the molecular ion can even be absent from the spectrum.

2.4.2. Chemical Ionization

For CI, a reagent gas, such as methane or ammonia, is introduced into the source of the mass spectrometer; protonated gas ions are produced as a result of collision of reagent gas with electrons from the electron beam. These ionize the analytes eluting from the column after vaporization into the ion source, provided the analyte is more basic than the protonated gas ion. Significantly less energy than in EI is transferred to the analytes, and as a result the dominant ion is usually the molecular ion. This means that the mass of the intact molecule can be easily obtained from the spectrum. Some fragmentation may be observed, but far less than in EI mode.

CI is essential for determining relative molecular masses of unknown analytes. Regardless of ionization mechanism structural elucidation and characterization of an analyte can be very complicated if the molecular mass cannot be determined.

3. DERIVATIZATION

3.1. The Need for Derivatization

Metabolites span a huge range of polarities: many metabolites in urine that have evolved to be highly soluble in order to be easily excreted are highly polar (e.g., acids and amines) and will tightly bind to typical GC columns, usually by forming hydrogen bonds with the polysiloxane-based column material. On the other hand, there are extremely nonpolar metabolites like lipids, usually found in tissues or serum, which can be analyzed in GC-MS without any derivatization.

Because the aim of GC-MS–based metabolomics is to cover as many metabolites as possible—ideally in a single GC-MS run—it is evident that it will be necessary to use a derivatization step in order to get maximal coverage of the metabolome. If the focus is on analyzing a subclass of metabolites, more targeted derivatization methods can be used. A number of good generic derivatization methods are available that can be used to get broad coverage of the metabolome. Derivatization offers several advantages: (i) the polarity of many analytes is reduced, increasing their volatility. This increases the range of components detectable by GC-MS. (ii) Chemical modification can considerably increase the thermal stability of certain analytes. A number of metabolites are known to be thermally labile and would be modified/degraded during thermal elution from the GC column if they were not derivatized prior to analysis. (iii) Derivatization can give significant improvement of chromatographic peak shape and resolution. This can be particularly important for metabolite classes that contain many structurally closely related molecules, (e.g., sugars). (iv) Derivatization can increase the intensity of observed molecular ion peaks or characteristic fragment ions, which is essential for metabolite identification.

The ideal derivatization method will be fast and reproducible, it will quantitatively convert each polar metabolite to a single derivatized species, and (considering the number of samples in typical metabolomic studies) has to be relatively inexpensive and fast. In the case of metabolomic analysis, the most commonly used generic method for derivatization is silylation. It is particularly suited for metabolite profiling studies as it will modify a wide range of functional groups commonly found in metabolites.

3.2. *Reagents for Derivatization*

In the process of silylation, a silyl group [$-Si(CH_3)_3$] replaces an active hydrogen within a molecule thus making the molecule less polar and more volatile. Active hydrogens are found in many different functional groups such as hydroxyls, thiols, amines, amides, imines, and carboxylic acids. Aldehydes and ketones exist in equilibrium with their enols and can also be derivatized (discussed later).Within a mixture of such components, the reactivity of the functional groups towards silylation is

$$\text{Alcohols} > \text{Phenols} > \text{Carboxylic Acids} \sim \text{Amines} > \text{Amides}$$

The enol forms of aldehydes and ketones have an acidic hydrogen and can therefore be derivatized to trimethylsilyl (TMS) ethers. These can be thermally and hydrolytically unstable, and it is therefore advisable to use methoxymation prior to the addition of the silylating agent to convert these functional groups to oximes or alkyloximes. If methoxymation is not carried out, incomplete derivatization can occur, resulting in multiple peaks for such components. Incomplete derivatization can make subsequent quantitation very difficult.

The most commonly used methoxymation reagent for metabolomic applications is *O*-methoxylamine hydrochloride, which is used as a pyridine solution. For silylation, the preferred reagents are BSTFA [*N,O*-bis(trimethylsilyl trifluoroacetamide] and MSTFA [*N*-methyl-*N*-(trimethylsilyl) trifluoroacetamide]. A small amount (1%) of a chlorinated silane (TMCS; trimethylchlorosilane) can be added to act as a catalyst for the silylation reaction. Derivatization conditions that are commonly used for different types of metabolomics samples will be given in **Section 4**.

In all cases, the samples must have been thoroughly dried prior to derivatization. Traces of any solvent containing active hydrogens such as water or methanol will derivatize along with the analytes of interest. During subsequent GC-MS analysis, most of these derivatized solvent components will elute at the very start of the chromatogram and can be dismissed once they have been identified and characterized. Adding an excess of the silylating agent will ensure that the analytes of interest will be quantitatively derivatized. This is also important to make sure that metabolite molecules containing more than one derivatizable group (e.g., sugars) will be converted to their fully derivatized form.

Solvents that can be used for sample derivatization include pyridine, tetrahydrofuran, acetonitrile, and dimethylformamide and are only required if the sample is insoluble in the derivatizing agent. Some derivatized components, such as amines, are readily hydrolyzed on exposure to air. To minimize hydrolysis, it is recommended to perform the derivatization reaction in sealed vials (crimp caps) from which the samples can be directly injected after derivatization.

More than one product can be produced from a single metabolite as a result of the derivatization process; *E* and *Z* geometrical isomers can be formed by methoxymation of sugars. Different metabolites may also be chemically transformed into the same analyte structure ***(9,17)***.

Only a very brief overview of the most commonly used derivatization reagents and methods has been given in this section. An excellent review on derivatization has been published by Halket and Zaikin ***(18)***. Many publications are available that describe derivatization methods that are tailored for particular classes of compounds; for a good overview, see Ref. ***(19)***. Once certain metabolites of interest have been identified in a profiling study, the use of more specific derivatization methods should be considered for potential targeted follow-up studies.

4. EXPERIMENTAL METHODS

Many metabolomic studies will be used to look for biomarker candidates for a particular disease state or a change in a metabolic profile due to a drug treatment/toxicologic effect. Two general approaches can be taken for profiling aiming at biomarker discovery: Ideally, a biomarker or toxicology marker can be measured in a matrix that is easily accessible both preclinically and clinically, plasma being the biofluid of choice. The seemingly most straightforward way to discover marker candidates is therefore the metabolic profiling of plasma. However, because plasma is a circulating representation of all body tissues and their physiologic and pathologic processes, the concentration of potential markers can be extremely low. An alternative approach is therefore to use a tissue that is most relevant for a given disease or a biofluid that is proximal to such a tissue; for example, cerebrospinal fluid (CSF) for brain-related diseases. Starting with tissue is the preferred approach if the tissue will be available later on (e.g., from biopsies) for disease diagnosis and classification; for progression to the development of plasma-based assays, it will need to be tested whether these markers can also be measured in plasma. In the following, generic protocols will be given that can be used for metabolic profiling of plasma, urine, and tissue samples. References for analysis of less commonly used biological samples will be given in a later section.

4.1. Sample Preparation

An excellent and very detailed protocol for metabolomic analysis of plasma can be found in a publication by Fiehn and Kind ***(20)***. In addition to step-by-step procedures covering the whole analysis process from sample collection to statistical analysis, it contains many useful hints and tips and is a highly recommended general source of information.

4.1.1. Plasma Samples

The first step in the profiling of plasma metabolites is the precipitation of plasma proteins by addition of organic solvent. The most commonly used methods for plasma profiling are based on a very thorough multiparameter optimization by A et al. ***(21)***. The authors investigated five different organic solvents, singly and in combination, for optimization of metabolite extraction. Derivatization conditions were also optimized. A key parameter that influences the nature of the detected metabolites is the concentration of methanol that is used in the initial extraction step. A methanol concentration of 80% was found to be an acceptable compromise to give reasonable extraction yields for both polar and nonpolar metabolites. The optimized protocol is as follows:

1. Thaw the plasma by incubation at 37°C and vortex-mix before use.
2. Add 800 μL of methanol to a mixture of 100 μL of plasma and 100 μL of water
3. Vortex-mix the solution for 10 s, keep on ice for 10 min, and then vigorously extract for 2 min.
4. Keep on ice for 120 min and centrifuge for 10 min at 4°C (19,600 × *g*).
5. Transfer a 200 μL aliquot of the supernatant to a GC vial and evaporate to dryness in a vacuum concentrator.
6. Add 30 μL of a solution of 15 mg/mL methoxylamine hydrochloride in pyridine and vortex vigorously for 10 min.
7. Incubate for 16 h at room temperature, vortex-mix again for 10 min, and then add 30 μL of MSTFA containing 1% TMCS as a catalyst.
8. Incubate for 1 h at room temperature, add 40 μL of heptane ,and use 1 μL for GC-MS analysis.

A very detailed step-by-step protocol that uses a mixture of acetone and isopropanol for precipitation can be found in Fiehn and Kind ***(20)***. Detailed instructions on the use of direct thermodesorption (DTD) in combination with automated liner exchange, which is an excellent method to ensure good data quality, particularly for lipids, can also be found in this publication. A method for profiling of serum samples can be found in Ref. ***(22)***.

4.1.2. Urine Samples

Some methods that have been used for metabolite profiling of urine [c.f. ***(23,24)***] are based on the publication by Shoemaker and Elliott ***(5)***. It contains a urease incubation step that removes urea. Urea is highly abundant in urine and can interfere with GC-MS analysis. Subsequently, proteins are precipitated with acetonitrile.

However, Kind et al. have recently published data showing that the metabolome can be altered by urease treatment ***(25)***. The authors found that the concentration of several metabolites was severely diminished by the urease treatment.

The following protocol, which omits the urease treatment and protein precipitation steps, has been adapted from Kind et al. ***(25)***:

1. Dry down 100 μL of urine (after addition of internal standards if required) in a vacuum concentrator.

2. Add 20 μL of 40 mg/mL methoxylamine hydrochloride in pyridine to the dried sample and agitate at 30°C for 30 min.
3. Add 180 μL of MSTFA and agitate the samples at 37°C for 30 min. Use a 1 μL aliquot for GC-MS analysis.

4.1.3. Tissue Samples

Extraction methods for tissue can be specifically optimized for a given target tissue. A generic method that works well in a range of tissues—heart, liver, skeletal muscle (gastrocnemius and soleus), diaphragm, and white adipose tissue from mouse—has been published by Atherton et al. ***(26)***. One hundred milligrams of tissue was used for tissues of which sufficient sample was available. It was shown that less than a quarter of metabolites could be identified in soleus tissue for which only 20 mg was available for the analysis.

The protocol is as follows:

1. Pulverize the tissue with dry ice and 600μL of a methanol-chloroform mixture (2:1).
2. Sonicate for 15 min and add 400 μL of chloroform:water (1:1).
3. Centrifuge the mixture (13,500 rpm for 20 min), collect the aqueous layer, and dry overnight in a vacuum concentrator. (At this point, the sample can be reconstituted in a larger volume of 600 μL D_2O for NMR analysis if required, and after analysis an aliquot can be dried down and used for GC-MS.)
4. If only GC-MS is to be performed, dissolve the dried sample in 120 μL of methoxylamine hydrochloride (20 mg/mL in pyridine), vortex-mix for 1 min, and leave at room temperature for 17 h.
5. Add 120 μL of MSTFA and incubate for 1 h at room temperature.
6. Dilute the derivatized sample with hexane (1:10) before GC-MS analysis.

A method for GC-MS analysis of the metabolites that were extracted into the organic phase is also given in Ref. ***(26)***. Metabolite extraction from spleen tissue is described in Ref. ***(27)***.

4.2. GC-MS Analysis

Samples should be left for at least 2 h at room temperature prior to GC-MS analysis but not be stored for longer than 2 days before analysis. Long periods of storage can result in significant amounts of polysiloxanes forming by reaction of the silylation reagent with traces of water vapor that enters the vials over time. It is also recommended to store the samples in the dark at room temperature. Make sure that the GC-MS system is in good condition, and run mass calibrations before analyzing a new batch of samples.

A good GC column for metabolomics analysis is a 35% phenyl-coated fused silica capillary column of 30 m length, 0.32 mm i.d., and 0.25-μm film thickness. It is recommended to run several QC samples at the beginning of each analysis batch to condition the column ***(12)***. Care also needs to be taken to randomize the injection sequence in order not to compromise subsequent statistical analysis.

It is very important to avoid cross-contamination of samples; if standard split/splitless liners are used, it is almost impossible to avoid this problem. The liner should be cleaned with every sample or ideally an automatic liner exchange device should be used ***(1)***.

1. Inject 1 μL onto the column with an injection temperature of 230ºC.
2. Start with a column temperature of 80ºC, 2 min isothermal, and ramp with 5ºC/min up to 330ºC, 5 min isothermal, and then cool back down to the starting conditions.
3. Detect analytes at an ion source filament energy of 70eV and a source temperature of 250ºC.
4. Scan a mass range of 83 to 500 Da; if low mass-range fragments are to be observed, use 40 to 500 Da.

5. DATA ANALYSIS

5.1. Data Processing

Once the raw data have been acquired, processing is necessary to ensure that subsequent data analysis can be performed as accurately as possible. In targeted metabolomics studies where usually just a few metabolites are analyzed, only certain regions of the chromatograms or certain m/z values in the data will be of interest. For nontargeted metabolomics analyses, the entire chromatogram is relevant, and it is important to extract information for as many peaks as possible. A number of peaks in each chromatogram are derived from compounds that are produced during sample derivatization—identifiable from analyzing blank derivatization samples—and areas of the chromatograms that contain these peaks should be excluded from the subsequent analysis. A typical data processing routine consists of several steps:

1. The first step in data processing is usually peak enumeration, which also includes extraction of lower intensity peaks from spectral noise.
2. Frequently, peaks are not made up of individual analytes but are mixtures of coeluting components. Spectra of putative pure components can be obtained from overlapping peaks by applying spectral deconvolution methods. These methods are now quite commonly used in chromatography-based metabolomics studies, and mass spectral deconvolution has been shown to increase the number of detectable peaks by a factor of three ***(28)***.
3. Retention times for the same analytes can drift between different runs. An important step in data preprocessing is therefore to match and align peaks that represent the same analyte from different samples.

4. After peak detection and alignment, some of the peaks will have none or only a few matches in other samples. Reasons for this are that a peak might not be present in a specific sample, peak detection may have failed because of noisy raw data, or that inaccurate parameter settings have been used for peak detection and chromatographic alignment ***(29)***.
5. Normalization is used to remove unwanted systematic variation in the data. It can be introduced by a change in instrument response during the course of the analysis or can be caused by the fact that some samples are more dilute than others. The latter is quite commonly found in the analysis of urine samples. The concentrations of all metabolites in a sample can be normalized to one component (either an endogenous metabolite that is estimated to be present at a relatively constant level or a compound that has been spiked into the samples). Metabolite concentrations can also be normalized to the total concentration of all endogenous metabolites.

A number of software solutions exist that can perform some or all of the above-mentioned data-processing steps. A freely available program for deconvolution is AMDIS (www.nist.gov). Practically all the major vendors of mass spectrometers offer software for data processing; however, these software packages can usually only process data that have been acquired on instruments from this vendor. As an alternative, there are other proprietary programs available that work for a variety of manufacturers' hardware, for example MetAlign (Plant Research International, Wageningen, The Netherlands). During the past few years, several public-domain programs have emerged that can process data irrespective of what type and make of instrument they have been acquired on. The most commonly used programs are XCMS ***(30)*** and MZmine ***(29)***; both programs are freely available from the Web sites mentioned in these two references. They allow data import and export irrespective of the instrumental platform that is being used. Program documentation can also be found on these Web sites. A fairly detailed description of the data-processing workflow in both XCMS and MZmine (including detailed parameter settings) has been described by Kind et al. ***(25)***. A comprehensive list of public-domain software for mass spectrometry data processing can be found at the following Web site: http://fiehnlab.ucdavis.edu/staff/kind/Metabolomics/Peak_Alignment/.

5.2. Statistical Analysis

In contrast with targeted studies where one or a few metabolites are analyzed and conventional statistical methods like the Student's *t*-test can be used for data analysis, multivariate statistical methods will need to be used for data analysis in metabolomics studies. Initially, unsupervised methods should be applied to find features that can be useful for categorization. The most commonly used approach is principal components analysis (PCA) ***(31)***. No information about the structure of the data (i.e., which group a sample belongs to) is used at the outset. PCA will look for linear combinations of individual variables that explain the most variation in the data set.

To find out which individual analytes—in our case peaks in the mass chromatogram—contribute to this separation, a loadings plot can be constructed. Unsupervised methods will find any natural partitions in a given data set; this can be parameters that are in no way related to the groups of interest (control–diseased, etc.); for example, they might find that the biggest separation in the data set is driven by the age of the subject. Therefore, it is very important wherever possible to control these factors in the design of the study (e.g., sex, age) and to collect as much information as possible on any remaining confounding factors (e.g., diet, medication).

Once the data have been analyzed by unsupervised methods, supervised methods [e.g., partial least squares (PLS) analysis ***(32)***], should be used for further evaluation. This will give an indication as to whether candidates that have been identified using an unsupervised method can be used to classify samples. The preferred approach is to split the data set into an analysis and a test set, given that the data set is big enough. Once PLS analysis has been performed on the analysis set, it can be checked whether the classifiers can correctly classify the samples in the test set. A study investigating metabolic effects of exercise in human serum with a strong focus on statistical analysis has recently been published ***(22)***.

At this stage, one will hopefully be in a situation where a number of biomarker candidates have been identified. Two major workstreams will have to be followed from this point:

(i) Validation studies, usually with much larger sample numbers, will have to be performed, and more targeted GC-MS methods can be developed to focus on the metabolites of interest.
(ii) More work usually has to be done to increase confidence in the biomarker candidates. Although peaks in a chromatogram could in principle be used as biomarkers, it is highly desirable to identify the metabolite(s) in question and to rationalize their connection to the condition under study.

5.3. Metabolite Identification and Biomarker Candidate Qualification

Only a minority of all metabolites have been identified and structurally characterized to date. One way for the identification of unknowns is to perform a database search using EI spectra. M/z values and relative ion intensities in a spectrum are matched against spectra in a reference library. The most commonly used database of EI spectra is the NIST (U.S. National Institute of Standards and Technology) database (http://www.nist.gov/srd/nist1.htm), which is commercially available from this site. A spectral library of metabolites is also maintained by the Max Planck Institute for Molecular Plant Physiology in Golm, Germany (http://csbdb.mpimp-golm.mpg.de/csbdb/gmd/gmd.html).

A database search will usually return a list of possible hits, ranked by the probability of the match. Even if a match is seemingly very good, the metabolite should not be considered as identified at this point. To aid further analysis high-resolution EI and CI spectra could be acquired—ideally on an FT-MS instrument—and the high mass accuracy (and hence empirical formula) combined with isotope abundance information can be very useful to further confirm the identity of a metabolite *(15)*. The final proof of identity should always be the comparison to EI/CI spectra and retention time obtained from analyzing a reference compound under identical analytical conditions.

Once the structures of the metabolites of interest are known, an important step is to rationalize how a change in their levels is connected to the condition under study ("qualification"). Consulting databases like the Kyoto Encyclopedia of Genes and Genomes (KEGG; available at http://www.genome.ad.jp/kegg/) can give valuable insights into how the changes might be connected to different metabolic pathways.

6. RECENT DEVELOPMENTS IN ANALYTICAL TECHNOLOGY AND SAMPLE PREPARATION

6.1. GCxGC-MS

Comprehensive two-dimensional gas chromatography–mass spectrometry (GCxGC-MS) was developed in the 1990s *(33)* and uses two consecutive GC separations to resolve significantly more analytes than GC. Two columns with different polarity—usually a nonpolar column is chosen for the first dimension and moderately polar column for the second one—are run in series; the majority of the eluent from the first column is passed on to the second column and only a small fraction is routed to a first-dimension detector. Peaks eluting from the second column are detected by MS.

The challenge is to avoid continuously transmitting analyte onto the second column, which would lead to a loss of resolution. A solution to this problem is to make the separation on the second column much faster than the separation on the first column. So-called heartcuts from the first column are trapped in a cryogenic modulation interface between the two columns and then released into the second column *(33)*; the second dimension separation is then started by initiating the temperature gradient. The number of heartcuts from the first column that can be separated on the second column is obviously limited by the analysis and recovery times of the second column. In order to get ready for trapping of the next heartcut, the entry area of the second column is rapidly cooled down again once the previous heartcut is on its way.

Typically, secondary column separations are of the order 0.1 min or less; this is achieved by using very short, narrow-bore columns and linear velocities that are several times higher than the optimum. Hence the separation in the second dimension is not optimal, but given that it only needs to handle heartcuts with relatively low complexity, it is usually sufficient. The first column is being run at normal speed.

The first metabolomics study using GCxGC-MS was published in 2005 *(27)*. The three main benefits of the GCxGC-TOF approach were found to be (i) maximum comprehensive separation obtainable with a single chromatographic run, (ii) increased average mass spectral purities, and (iii) increased sensitivity. Twelve hundred compounds could be detected in the analysis.

6.2. Analysis of Less Common and Lower Abundance Samples

GC-MS is a technology that allows for highly sensitive detection of metabolites; in typical plasma profiling experiments, the equivalent of 0.2 to 0.5 μL of plasma is used per injection, whereas in a typical LC/MS-based profiling experiment, 10 μL of plasma is required for the analysis. The inherent very high sensitivity of GC-MS makes it possible to analyze biological samples for which only small amounts can be obtained—fluids like saliva, tears, or sweat come to mind. This type of samples could be attractive because they are easily accessible. On the other hand, sample variability might be of concern because unlike plasma, the fluids mentioned above are in direct contact with the environment, and their composition can be strongly dependent on external factors; diet and the oral microflora are likely to have a strong influence on the saliva metabolome. Very interesting results have been published recently on the use of GC-MS for metabolomics analysis of human sweat *(34)*. The authors analyzed axillary sweat, urine, and saliva from 197 adults (5 samples per subject taken over 10 weeks) using a novel sampling technology, sorptive stir bar extraction *(35)*. They could show that more volatile compounds were present in sweat than in urine or saliva. Both individually distinct as well as gender-specific components could be reproducibly detected.

The inherent high sensitivity of GC-MS also allows you to work with minute amounts of sample. As mentioned before, an alternative to starting a biomarker discovery approach at the plasma level is to use tissue that is most relevant for a given disease; being able to focus on subsections of tissues directly affected by the disease should increase the chances for finding disease-specific markers even more. If one can embark on a metabolomic study aimed at biomarker discovery for a certain carcinoma using isolated cancer cells instead of a larger sample from the affected tissue, the likelihood of finding relevant markers can be significantly higher. Over the past few years, a technique called laser capture microdissection (LCM) ***(36)*** has found more and more widespread use; in LCM, tissue sections that can be stained to better visualize cells of interest are covered with a Perspex cap, which is coated with a thermolabile film. The sections are observed through a microscope, and cells of interest can be excised by tracing their outline with a laser beam. The film melts and fuses with the underlying cells so that they are retained on the Perspex cap from where they can be solubilized or extracted with appropriate solvents. LCM-derived samples are now commonly used for transcript profiling, and a number of publications have described proteomic analysis of laser capture microdissected samples; for example, a study on ~3000 laser-microdissected breast carcinoma cells ***(37)***. In a plant metabolomics study, laser capture microdissected vascular bundles of *Arabidopsis thaliana* have been analyzed, and it was shown that 66 metabolites could be identified from as few as 5000 pooled single plant cells ***(38)***.

6.3. Additional Sampling and Sample Processing Techniques for GC-MS

The sample preparation protocols that were described in **Section 6** have been developed for profiling of a broad range of metabolites from biological fluids and tissues. Because GC-MS is ideally suited for the analysis of volatile and hydrophobic components, sampling technologies specifically developed for these analytes can be used.

6.3.1. Headspace Gas Analysis

In headspace gas analysis (HGA), volatiles in the headspace above a liquid or solid are directly sampled for GC-MS analysis. In most cases, no further sample preparation is necessary because the sample can be placed neat into the sample vial for subsequent heating/shaking to promote the extraction of the volatiles of interest. It is important to stabilize the sample during the period of extraction and to make sure that all volatiles are released into the headspace for collection and analysis. Grinding or homogenizing material to increase surface area can improve yields of volatiles observed. To preserve the sample and minimize losses of volatiles, this must be carried out with liquid nitrogen, and ideally the samples will be analyzed immediately after homogenization. Disadvantages of HGA are (i) only volatile or semivolatile components will be observed giving only limited coverage of the metabolome; (ii) because the analytes of interest are volatile, sampling and analysis must be carried out as quickly and efficiently as possible to minimize losses; (iii) the analytes may only be present in very small quantities, so a preconcentration step may be required to trap sufficient material for detection.

HGA has mainly been used in plant metabolomics studies [see, e.g., ***(39)***] but has also been used for analysis of volatiles in human urine ***(40)***. A recent review on its use for screening of various biological fluids can be found in Ref. ***(41)***.

6.3.2. Solid-Phase Microextraction

Solid-phase microextraction (SPME) uses fibres that are coated with a liquid or solid extraction phase. Liquid or gaseous analytes are brought into contact with the SPME device, and after a sufficient equilibration time that needs to be determined experimentally, the device is inserted into the injection port of the GC-MS system for desorption and analysis. Samples adsorbed to SPME devices do not have to be analyzed immediately after adsorption, given they are stored under conditions where they are stable. The extraction phase can be tailored to certain classes of analytes, and the technology can be used for more targeted analyses or for sampling subclasses of a metabolome. A review on the use of SPME for analysis of biological samples can be found in Ref. ***(42)***.

7. CONCLUSION

It has been suggested that metabolomics provides the most "functional" information of all the "omics" technologies ***(43)***. As has been described above, GC-MS–based metabolomics is an excellent tool to separate, detect, and quantify large numbers of metabolites. Fairly generic and robust protocols have been developed over the past few years, and some examples for analysis of the most commonly used types of samples have been given in **Section 4**. Some limitations and advantages of GC-MS will be briefly discussed in the following.

Coverage of the metabolome: As described, the disadvantage of GC being able to separate only nonpolar analytes can be turned into an advantage by introducing an upfront derivatization step. This makes compounds with a very large range of polarities amenable to analysis by GC. In fact, a number of analytes that can be very difficult to analyze by LC-MS–based methods like amino acids can be easily measured by GC-MS. During the past few years, derivatization methods that are reasonably robust and quick and give good broad coverage of the metabolome have been developed.

For biomarker discovery, it is desirable to be able to detect as many metabolites as possible. In typical GC-MS studies, several hundred features can usually be detected. As mentioned above, the application of GCxGC technology has great potential to give better coverage of the metabolome, and the number of features that can be detected in a single run has been shown to be well over 1000 *(27)*. Considering that this technology is still far from being mainstream for metabolome analysis, considerable improvements in the ease of use and number of metabolites detected can be expected over the next few years.

Although the number of features that can be detected by LC-MS–based methods is almost twofold higher *(44)*—one of the reasons being that compounds with a molecular weight >600 are usually not volatile enough to be amenable to GC-MS analysis—the overlap between LC-MS and GC-MS with respect to the range of detectable compounds is limited due to the differences in separation and ionization mechanisms *(27)*. For maximal coverage of the metabolome, the application of both methods in parallel should therefore be considered.

Sensitivity: Very small amounts of analytes can be separated and characterized by GC-MS. This will open up interesting possibilities for research with samples that have high relevance with respect to a disease under study; for example, low milligram amounts of tumor tissue samples *(1)* or laser microdissected tumor cells.

Metabolite identification still remains a major problem: Although EI is an excellent method to generate highly reproducible fragmentation spectra, only a relatively small fraction of metabolites can be identified by searching databases like NIST, a reason being that this database has traditionally been a repository of EI spectra of synthetic organic compounds. Only over the past few years has the number of metabolite spectra started to increase. The gold standard for definitive metabolite identification is still to compare mass spectra and retention indices with those of reference compounds; if they are not commercially available, metabolite identification can turn into a major task.

Assay development and applicability in clinical settings: Once biomarker candidates have been discovered, the next steps are validation of the candidates, assay development, and ultimately its application in a clinical environment. GC-MS is robust in terms of automation and in terms of cost compares favorably with other analytical tools that are being used in the clinic; a nice example on how a GC-MS–based assay can be developed for a clinical setting can be found in Ref. *(45)*.

The scope of metabolome analysis goes far beyond the discovery of biomarkers. GC-MS–based data can be integrated with data obtained from other metabolome profiling technologies like NMR and LC-MS to try to get a deeper understanding of the molecular mechanisms underlying the changes seen in a diseased state or as a response to drug treatment. An example for such a more holistic approach can be found in Ref. ***(26)***. Scientists are even going one step further and are trying to combine the information from transcriptomics, proteomics, and metabolomics studies in a "systems biology" or "high-dimensional biology" approach to get more insights into how an organism functions and responds to external stimuli. Significant efforts in the development of new data analysis and mining tools will be necessary to ensure that as much information as possible can be extracted from the multidimensional data sets ***(46)***.

REFERENCES

1. Denkert C, Budczies J, Kind T, et al. Mass spectrometry-based metabolic profiling reveals different metabolite patterns in invasive ovarian carcinomas and ovarian borderline tumours. Cancer Res 2006;66:10795–10804.
2. Förster J, Famili I, Fu P, Palsson B, Nielsen J. Genome-scale reconstruction of the Saccharomyces cerevisiae metabolic network. Genome Res 2003;13: 244–253.
3. Griffin J. The Cinderella story of metabolic profiling: does metabolomics get to go to the functional genomics ball? Philos Trans R Soc Lond B Biol Sci 2006;361:147–161.
4. Jellum E, Stokke O, Eldjarn L. Application of gas chromatography, mass spectrometry, and computer methodsin clinical biochemistry. Anal Chem 1973;45:1099–1106.
5. Shoemaker J, Elliott W. Automated screening of urine samples for carbohydrates, organic and amino acids after treatment with urease. J Chromatogr 1991;562:125–138.
6. Kuhara T. Gas chromatographic-mass spectrometric urinary metabolome analysis to study mutations of inborn errors of metabolism. Mass Spectrom Rev 2005;24:814–827.
7. Lisec J, Schauer N, Kopka J, Willmitzer L, Fernie A. Gas chromatography mass spectrometry-based metabolite profiling in plants. Nature Protocols 2006;1: 387–396.
8. James A, Martin A. Gas-liquid partition chromatography: the separation and micro-estimation of volatile fatty acids from formic acid to dodecanoic acid. Biochem J 1952;50:679–690.
9. Kopka J. Gas chromatography mass spectrometry. In: Saito K, Dixon R, Willmitzer L, eds. Biotechnology in Agriculture and Forestry. Berlin, Heidelberg: Springer-Verlag, 2006.

10. Grob R, Barry E. Modern Practice of Gas Chromatography. 4th ed. New York: John Wiley & Sons, 2004.
11. Skoog D, Holler F, Nieman T. Gas chromatography. In: Principles of Instrumental Analysis. 5th ed. Philadelphia: Saunders College Publishing, 1998: 702–722.
12. Sangster T, Major H, Plumb R, Wilson A, Wilson I. A pragmatic and readily implemented quality control strategy for HPLC-MS and GC-MS based metabanomic analysis. The Analyst 2006;131:1075–1078.
13. De Hoffmann E, Stroobant V. Mass Spectrometry—Principles and Applications. 3rd ed. New York: John Wiley & Sons, 2007.
14. Williamson L, Bartlett M. Quantitative gas chromatography/time-of-flight mass spectrometry: a review. Biomed Chromatogr 2007;21:664–669.
15. Kind T, Fiehn O. Metabolomic database annotations via query of elemental compositions: mass accuracy is insufficient even at less than 1 ppm. BMC Bioinformatics 2006;7:234.
16. Brown S, Kruppa G, Dasseux JL. Metabolomics applications of FT-ICR mass spectrometry. Mass Spectrom Rev 2005;24:223–231.
17. Kopka J. Current challenges and developments in GC-MS based metabolite profiling technology. J Biotechnol 2006;124:312–322.
18. Halket J, Zaikin V. Derivatization in mass spectrometry – 1. Silylation. Eur J Mass Spectrom 2003;9:1–21.
19. Blau K, Halket J. Handbook of Derivatives for Chromatography. New York: John Wiley & Sons, 1993.
20. Fiehn O, Kind T. Metabolite profiling in blood plasma. In: Methods in Molecular Biology 358 (Metabolomics). Totowa, NJ: Humana Press Inc., 2007:3–17.
21. A Trygg J, Gullberg J, et al. Extraction and GC/MS analysis of the human blood plasma metabolome. Anal Chem 2005;77:8086–8094.
22. Pohjanen E, Thysell E, Jonsson P, et al. A multivariate screening strategy for investigating metabolic effects of strenuous physical exercise in human serum. J Proteome Res 2007;6:2113–2120.
23. Zhang Q, Wang G, Du Y, Zhu L, A. GC/MS analysis of rat urine for metabonomic research. J Chromatogr B 2007;854:20–25.
24. Fancy S, Beckonert O, Darbon G, et al. Gas chromatography/flame ionisation detection mass spectrometry for the detection of endogenous urine metabolites for metabonomics studies and its use as a complementary tool to nuclear magnetic resonance spectroscopy. Rapid Commun Mass Spectrom 2006;20:2271–2280.
25. Kind T, Tolstikov V, Fiehn O, Weiss R. A comprehensive urinary metabolomic approach for identifying kidney cancer. Anal Biochem 2007;363:185–195.
26. Atherton H, Bailey N, Zhang W, et al. A combined ^{1}H-NMR spectroscopy- and mass spectrometry-based metabolomic study of the PPAR-α null mutant mouse defines profound systemic changes in metabolism linked to the metabolic syndrome. Physiol Genomics 2006;27:178–186.

27. Welthagen W, Shellie R, Spranger J, Ristow M, Zimmermann R, Fiehn O. Comprehensive two-dimensional gas chromatography–time-of-flight mass spectrometry (GCxGC-TOF) for high resolution metabolomics: biomarker discovery on spleen tissue extracts of obese NZO compared to lean C57BL/6 mice. Metabolomics 2005;1:65–73.
28. Weckwerth W, Loureiro M, Wenzel K, Fiehn O. Differential metabolic networks unravel the effects of silent plant phenotypes. Proc Natl Acad Sci U S A 2004;101:7809–7814.
29. Katajamaa M, Miettinen J, Orešic M. MZmine: toolbox for processing and visualization of mass spectrometry based molecular profile data. Bioinformatics 2006;22:634–636.
30. Smith C, Want E, O'Maille G, Abagyan R, Siuzdak G. XCMS: Processing mass spectrometry data for metabolite profiling using nonlinear peak alignment, matching, and identification. Anal Chem 2006;78:779–787.
31. Wold S, Esbensen K, Geladi P. Principal components analysis. Chemom Intell Lab Syst 1987;2:37–52.
32. Wold S, Albano C, Dunn W, et al. Multivariate analysis in chemometrics. In: Kowalsi B, ed. Chemometrics: Mathematics and Statistics in Chemistry. Dordrecht, The Netherlands: D Reidel Publishing Company, 1984.
33. Phillips J, Beens J. Comprehensive two-dimensional gas chromatography: a hyphenated method with strong coupling between two dimensions. J Chromatogr A 1999;856:327–334.
34. Penn D, Oberzaucher E, Grammer K, et al. Individual and gender fingerprints in human body odour. J R Soc Interface 2007;4:331–340.
35. Soini H, Bruce K, Wiesler D, David F, Sandra P, Novotny M. Stir bar sorptive extraction: a new quantitative and comprehensive sampling technique for determination of chemical signal profiles from biological media. J Chem Ecol 2005;31:377–392.
36. Emmert-Buck M, Bonner R, Smith P, et al. Laser capture microdissection. Science 1996;274:998–1001.
37. Umar A, Luider T, Foekens J, Paša-Tolic L. NanoLC-FT-ICR MS improves proteome coverage attainable for ~3000 laser-microdissected breast carcinoma cells. Proteomics 2007;7:323–329.
38. Schad M, Mungur R, Fiehn O, Kehr J. Metabolic profiling of laser microdissected vascular bundles of Arabidopsis thaliana. Plant Methods 2005;1:1–10.
39. Tikunov Y, Verstappen F, Hall R. Metabolomic profiling of natural volatiles: headspace trapping: GC-MS. In: Weckwerth W, ed. Methods in Molecular Biology. Totowa, NJ: Humana Press Inc., 2007:39–53.
40. Wahl H, Hoffmann A, Luft D, Liebich H. Analysis of volatile organic compounds in human urine by headspace gas chromatography-mass spectrometry with a multipurpose sampler. J Chromatogr A 1999;847: 117–125.

41. Lechner M, Rieder J. Mass spectrometric profiling of low-molecular-weight volatile compounds—diagnostic potential and latest applications. Curr Med Chem 2007;14:987–995.
42. Musteata F, Pawliszyn J. In vivo sampling with solid phase microextraction. J Biochem Biophys Methods 2007;70:181–193.
43. Sumner L, Mendes P, Dixon R. Plant metabolomics: large-scale phytochemistry in the functional genomics era. Phytochemistry 2003;62:817–836.
44. Nordström A, ÓMaille G, Qin C, Siuzdak G. Nonlinear data alignment for UPLC-MS and HPLC-MS based metabolomics: Qunatitative analysis of endogenous and exogenous metabolites in human serum. Anal Chem 2006;78:3289–3295.
45. Masood A, Stark K, Salem N. A simplified and efficient method for the analysis of fatty acid methyl esters suitable for large clincial studies. J Lipid Res 2005;46:2299–3305.
46. Ekins S, Nikolsky Y, Burgrim A, Kirillov E, Nikolskaya T. Pathway mapping tools for analysis of high content data. Methods Mol Biol 2007;356:319–350.

16

NMR-Based Metabolomics for Biomarker Discovery

Narasimhamurthy Shanaiah, Shucha Zhang, M. Aruni Desilva, and Daniel Raftery

Summary

Metabolomics provides a powerful set of tools for pharmaceutical and clinical research in a number of important areas that include drug development, early disease detection, patient stratification for treatment, and information on disease processes. With its ability to discover new metabolic markers, metabolomics (as well as metabolic profiling and metabonomics) is highly effective for drug development by providing early preclinical indications of efficacy and toxicity. The most information-rich techniques currently employed in metabolomics-based studies today are nuclear magnetic resonance (NMR) spectroscopy and mass spectrometry (MS). NMR spectra of untreated biosamples provide an overview of all metabolites present, and the complete spectrum can be used as a fingerprint of metabolic status. Analysis by multivariate statistical methods is used to identify potential biomarkers of altered metabolism that can improve the understanding of the health and disease processes. Current trends and recent advances in NMR-based metabolomics are focused on the development of advanced NMR methods, improved multivariate statistical data analysis, and a number of efforts to identify altered metabolites and pathways. Applications in the areas of toxicology, inborn errors of metabolism, cardiovascular disease, and cancer detection are described, and the prospects and the future directions of the technology are highlighted.

Key Words: cancer; coronary heart disease; inborn errors of metabolism; metabolic profiling; metabolomics; metabonomics; multivariate statistical analysis; NMR; toxicology

1. INTRODUCTION

Metabolomics, and the closely related fields of metabonomics and metabolite profiling, which are commonly understood as the study of essentially all detectable small molecules contained in cells, tissue, organs, or biological fluids involved in primary or secondary metabolism, has risen

From: *Methods in Pharmacology and Toxicology: Biomarker Methods in Drug Discovery and Development*
Edited by: F. Wang © Humana Press, Totowa, NJ

in prominence over the past several years *(1–6)*. Despite the vast developments in genomics in the past two decades and more recently in proteomics, additional evidence of end-point biomarkers for disease diagnosis and more rapid evaluation of beneficial or adverse drug effects are highly desired. The parallel detection of up to hundreds of metabolites provides a key and efficient method to monitor altered biochemistry and provide actionable diagnostic information. Thus, it is reasonable to expect that biochemical profiling or metabolomics strategies will reveal information that is closely related to the current disease or therapy status. Numerous factors can affect the biology of an organism in addition to changes in gene expression or single nucleotide polymorphisms. Environmental factors such as diet, age, ethnicity, lifestyle, and gut microfloral populations have a large influence on biology, and these various factors need to be deconvoluted *(7,8)*. In addition, the concentrations of metabolites are often altered more significantly than gene expression or protein levels, making the detection of metabolite profiles a relatively sensitive measure of biological status. As has been demonstrated in numerous studies, metabolomics-based approaches are successful because disease, drugs, or toxins cause perturbations in the concentrations and fluxes of endogenous metabolites involved in a number of key cellular pathways. A number of review articles providing information on the background of the metabolomics field, its various applications and technologies, and the advantages and limitations of the metabolomics approach have appeared *(9–15)*.

The two main analytical methods used for metabolomics are nuclear magnetic resonance (NMR) spectroscopy and mass spectrometry (MS), and a number of methodologies within these approaches are currently being developed. For MS-based metabolomics, it is generally necessary to carry out a separation step, usually using liquid chromatography (LC) or chemical derivatization and gas chromatography (GC) before the MS detection *(16)*. The use of Fourier transform MS with its exceptional resolution may reduce the need for the separation step *(17)*, and newer approaches that do not require chromatography have been reported *(18)*. Moreover, MS is highly sensitive, with detection limits in the picogram range, which allows the detection of up to several thousand metabolites in serum or urine samples. However, challenges exist in MS such as the nonuniform detection caused by variable ionization efficiency. The main alternative to MS-based approaches to metabolic profiling is provided by NMR. NMR spectroscopy is especially suitable for metabolomics as it requires little or no sample preparation; is rapid, nondestructive, and noninvasive; and provides highly reproducible results. These features can provide important advantages in a number of situations. NMR spectroscopy typically have lower sensitivity compared with MS methodologies. However, with the introduction of higher field magnets (up to 900 MHz), cryogenically cooled probes (that reduce thermal noise), and microprobes equipped to handle very small (high nanoliter to low microliter) samples, methodologies that couple NMR to liquid chromatography and solid-phase extraction ***(19,20)***, sensitivity and resolution are being improved. More recently, high-resolution magic-angle spinning NMR techniques have created opportunities for the application of metabolomics to intact tissue samples ***(21)***.

Typically, ^{1}H NMR spectra of biofluids such as urine and plasma contain thousands of signals arising from hundreds of endogenous molecules representing many biochemical pathways. Conventional measurements of the major NMR signals can be used to detect biochemical changes, but the complexity of the spectra and the presence of natural biological variation across a set of samples make the use of data reduction and pattern recognition techniques highly advantageous. The use of statistical tools such as principal component analysis (PCA), hierarchical cluster analysis (HCA), stepwise linear regression (SLR), and partial least squares (PLS) in the interpretation of metabolic profiling data set is now widespread ***(7,22)***. These methodologies are extremely helpful tools for filtering the large amounts of data and for accessing the often-subtle biochemical perturbations latent in the spectra. In addition, these approaches are used to extract single biomarkers or sets of biomarkers with the best properties for the prediction of diseases, organ function, or drug toxicity and efficacy.

In this review, we focus on the recent developments in NMR-based metabolomics methods for biomarker discovery in the areas of drug development and early disease detection. We will also highlight the main methods of statistical analysis currently used for metabolic profiling. Finally, we will describe potential applications of metabolomics for identifying altered metabolites and metabolic pathways in the area of drug development (especially toxicology and drug development) and the diagnosis of human diseases such as cardiovascular disease, inborn errors of metabolism, and cancer.

2. SAMPLES: ANALYSIS OF BIOFLUIDS AND TISSUE

Metabolomics-based studies are usually conducted on biofluids such as urine, blood plasma, or serum, which can be obtained noninvasively and easily. Also when available, other natural fluids such as cerebrospinal fluid, bile or seminal fluid, and applied fluids such as those used in renal dialysis or lung aspiration can be used. It is also possible to use cell culture supernatants, tissue extracts, and similar preparations, and in special cases, as described below, intact tissue biopsy samples.

2.1. Analysis of Urine

Urine is a readily collected, information-rich biofluid that can provide insight into the metabolic state of an organism. Compared with other biofluids, the NMR analysis of urine has certain obvious advantages. The relatively low concentrations of proteins and high concentrations of low-molecular-weight compounds minimize sample preparation and result in high-quality measurements due to the narrow linewidths of the spectral peaks *(23)*. Many publications in the field of toxicology and urinary metabolite profiling using NMR allow for comparison of experimental data with archived data from the literature. This enhances the process of biomarker identification. As a result, urine is often a focus in metabolomics investigations using NMR spectroscopy in both diagnostic and monitoring applications.

2.2. Analysis of Serum/Plasma

Metabolite analysis of fluids from the circulatory system provides a view of the instantaneous metabolic state of an organism. Unlike urine analysis, which measures an organism's waste products, serum or plasma analysis measures homeostatic levels of metabolites throughout the organism. The ^{1}H NMR spectrum of a serum sample includes both sharp, narrow signals from small-molecule metabolites and broad signals from proteins and lipids. The most noticeable effect of the high protein content on the spectrum of a typical serum sample is baseline distortion. As a result, many methods of analyzing serum spectra rely on methods that can remove the effects of protein resonances, after which standard analysis techniques may be used. A common method for removing the contribution of the broad protein resonances is the use of the NMR pulse sequence CPMG (Carr-Purcell-Meiboom-Gill; see below for description) or others, which allow the resonances from larger molecules to relax before the longer-lived resonances from small molecules are detected. These CPMG or "relaxation-edited" spectra have improved baselines, and contributions from weaker signals are more clearly visible.

2.3. Analysis of Tissue

NMR data from small intact tissue samples with no pretreatment can be obtained using a technique called high-resolution ^{1}H magic angle spinning (HR-MAS) NMR spectroscopy *(24,25)*. Rapid spinning of the sample (typically ~4 to 6 kHz) at an angle of 54.7 degrees relative to the applied magnetic field will reduce the loss of information caused by line-broadening effects that result from sample heterogeneity and residual anisotropic NMR parameters. The resolution achieved by HR-MAS NMR tissue specimens and cultured cell line samples is comparable with the resolution obtained in conventional high-resolution liquid-state experiments. As the result, most pulse sequences used in HR-MAS are directly applied from liquid-state NMR without any modification but with careful consideration of the solid-like properties of some metabolites in cell or tissue samples, such as short transverse relaxation time and slow diffusion. A water suppression technique based on the combination of selective excitation pulses and pulsed field gradients as proposed by Chen et al. in the acquisition of HR-MAS NMR spectra of tissue specimens and cell samples enables efficient water suppression for intact cells and tissue samples and eliminates signal loss from cellular metabolites *(26)*.

3. NMR METHODS USED IN METABOLOMICS

3.1. Standard 1D NMR Methods

NMR metabolomics data are typically obtained from the single pulse or 1D NOESY (nuclear Overhauser spectroscopy) sequences because of their simplicity, reproducibility, reliability, and quantitative accuracy. In addition, due to the enormous signal intensity of the proton signal from water, most methods of acquiring NMR spectra of aqueous samples such as urine or serum involve some form of water suppression. A number of NMR pulse sequences are available to suppress the water signal. The most straightforward approach is to saturate the water signal using a long, low-power presaturation pulse, or PRESAT. PRESAT can be added to other NMR sequences, such as 1D NOESY, which tends to give better suppression and flatter spectral baselines. More advanced sequences, such as WATERGATE (water suppression by gradient tailored excitation), excitation sculpting, and others can improve suppression, although they often attenuate solute peaks as well *(27,28)*. As a result, estimating the concentration of a compound that has peaks close to the attenuated water signal can give artificially low results. An improved water saturation based on presaturation has recently been developed to reduce the residual water signal and to maintain good quantitation *(29)*.

3.2. CPMG Experiment

For serum or tissue samples, which contain large-molecular-weight species such as proteins, lipoproteins, or phospholipids, broad peaks will be detected, along with some narrow peaks from lower-molecular-weight metabolites. In such cases, the CPMG experiment *(30)* can be used to attenuate the broad signals resulting from macromolecules, which relax during the spin echo pulse sequence and are greatly diminished in intensity. Although the pulse sequence parameters should be properly optimized for each sample type, the sequence is generally robust and has been widely used in a number of studies to date. NMR spectra of serum, urine, and bile fluids are shown in **Fig. 1** obtained using the PRESAT and CPMG pulse sequences.

3.3. 2D NMR Methods

It is well-known that 2D NMR experiments improve the ability to interpret spectra because of the higher resolution provided by dispersing the spectral peaks. However, 2D NMR methods have not been widely used in metabolomics to date because of their increased acquisition time, data size, and complexity in data analysis. Nevertheless, a small but growing number of papers report using 2D approaches in metabolomics studies ***(31–34)***. The most commonly used experiments include correlation spectroscopy (COSY), total correlation spectroscopy (TOCSY), 2D-J spectroscopy, and heteronuclear single quantum coherence (HSQC) spectroscopy. With modern NMR spectrometers, these experiments are readily accessible.

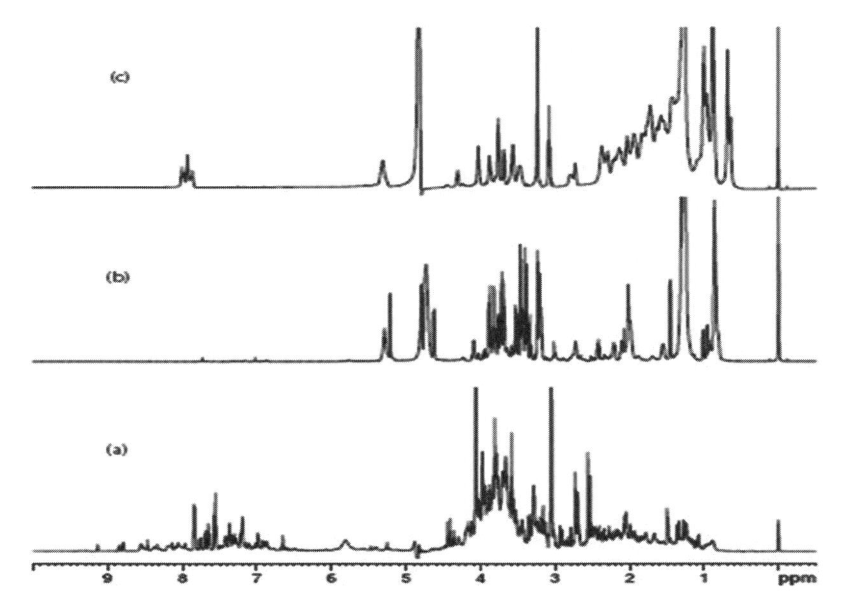

Fig. 1. ^{1}H NMR spectra of **(A)** control urine with water presaturation using 1D NOESY, **(B)** CPMG of control blood serum, and **(C)** control bile sample with water presaturation.

3.4. 2D J-Resolved Spectroscopy

J-resolved (JRES) spectroscopy is a simple homonuclear 2D experiment in which J-coupling information is removed from the first spectral dimension and separated into a second dimension. The 1D projection of the 2D spectrum retains only chemical shift information, leading to a substantial simplification of the spectra. This technique has been successfully applied to many body fluids including cerebrospinal fluid (CSF), seminal body fluid, blood plasma, and urine *(35)*. The integral of signals in chemical shift dimension is strongly influenced by T_2 relaxation and hence only a relative quantification of concentration of metabolites is possible. Recently, Viant et al. used ^{1}H-JRES spectra to monitor the metabolic developmental trajectory of embryogenesis and to succinctly describe changes in the NMR-visible metabolome for examining toxicity *(36)*.

3.5. Selective TOCSY and Its Application to Biofluid Analysis

Although the use of single pulse (with presaturation) or NOESY sequences is powerful and ubiquitous in NMR-based metabolomics, a key limitation is that such NMR experiments necessarily observe metabolite signals that are potentially highly overlapped. This problem is compounded because it is almost always the case that a given biofluid will contain a relatively few number of species present at high concentration levels that will dominate the statistical analysis. Scaling methods such as Pareto scaling or logarithmic scaling can be used to change the relative emphasis of large and small peaks *(37)*, however this can still be insufficient for very similar samples or those in which uninteresting metabolites have large changes in concentration. It therefore is important that NMR methods be developed to enhance the sensitivity of detection of low concentrated metabolites for metabolomics applications. With this view, Sandusky and Raftery *(38,39)* developed a 1D selective TOCSY approach that detects metabolites quantitatively even if they are found at concentrations 10 to 100 times below those of the major components (**Fig. 2**). Using this approach, a selected subset of potentially important biomarkers in body fluids can be targeted to improve their sensitivity dramatically. For example, urine samples spiked with 250 μM isoleucine (an amount similar to that found in newborn babies with maple syrup urine disease) are not easily distinguished from normal urine (**Fig. 2A, B**). However, selective TOCSY easily differentiates these spectra as seen in **Fig. 2C, D**.

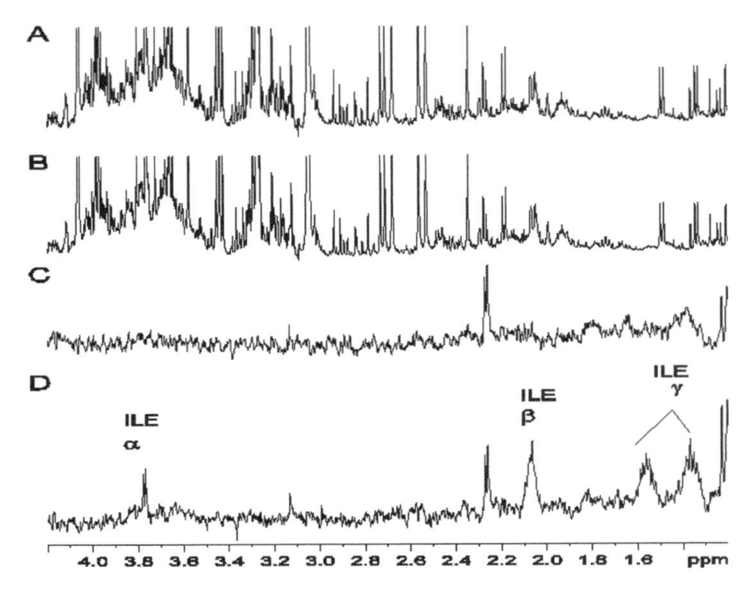

Fig. 2. ^{1}H NMR of (**A**) normal urine and (**B**) urine spiked with 0.25 mM isoleucine. ^{1}H selective TOCSY NMR of (**C**) normal urine and (**D**) urine spiked with 0.25 mM isoleucine. The selective TOCSY experiment is much more sensitive to the altered metabolite concentration.

3.6. Chemical Derivatization Methods and ^{13}C NMR

^{13}C NMR can also serve as a useful alternative to ^{1}H NMR for identifying and quantifying metabolites. However, ^{13}C NMR suffers from poor sensitivity due to the low natural abundance (~1.1%) and low gyromagnetic ratio of ^{13}C nuclei and therefore requires unacceptably long data acquisition times. Hence, the application of natural-abundance ^{13}C NMR to metabolic profiling has been limited ***(40,41)***. Although the use of cryogenically cooled probes that enhances the sensitivity of natural-abundance ^{13}C NMR may be potentially useful for metabolomics analysis on a timescale suitable for routine experiments ***(42)***, substantial gains in sensitivity are still needed for routine applications of ^{13}C NMR to metabolomics-based biomarker discovery. ^{13}C isotope labeling provides a potentially useful strategy to improve sensitivity and resolution for NMR-based metabolomics. Recently, Shanaiah et al. have shown that an isotope-labeled acetylation reaction can be carried out directly in aqueous solution at ambient temperature and thus provide an alternative approach to the analysis of complex mixtures ***(43)***. In general, this approach consists of labeling specific classes of metabolites with easily observed, isotopically enriched reactant species under physiologic pH. This is especially attractive for complex mixtures such as urine, serum, or other biofluids when combined with sensitivity-improved 2D inverse detected (^1H-^{13}C) heteronuclear experiment to yield spectra with good signal-to-noise ratios with reasonable acquisition times.

The derivatization approach results in a much simplified ^{13}C spectrum in 1D and an enhanced sensitivity for 2D spectra when compared with the normal 1D ^1H spectrum. Shown in **Fig. 3** is the 2D HSQC spectrum of derivatized urine from a patient with the inborn error of metabolism, phenylketonuria (PKU). Well-resolved peaks from a number of amine-containing metabolites are easily visible and differentiated from the spectrum of normal urine *(43)*. Hence, the use of metabolite derivatization and ^{13}C NMR spectroscopy produces data suitable for metabolite profiling analysis of biofluids on a timescale that allows routine use. The improved, quantitative detection of low-concentration metabolites by NMR creates

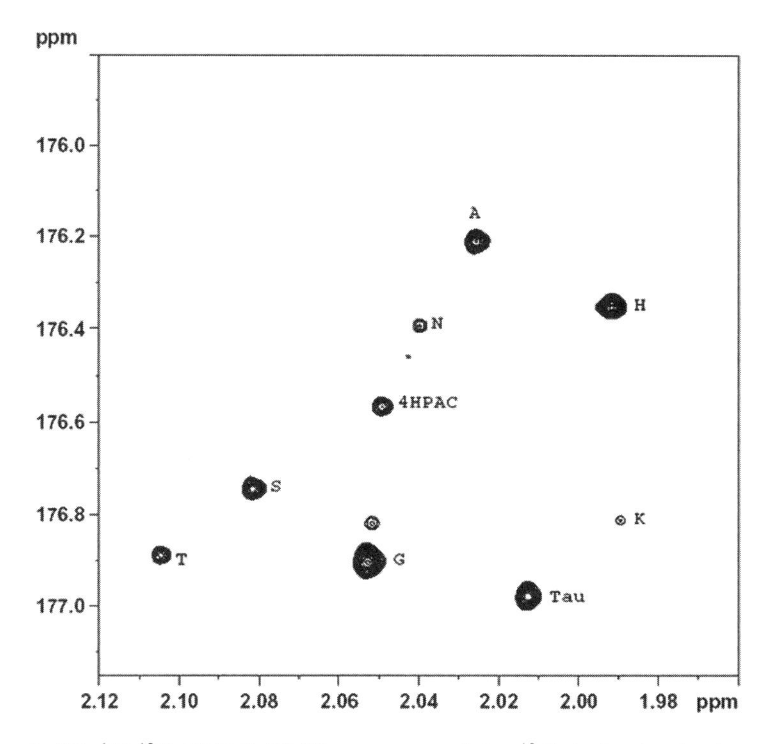

Fig. 3. 2D ^1H-^{13}C HSQC NMR spectrum of the ^{13}C derivatized urine sample from a patient with PKU metabolic disorder. The spectrum shows the labeled amino acids and other identified metabolite signals, typical for the disease. Tau, taurine; 4HPAC, 4-hydroxy-phenylacetate.

opportunities to improve the understanding of a number of biological processes. In particular, the ability to follow the quantitative changes of a class of metabolites such as amino acids across a variety of metabolic pathways will be highly useful for the metabolic network analysis. Combinations of derivatization, sensitivity-enhanced NMR experiments such as 2D HSQC, and pattern-recognition methods such as PCA would be potentially very useful in biomarker discovery for a number of diseases.

4. DATA PREPROCESSING

After data acquisition, the application of various preprocessing steps such as baseline correction, alignment and subsequent binning, scaling or data normalization is used to enhance the specific class-related information and minimize the effects of irrelevant factors.

Baseline correction is used to reduce the effect of any nonideal offsets in individual spectra. Typically, it is more difficult to correct the baseline around the water spectral region because water suppression readily distorts the baseline. In particular, low-abundance metabolites that have small peaks in the NMR spectrum are more prone to baseline artifacts than are high-abundance metabolites. Careful baseline correction can be made using a variety of commercially available software and ranges from constant and linear phase corrections to higher-order polynomial fitting. Good baseline correction is a necessary antecedent for the discovery of true biomarkers.

Statistical analysis of spectral data sets requires each spectral peak (or variable) be compared throughout all observations (samples). Misalignment will jeopardize the construction of an appropriate model, which readily produces incorrect metabolic patterns and erroneous biomarkers. Paralleled NMR spectra can be first aligned using a chemical shift reference such as TSP (trimethylsilyl-propionic acid-D_4, sodium salt) or similar compound to account for any frequency shift. Detailed alignment can be focused on individual peaks based on metabolite features such as NMR chemical shifts, peak shape, multiplicity or linewidth, and so forth. Autoalignment available in metabolomic data analysis packages such as KnowItAll (BioRad, Philadelphia, PA) allows rapid analysis of biosamples with improved alignment. However, even sophisticated alignment procedures are not perfect as small shifts due to pH, ion concentration, and complexation effects often remain. Alternative approaches, such as the addition of ethylenediamine tetraacetic acid (EDTA), which may reduce these dependencies further, are also being developed ***(44)***.

Frequency binning normally follows spectral alignment. Binning reduces the original high-resolution NMR spectral data to lower resolution by averaging all intensity values of each spectral region within a defined bucket (e.g., 0.04 ppm). Binning reduces the NMR spectral peak misalignments. At the extremes, binning size does affect statistical pattern recognition and thus biomarker exploration, although moderate binning is advantageous, and the full-resolution NMR spectrum is recoverable if necessary for metabolite identification. Historically, the bucket widths were set to equal values. This approach is often insufficient because biosamples provide a huge variety of metabolites showing diverse peak shapes and linewidths. Variable binning reduces the problems associated with peaks splitting into multiple frequency bins or buckets, thus each binned value is comparable with the integral of the corresponding peak.

Additional preprocessing steps include scaling, which allows the emphasis of smaller-concentration metabolites. A number of scaling parameters are used, including variance scaling (division by the standard deviation of the peak intensities across the set of spectra) and Pareto scaling (division by the square root of the standard deviations). Log scaling has been used to reduce the size of very large and dominant peaks. The data are then typically mean centered by subtracting the average of all the spectra and normalized to reduce any dependence on overall concentration differences. This last step is particularly important for urine samples, which can vary by a factor of 10 or more in overall intensity.

5. STATISTICAL ANALYSIS

Because of the complexity of the NMR spectral data, pattern recognition tools such as multivariate statistical analysis are used to reduce the dimensionality of the data. This combination of high-resolution NMR data with multivariate statistical analysis allows the identification of potential biomarkers in biological specimens and their associated and specific metabolic pathways or networks associated with disease ***(18,45–47)***. Both unsupervised and supervised techniques can be used to derive metabolic profiles. Unsupervised methods such as PCA and HCA require no prior classification information. Such analyses enable easy visualization of data classification in terms of clusters and identification of spectral features or potential biomarkers responsible for the classification.

Supervised techniques use the information of class membership, such as disease/healthy status, to classify a given data set. Normally, a supervised analysis is applied to a training set of samples, and a second set is tested to validate the statistical model developed using the training set. Cross-validation ***(48)*** is often used in this approach. Supervised techniques can be appropriate to either force classification (such as in determining which metabolites distinguish between groups) or regress a pattern against a trend (such as correlating a temporal progression with metabolic changes). Methods for supervised pattern recognition include prediction to latent structures through PLS, orthogonal signal correction (OSC), genetic programming, and neural networks. In general, it is extremely important to validate the findings of PCA, PLS, or other methods using extensive cross-validation and even a second set of samples (preferably blinded and from a second location). Ultimately, biological validation involving a disease hypothesis will likely be required before acceptance by a large community is achieved.

5.1. Principal Component Analysis

As an unsupervised method, PCA has been the most extensively used multivariate statistical method in the metabolomics field. Principal components (PCs) are a set of orthogonal vectors that describe the variance of the spectra data and are linear combinations of that original data. The first PC explains the largest amount of variance, and subsequent PCs explain progressively less variance. The original data matrix is expressed as the product of a scores matrix and the transpose of a loadings matrix. Scores are used for sample classification, and loadings locate variables accounting for maximal amounts of variation. Therefore, PCA allows one to differentiate samples based on scores and to seek key components from loadings. Potential biomarkers are normally proposed by focusing on resonance regions showing large loadings (high variance) on the PC along which classes are best separated. Independence of any prior knowledge of class identity makes PCA one of most preferable methods in metabolomics as any clusters observed in PCA score plots are in general considered as naturally separated classes. To date, PCA has been widely used in recognizing the differences among samples with variable features such as gender *(49)*, ethnicity, diet *(46,50)*, stress *(51)*, toxicity (see **Section 6.1**), and disease *(52)*. However, a large variance does not guarantee a relevant biomarker. Biological validation and additional statistical validation are normally required for correct biomarker identification.

5.2. Hierarchical Cluster Analysis

HCA aims to define natural clusters based on comparing distances between pairs of samples (or variables): small distances between samples imply that the samples (biological samples, urine, blood, and tissue) share similar metabolite contents representing similar physiologic properties, dietary habits or disease grades, and so forth. HCA represents analytical results in the format of a dendrogram and facilitates the visualization of different categories with a given similarity level *(53)*. HCA can also be conveniently used for both sample clustering and variable clustering, similar to all cluster analysis. In biomarker discovery, HCA is usually used as a supporting method to more powerful methods such as PCA in order to target key individual metabolites or spectral region that most correlate with the class membership. As an example proton NMR spectroscopy of sera coupled with PCA and HCA has been successfully used in discriminating 120 serum samples into three baseline clusters and two treatment clusters to detect variations in the metabolism of lipids resulting from statin treatments *(54)*.

For discriminative biomarker discovery, PCA and HCA can be combined to construct a robust and interpretable framework for unbiased mechanistic screening *(55)*.Clusters achieved by PCA can be cross-validated from the detailed description of clustering in HCA *(55)*. Both PCA and HCA are sensitive enough to recognize spectral differences detected by NMR.

5.3. Partial Least Squares

As an extension to PCA regression, PLS seeks to derive latent variables (analogous to PCs) by maximizing the covariation between the measured data (X), such as spectral intensity values, and the response variables (Y), such as measurement response (e.g., toxic effects) and disease presence or grade. In the case of diseases (cancer, heart disease, etc.) studied by metabolomics-based approaches, a special form of PLS, partial least squares–discriminant analysis (PLS-DA), is often applied, in which each class is assigned a dummy value 1 or 0, representing disease or normal. PLS-DA has potential advantages over PCA for the purpose of identifying significant biomarkers as the coordinates of its defined latent variables can point more directly to the desired class separation than those found by PCA.

5.4. Orthogonal Signal Correction

A preprocessing step to PLS-DA, OSC works to remove orthogonal variations to the class of interest. Because of its powerful attributes, it has received increased emphasis in metabolomics studies. OSC can filter essentially any confounding factors that obscure interesting biological variation. For example, morning urine contains higher levels of creatine, hippurate, trimethylamine, succinate, citrate, and 2-oxo-glutarate, and evening urine contains lower levels of trimethylamine *N*-oxide, taurine, spermine, and 3-hydroxy-iso-valerate *(56)*. The diurnal-related metabolic variation can be removed by OSC when the purpose is to focus on animal strain differences *(56)*. Variations in gender, diet, strain, age, and weight, or analytical variation such as instrumental drift, are confusing parameters in any particular disease biomarker discovery process. Because NMR is very reproducible, many confusing factors can be removed by conducting very careful experiments. Nevertheless, it is usually hard to obtain an absolutely "clean" sample set due to the extreme complexity of biological samples, especially in the case of human subjects. OSC, used to enhance the focus on specific class-related information while minimizing all other unrelated biological variation, is thus attractive *(57)*. In principle, data analysis using OSC-edited PCA or PLS-DA facilitates more targeted biomarker discovery than solely using PCA or PLS-DA. Again, care must be taken when using powerful statistical methods such as OSC.

In parallel with the development of these highly useful and flexible statistical tools, the field of metabolomics has benefited an increasing number of options for statistical software. Some software packages such as Pirouette (Infometrix Inc., Woodinville, WA), Minitab (Minitab Inc., State College, PA), MATLAB (The Math Works Inc., Natick, MA), and KnowitAll (BioRad) have improved to a great extent such that only a minimal training is needed to run advanced multivariate statistical analyses on the data. Nevertheless, the data interpretation, for example the identification of interesting metabolites as putative biomarkers, often requires a deep background knowledge and sophisticated skills in both statistics and analytical techniques. Data interpretation can be biased by many factors including experimental design and preprocessing (types of baseline correction, binning, scaling, and normalization chosen) as well as the type of statistical model chosen. More advanced statistical methods and easily accessible software packages are being produced to meet the demands of challenging metabolomics studies involving very complex biological system. The expansion to the techniques of PLS and O-PLS from PCA exemplifies such an evolution in the field.

5.5. Statistical Methods to Improve NMR Spectra

Recently, a highly useful method for identifying multiple NMR peaks from the same molecule in a complex mixture based on the concept of statistical total correlation spectroscopy (STOCSY) has been introduced. Statistical TOCSY converts a set of ^{1}H NMR spectra to a pseudo two-dimensional NMR spectrum by displaying the correlations among various spectral peak intensities. Similar to traditional TOCSY, statistical TOCSY is capable of finding multiple peaks from the same metabolite. Additionally, it can find two or more metabolites involved in the same metabolic pathway, showing correlation or anticorrelation. For example, a constructed statistical TOCSY spectrum using urine samples from three mouse strains successfully displayed metabolites (e.g., 2-oxoglutarate, citrate, 3-hydroxyphenylpropionate) identified from the traditional 2D

TOCSY NMR spectrum of a single urine sample ***(58)***. The statistical TOCSY spectrum also displayed the correlation between methylamine and dimethylamine and that between dimethylamine and trimethylamine.

The approach is not limited to NMR spectra alone and can be extended to other forms of data. The co-analysis of both NMR and mass spectra by this method, known as statistical heterospectroscopy (SHY), is a valuable tool for metabonomic studies ***(59)***. In contrast with statistical TOCSY, SHY produces a two-dimensional spectrum with one axis corresponding with the NMR chemical shift and the second axis corresponding with mass to charge ratio from MS. An application of NMR and ultra performance liquid chromatography (UPLC)-MS–based SHY has efficiently identified key metabolites such as creatine, 2-aminoadipate, hippurate, and spermine from hydrazine-treated urine sample ***(59)***. The SHY approach is of general applicability to complex mixture analysis if two or more independent spectroscopic data sets are available for any sample cohort.

6. APPLICATIONS

Given the wealth of information accessible by metabolomics-based methods and the powerful spectroscopic methods and advanced statistical packages that are currently available, there has been an explosion of applications in the field. Several important examples are chosen to represent some applications of current interest.

6.1. Toxicology

With increasing demands to reduce the time and costs of drug development, one of the most important aims of research and development in the pharmaceutical industry is the selection of robust drug candidate based on the minimization of adverse effects. NMR-based metabonomics promises to provide information on *in vivo* toxicity and efficacy in all stages of drug discovery and development and has been well documented by several reviews ***(5,6,10,11,60–62)***.

NMR-based metabonomic toxicology studies have employed ^{1}H NMR of biofluids, especially urine before and after the administration of a wide range of toxins. The toxin-induced deviation from normal metabolite profile can be measured efficiently using multivariate statistical methods. Predictive statistical models have been constructed to assess toxicologic effects on three levels. The first level is simply to distinguish whether a sample is normal (i.e., whether it belongs to a control population). The second level involves classifying samples by target organ toxicity, regions with organs or biochemical mechanisms, with a view to predicting the toxicity of novel pharmacological compounds. The final level is to identify the spectral regions that are responsible for the deviation from the normal profile and to determine the biomarkers of toxicity within those regions, as this might help to elucidate mechanisms of toxicity *(7)*.

Because the response to toxic insult is dynamic, biofluid profiles are in a constant state of flux, and the timescale of metabolic responses is also characteristic for specific toxins. Because of the noninvasive nature of the NMR biofluid analysis, time course evaluations can be readily conducted, providing an assessment of toxic change from the onset, through evolution and regression. The information that is derived from databases of NMR spectra can be maximized using appropriate chemometric and multivariate analytical strategies. In a study of organ-specific toxicity, Robertson et al. ***(63)*** have investigated two hepatotoxins (carbon tetrachloride and α-naphthylisothiocyanate) and two nephrotoxins (2-bromo-ethylamine and 4-aminophenol). They analyzed ^{1}H NMR data of urine samples from rats before treatment and 1 to 4 days after treatment by PCA and compared their data with clinical chemistry indices and histology. A 3D map of the first three PCs depicts clusters of data points in localized regions for each of the four toxins; these are clearly displaced from the data for untreated control rats. PCA data were also more consistent with regard to temporal toxicity than either clinical chemistry or pathology. These studies demonstrate how a 3D model of metabolic space (a metabolic map) can be devised and used to predict target organ toxicity. Such techniques may serve as an *in vivo* screening method for assessment of toxicity.

The usefulness of NMR-based metabonomics for the evaluation of xenobiotic toxicity effects has recently been comprehensively explored by the Consortium for Metabonomic Toxicology (COMET) project, a union of several pharmaceutical companies and Imperial College (London, UK) ***(64)***. This group developed new methodologies for analyzing and classifying the complex data sets and constructed predictive and informative models of toxicity using NMR-based metabonomic data in order to delineate the time course of toxicity. The group curated databases of spectral (~35,000 NMR spectra) and conventional results (clinical chemistry, histopathology, etc.) for 147 model toxins and treatments that serve as the basis for computer-based expert systems for toxicity prediction.

6.2. Clinical Biomarkers

Many examples exist in the literature on the use of NMR-based metabolic profiling to aid human disease diagnosis, such as the investigation of diabetes using plasma and urine, neurologic conditions such as Alzheimer's disease using cerebrospinal fluid, arthritis using synovial fluid, and male infertility using seminal fluid. In addition, analysis of urine has been used in the investigation of drug overdose, renal transplantation, and various renal diseases. A number of clinical applications of metabonomics have been reviewed previously *(65–67)*, and NMR spectroscopy of urine and plasma has been used extensively for the diagnosis of inborn errors of metabolism in children *(68)* as discussed in more detail in **Section 6.3**. More recent studies include cerebrospinal fluid sample analysis using NMR spectroscopy to distinguish various types of meningitis infection (bacterial, viral and fungal) *(69)* and an investigation of subarachnoid hemorrhage *(70)*.

Tissue samples can be studied using metabolomics through MAS techniques, and examples of this approach include the study of prostate cancer *(71)*, breast cancer *(72)*, and various brain tumors. One recent study has investigated mouse models of various cardiac diseases [including Duchenne muscular dystrophy (DMD), cardiac arrhythmia, and cardiac hypertrophy] in which metabolic profiling of cardiac tissue through HR-MAS ^{1}H NMR spectroscopy was combined with multivariate statistical analysis. It was shown that although the mouse strain was a major component of the mouse phenotype, it was possible to discover underlying profiles characteristic of each abnormality *(73)*. A study for assessing intact liver tissue during the whole process of human liver transplantation using HR-MAS ^{1}H NMR spectroscopy by Durate et al. enabled a determination of liver metabolic profiles before removal from donors and after implant into recipients. Glycero-phosphocholine (GPC) was proposed as new biomarker for liver function *(74)*.

An interesting study by Sharma et al. used NMR spectroscopy to quantify metabolites in seminal plasma from subjects injected with a new male contraceptive called RISUG (a copolymer of styrene maleic anhydride dissolved in DMSO), and in seminal plasma *(75)*. No significant difference in the concentration of citrate was observed between the groups, indicating that the prostate was not affected by RISUG. The citrate:lactate and GPC:choline ratios were significantly lower ($p < 0.01$) in subjects injected with RISUG compared with controls, which was interpreted as indicating the occurrence of partial obstructive azoospermia. The study reported that intervention with RISUG in vas deferens for as long as 8 years was safe and did not lead to prostatic diseases.

For personalized health care, an individual's drug treatments must be tailored so as to achieve maximal efficacy and avoid adverse drug reactions. One of the approaches has been to understand the genetic makeup of different individuals (pharmacogenomics) and to relate these differences to their varying abilities to handle pharmaceuticals both for their beneficial effects and for identifying adverse effects. Very recently, an alternative approach to understanding such intersubject variability in response to drug treatment using a combination of predose metabolite profiling and chemometrics has been developed *(76)* and termed *pharamacometabonomics*. Unlike pharmacogenomics, this approach is sensitive to both the genetic and modifying environmental influences that determine the metabolic fingerprint of an individual. This new approach has been illustrated with studies of the toxicity and metabolism of compounds with very different modes of action (allyl alcohol, galactosamine, and acetaminophen) that were administered to rats.

6.3. Inborn Errors of Metabolism

Inborn errors of metabolism (IEM) form a considerable group of genetic diseases. The majority are due to defects of single genes that code for specific enzymes. In most of the disorders, problems arise due to accumulation of a substances that are toxic or interfere with normal function. The diagnosis of IEM may rely on the detection of abnormal accumulation of specific metabolite, new metabolites that are normally not present, or the decreased concentration of a metabolite that is always present in body fluids such as urine, plasma, serum, or cerebrospinal fluid. ^{1}H NMR spectroscopy has been successfully applied to many IEM *(68,77)*. Compared with existing GC-MS, ion exchange chromatography, and HPLC methods, ^{1}H NMR spectroscopy is essentially nonselective and provides an overview of proton-containing metabolites and allows simultaneous quantification of many metabolites over a large concentration range in a short time frame. NMR can therefore be considered as an alternative analytical approach for diagnosing known and unknown IEM. Recently, metabolomics-based approaches have been applied to IEM detection *(78–80)*. In a paper by Constantinou et al., IEMs were detected using ^{1}H NMR of urine *(80)*. The authors detected PKU and maple syrup urine disease (MSUD). A second paper by the same group showed that the NMR approach could be applied to the analysis of blood spots, as is currently used for newborn screening *(78)*. Pan et al. studied six different IEMs using NMR and MS-based metabolomics and showed how the two techniques could be combined to improved biomarker identification *(47)*. NMR PCA data from this study are shown in **Fig. 4**. The metabolomics-based approach appears to be useful in identifying additional potential biomarkers of IEMs that could be used to identify borderline cases or for subclassifying the disease. The IEMs consisting of amino acid disorders, carbohydrate-related disorders, lysosomal diseases, organic acidurias, lipid metabolism disorders, purine/pyrimidines disorders, vitamin-related disorders, and other studied using NMR spectroscopy have been reviewed extensively *(81,82)*.

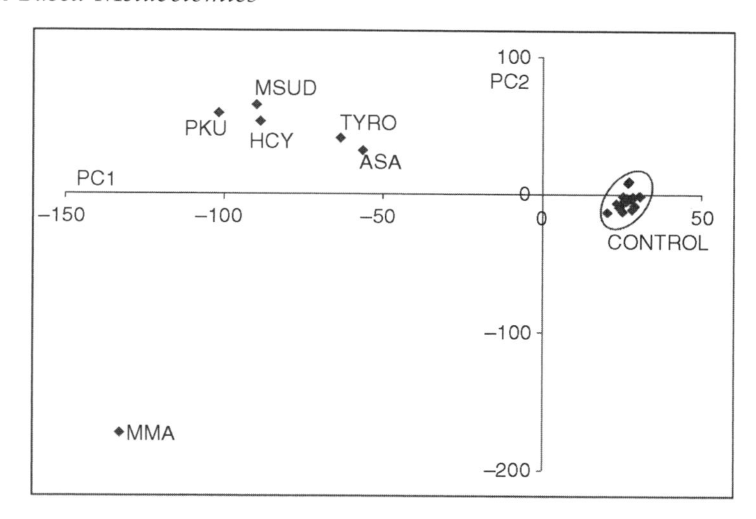

Fig. 4. PCA of ^{1}H NMR spectra of urine samples from patients with six different IEMs along with healthy control urine samples.

6.4. Cardiovascular Disease

Coronary heart disease (CHD) is the biggest cause of morbidity and mortality in developed countries, affecting as many as one in three individuals before the age of 70 years *(83)*. Currently the accurate diagnosis of CHD can only be made through application of angiography, which is both expensive and invasive. Identifying the individuals who will suffer a myocardial infarction as a result of CHD remains an important limiting factor in the delivery of clinical care in the cardiovascular arena. As a result, the early diagnosis of CHD is an appealing target for the development of new metabolic profiling diagnostics. Currently, C-reactive protein as well as the NMR-based lipoprofile *(84,85)* are used to help diagnose patients at risk for CHD. Brindle et al. have performed a pilot study to investigate the diagnostic utility of a metabolic profile obtained by using ^{1}H NMR spectroscopy *(86)*. In this study, they applied multivariate statistical analysis to 600-MHz NMR spectra of serum from individuals with no evidence of stenosis (normal coronary arteries; NCA), or severe CHD defined as at least 50% of stenosis (triple vessel disease; TVD), that were extensively characterized by angiography and for a wide range of conventional risk factors. Supervised PLS-DA following application of a single round of OSC revealed a signature completely separating the NCA and TVD groups. Although the peaks around 1.30 ppm are known to result from lipid CH_2 resonances and to correlate with the levels of low-density lipoprotein (LDL)-cholesterol, subtle chemical differences in lipid composition between TVD and NCA individuals seems to be particularly important *(86)*. It is suggestive of the degree of fatty acid side-chain unsaturation, which is largely not distinguishable by 1D NMR spectroscopy. These findings are also consistent with those of Otvos and co-workers, who have used NMR to characterize lipoproteins profiles in individuals with and without CHD *(84,85)*. However, a follow-up study by Kirschenlohr et al. reported a lower diagnostic accuracy, indicating again that metabolomic studies need to be made on large numbers of patients and with extreme caution as to the sample treatment and data analysis *(87)*. In another study, Brindle et al. have used ^{1}H NMR–derived human serum metabolic profiles to search for signatures associated with hypertension *(88)*. This analysis can clearly distinguish low/normal systolic blood pressure (SBP) serum samples from borderline and high SBP samples. However, borderline and high SBP samples could not be distinguished from one another.

6.5. Cancer

The clinical procedure in modern oncology can be characterized by one of the following categories of disease management: screening, diagnosis, or therapy. As cancer is typically detected radiographically and often in late stage in which therapy options are limited, there is high demand for alternative, earlier, and chemically based detection modalities. With the creation of new testing tools like magnetic resonance spectroscopy (MRS), HR-MAS, and ^{1}H NMR spectroscopy, there is a high potential for early diagnosis and even therapy management using highly accurate, chemically based disease markers, if they can be developed.

Early indications are that such signals exist and can be detected. HR-MAS and MRS studies show that increased levels of phospholipids (phosphocholine) and decreased levels of citrate are the hallmarks of prostate cancer *(89)*. A NMR method that can be used for metabolomics-based cancer pathology measurement with intact tissue is high-resolution MRS. In general, this approach is capable of analyzing cellular metabolites and tissue histopathology from the same sample. A wide range of metabolites have been shown to be useful in distinguishing tumors from healthy tissue and in monitoring cellular activities such as cell-cycle progression or apoptosis by MRS *(90)*. One of the first applications of HR-MAS ^{1}H NMR spectroscopy was to distinguish between normal lymph nodes and those that contained malignant cells *(91)*. Information about the metabolic environment of the tumor can also be obtained by this technique, which can be used to identify metabolites with a range of physical properties. These approaches have also been used to follow the effects of various therapeutic agents on tumor cells *in vitro* and *in vivo*. Metabolomics-based approaches have been used to study the function of hypoxia-inducible factor 1β in tumor growth and shown that this transcription factor is involved in increasing glucose metabolism, rather than inducing angiogenesis, in hepatomas ***(92)***. *In vivo* studies have shown that MRS can be used to identify tumor types, especially brain tumors, by their metabolic profiles ***(93)***. However, MRS is not feasible for large-scale population screening because of its high cost. Therefore, it is desirable to develop biofluid metabolomics-based methods.

In fact, the early diagnosis of cancer through analysis of biofluids such as blood plasma by NMR-based metabolomics appears promising. Odunsi and colleagues used ^{1}H NMR spectroscopy coupled with multivariate statistical analysis to distinguish women with epithelial ovarian cancer (EOC) from healthy women ***(52)***. In a study of 103 patients, ^{1}H NMR spectra of blood serum from 38 patients with EOC prior to surgery, 12 patients with benign ovarian cysts, and 53 healthy women (both pre- and postmenopausal) were obtained. After data reduction, the authors used both PCA and soft independent modeling of class analogy (SIMCA) for pattern recognition. In addition, they analyzed the regions of NMR spectra that most strongly influence separation of sera of EOC patients from healthy controls. The PCA analysis produced 100% separation of all 38 EOC patient serum samples from the premenopausal normal samples and from patients with benign ovarian disease. In addition, it was possible to separate cancer samples from postmenopausal control sera with 97% accuracy. However, the putative biomarkers consisted of a nonspecific signal emanating from the lipid region and 3-hydroxybutarate, which is often seen in the serum from control patients as well. Additional studies on larger numbers of patients will have to be made to verify these findings.

7. CONCLUSION

NMR-based metabolomics offers exciting opportunities in biomarker discovery and disease diagnostics and clearly has the potential to make a significant impact in toxicology and pharmaceutical development. In both screening and mechanistic exploration, NMR-based metabolic profiling can offer rapid, noninvasive toxicologic information that is robust and reproducible with little or no added technical resources to existing *in vivo* studies. New technologic advances will also improve the information that can be obtained using NMR. Increased automation will allow the rapid generation of metabolomic databases to assist in patient screening. One significant challenge is to validate metabolomics findings in large and prospective, well-controlled clinical studies of diverse patients across multiple institutions. Another challenge is the integration of biochemical, genetic, clinical, and various "omics" data to better understand organisms and disease states. This integrative, systems biology approach is perhaps the most exciting prospect of the field for the future. The final challenge is how to implement these data into clinical practice and make early detection, therapy monitoring, and faster drug development a reality.

REFERENCES

1. Nicholson JK, Wilson ID. High resolution proton magnetic resonance spectroscopy of biological fluids. Prog NMR Spectrosc 1989;21:449–501.
2. Shockcor JP, Holmes E. Metabonomic applications in toxicity screening and disease diagnosis. Curr Topics Med Chem 2002;2:35–51.
3. Nicholson JK, Wilson ID. Understanding 'global' systems biology: metabonomics and the continuum of metabolism. Nat Rev Drug Discov 2003;2: 668–676.
4. Lindon JC, Holmes E, Nicholson JK. Metabonomics and its role in drug development and disease diagnosis. Expert Rev Mol Diagn 2004;4:189–199.
5. Robertson DG. Metabonomics in toxicology: a review. Toxicol Sci. 2005;85; 809–822.
6. Lindon JC, Holmes E, Nicholson JK. Metabonomics in pharmaceutical R & D. FEBS J 2007;274:1140–1151.
7. Nicholson JK, Connelly J, Lindon JC, Holmes E. Metabonomics:a platform for studying drug toxicity and gene function. Nat Rev Drug Discov 2002;1: 153–161.
8. Nicholson JK, Lindon JC, Holmes E. 'Metabonomics': understanding the metabolic responses of living systems to pathophysiological stimuli via multivariate statistical analysis of biological NMR spectroscopic data. Xenobiotica 1999;29:1181–1189.
9. Lindon JC, Holmes E, Nicholson JK. Pattern recognition methods and application in biomedical magnetic resonance. Prog NMR Spectrosc 2001;39:1–40
10. Griffin JL. Metabonomics:NMR spectroscopy and pattern recognition analysis of body fluids and tissues for characterisation of xenobiotic toxicity and disease diagnosis. Curr Opin Chem Biol 2003;7:648–654.
11. Lindon JC, Holmes E, Nicholson JK. Toxicological applications of magnetic resonance. Prog NMR Spectrosc 2004;45:109–143.
12. Harrigan GG, Goodacre R, eds. Metabolic Profiling: Its Role in Biomarker Discovery and Gene Functional Analysis. Boston: Kluwer Academic, 2003:1–8.
13. Fiehn O. Metabolomics-the link between genotype and phenotype. Plant Mol Biol 2002;48:155–171
14. Kell DB. Metabolomics and systems biology: Making sense of the soup. Curr Opin Microbiol 2004;7:296–307.
15. Robertson DG, Reily MD, Baker JD. Metabonomics in pharmaceutical discovery and development. J Proteome Res 2007;6:526–539.
16. Nissen WMA. Liquid Chromatography-Mass Spectrometry. New York: Marcel Dekker Inc., 1999.

17. Asamoto B. FT-ICR/MS: Analytical Applications of Fourier Transform Ion Cyclotron Resonance Mass Spectroscopy. New York: VCH Weinheim, 1991.
18. Chen HW, Pan Z, Talaty N, Cooks RG, Raftery D. Combining desorption electrospray ionization mass spectrometry and nuclear magnetic resonance for differential metabolomics without sample preparation. Rapid Commun Mass Spectrom 2006;20:1577–1584.
19. Wasim M, Hassan MS, Brereton RG. Evaluation of chemometric methods for determining the number and position of components in high-performance liquid chromatography detected by diode array detector by diode array detector and on-flow ^{1}H nuclear magnetic resonance spectroscopy. Analyst 2003;128: 1082–1090.
20. Lindon JC. HPLC-NMR-MS:past, present and future. Drug Discov Today 2003;8:1021–1022.
21. Moka D, Vorreuther R, Schicha H, Spraul M, Humpfer E, Lipinski M, et al. Biochemical classification of kidney carcinoma biopsy samples using magic angle spinning ^{1}H nuclear magnetic resonance spectroscopy. J Pharm Biomed Anal 1998;17:125–132.
22. Fernie AR, Trethewey RN, Krotzky AJ, Willmitzer L. Metabolite profiling:from diagnostics to systems biology. Nat Rev Mol Cell Biol 2004;5:763–769.
23. Neild GH, Foxall PJ, Lindon JC, Holmes EC, Nicholson JK. Uroscopy in the 21st century: high-field NMR spectroscopy. Nephrol Dial Transplant 1997;12: 404–417.
24. Tomlins AM, Foxall PJD, Lindon JC, Nicholson JK, Lynch MJ, Spraul M, Everett JR. High-resolution magic angle spinning ^{1}H nuclear magnetic resonance analysis of intact prostatic hyperplastic and tumour tissues. Anal Commun 1998;35:113–115.
25. Cheng LL, Chang IW, Louis DN, Gonzalez RZ. Correlation of high-resolution magic angle spinning proton magnetic resonance spectroscopy with histopathology of intact human brain tumor specimens. Cancer Res 1998;58:1825–1832.
26. Chen J-H, Sambol EB, Kennealey PT, O'Connor RB, DeCarolis PL, Cory DG, Singer S. Water suppression without signal loss in HR-MAS ^{1}H NMR of cells and tissues. J Magn Reson 2004;171:143–150.
27. Liu M, Mao X-a, Ye C, Huang H, Nicholson JK, Lindon JC. Improved WATERGATE pulse sequences for solvent suppression in NMR spectroscopy. J. Magn Reson 1998;132:125–129.
28. Hwang TL, Shaka AJ. Water suppression that works—excitation sculpting using arbitrary wave-forms and pulsed-field gradients. J Magn Reson A 1995;112:275–279.
29. Mo H, Raftery D. Pre-SAT180, a simple and effective faraway water suppression method. J Magn Reson 2008;190:1–6.
30. Meiboom S, Gill D. Modified spin-echo method for measuring nuclear relaxation time. Rev Sci Instrum 1958;29:688–691.

31. Dumas M-E, Canlet C, Andre F, Vercauteren J, Paris A. Metabonomic assessment of physiological disruptions using ^{1}H-^{13}C HMBC NMR spectroscopy combined with pattern recognition procedures performed on filtered variables. Anal Chem 2002;74:2261–2273.
32. Tang H, Wang Y, Nicholson JK, Lindon JC. Use of relaxation-edited one-dimensional and two-dimensional nuclear magnetic resonance spectroscopy to improve detection of small metabolites in blood plasma. Anal Biochem 2004;325:260–272.
33. Dumas M-E, Canlet C, Vercauteren J, Andre F, Paris A. Homeostatic signature of anabolic steroids in cattle using ^{1}H-^{13}C HMBC NMR metabonomics. J Proteome Res 2005;4:1493–1502.
34. Xi Y, de Ropp JS, Viant MR, Woodruff DL, Yu P. Automated screening for metabolites in complex mixtures using 2D COSY NMR spectroscopy. Metabolomics 2006;2:221–233.
35. Holmes E, Foxall PJD, Spraul M, Farrant RD, Nicholson JK, Lindon JC. 750 MHz ^{1}H NMR spectroscopy characterisation of the complex metabolic pattern of urine from patients with inborn errors of metabolism: 2-hydroxyglutaric aciduria and maple syrup urine disease. J Pharm Biomed Anal 1997;15: 1647–1659
36. Viant MR. Improved methods for the acquisition and interpretation of NMR metabolomic data. Biochem Biophys Res Commun 2003;310:943–948.
37. Ramadan Z, Jacobs D, Grigorov M, Kochhar S. Metabolic profiling using principal component analysis, discriminant partial least squares, and genetic algorithms. Talanta 2006;68:1683–1691.
38. Sandusky P, Raftery D. Use of selective TOCSY NMR experiments for quantifying minor components in complex mixtures: application to the metabonomics of amino acids in honey. Anal Chem 2005;77:2455–2463.
39. Sandusky P, Raftery D. Use of semiselective TOCSY and the Pearson Correlation for the metabonomic analysis of biofluid mixtures: application to urine. Anal Chem 2005;77:7717–7723.
40. Moolenaar SH, Poggi-Bach J, Engelke UFH, Corstiaensen JMB, Heerschap A, de Jong JGN, et al. Defect in dimethylglycine dehydrogenase, a new inborn error of metabolism: NMR spectroscopy study. Clin Chem 1999;45:459–464.
41. Krawczyk H, Gryff-Keller A, Gradowska W, Duran M, Pronicka E. ^{13}C NMR spectroscopy:a convenient tool for detection of argininosuccinic aciduria. J Pharmaceut Biomed Anal 2001;26:401–408.
42. Keun HC, Beckonert O, Griffin JL, Richter C, Moskau D, Lindon JC, Nicholson JK. Cryogenic probe ^{13}C NMR spectroscopy of urine for metabonomic studies. Anal Chem 2002;74:4588–4593.
43. Shanaiah N, Desilva MA, Gowda GAN, Raftery MA, Hainline BE, Raftery D. Class selection of amino acid metabolites in body fluids using chemical derivatization and their enhanced ^{13}C NMR. Natl Acad Sci USA 2007;104:11540–11544.
44. Asiago V, Gowda GAN, Zhang S, Clark J, Raftery D. Minimizing ionic strength and pH dependent frequency shifts in ^{1}H NMR spectra of urine unpublished results.

45. Crockford DJ, Holmes E, Lindon JC, Plumb RS, Zirah S, Bruce SJ, et al. Statistical heterospectroscopy, an approach to the integrated analysis of NMR and UPLC-MS data sets: application in metabonomic toxicology studies. Anal Chem 2006;78:363–371.
46. Gu H, Chen H, Pan Z, Jackson AU, Talaty N, Xi B, et al. Monitoring diet effects via biofluids and their implications for metabolomics studies. Anal Chem 2007;79:89–97.
47. Pan Z, Gu H, Talaty N, Chen H, Shanaiah N, Hainline BE, et al. Principal component analysis of urine metabolites detected by NMR and DESI–MS in patients with inborn errors of metabolism. Anal Bioanal Chem 2007;387:539–549.
48. Johnson RA, Wichern DW. Applied Multivariate Statistical Analysis. 4th ed., New Jersey: Prentice Hall, 1998.
49. Hodson MP, Dear GJ, Roberts AD, Haylock CL, Ball RJ, Plumb RS, et al. A gender-specific discriminator in Sprague–Dawley rat urine: the deployment of a metabolic profiling strategy for biomarker discovery and identification. Anal Biochem 2007;362:182–192.
50. Stella C, Beckwith-Hall B, Cloarec O, Holmes E, Lindon JC, Powell J, et al. Susceptibility of human metabolic phenotypes to dietary modulation. J Proteome Res 2006;5:2780–2788.
51. Wang Y, Holmes E, Tang H, Lindon JC, Sprenger N, Turini ME, et al. Experimental metabonomic model of dietary variation and stress interactions. J Proteome Res 2006;5:1535–1542.
52. Odunsi K, Wollman RM, Ambrosone CB, Hutson A, McCann SE, Tammela J, et al. Detection of epithelial ovarian cancer using ^{1}HNMR-based metabonomics. Int J Cancer 2005;113:782–788.
53. Beckonert O, Bollard ME, Ebbels TMD, Keun HC, Antti H, Holmes E, et al. NMR-based metabonomic toxicity classification:hierarchical cluster analysis and k-nearest-neighbour approaches. Anal Chim Acta 2003;490:3–15.
54. Le Moyec L, Valensi P, Charniot JC, Hantz E, Albertini JP. Serum ^{1}H nuclear magnetic spectroscopy followed by principal component analysis and hierarchical cluster analysis to demonstrate effects of statins on hyperlipidemic patients. NMR Biomed 2005;18:421–429.
55. Pohjanen E, Thysell E, Lindberg J, Schuppe-Koistinen I, Moritz T, Jonsson P, Antti H. Statistical multivariate metabolite profiling for aiding biomarker pattern detection and mechanistic interpretations in GC/MS based metabolomics. Metabolomics 2006;2:257–268.
56. Gavaghan CL, Wilson ID, Nicholson JK. Physiological variation in metabolic phenotyping and functional genomic studies:use of orthogonal signal correction and PLS-DA. FEBS Lett 2002;530:191–196.
57. Dumas M-E, Barton RH, Toye A, Cloarec O, Blancher C, Rothwell A, et al. Metabolic profiling reveals a contribution of gut microbiota to fatty liver phenotype in insulin-resistant mice. Proc Natl Acad Sci USA 2006;103: 12511–12516.
58. Cloarec O, Dumas M-E, Craig A, Barton RH, Trygg J, Hudson J, et al. Statistical total correlation spectroscopy: an exploratory approach for latent biomarker

identification from metabolic ^{1}H NMR data sets. Anal Chem 2005;77: 1282–1289.

59. Crockford DJ, Holmes E, Lindon JC, Plumb RS, Zirah S, Bruce SJ, et al. Statistical heterospectroscopy, an approach to the integrated analysis of NMR and UPLC-MS data sets: Application in metabonomic toxicology studies. Anal Chem 2006;78:363–371.
60. Holmes E, Nicholls AW, Lindon JC, Connor SC, Connelly JC, Haselden JN, et al. Chemometric models for toxicity classification based on NMR spectra of biofluids. Chem Res Toxicol 2000;13:471–478.
61. Reo NV. NMR based metabolomics. Drug Chem Toxicol 2002;25:375–382.
62. Griffin JL, Bollard ME. Metabonomics: its potential as a tool in toxicology for safety assessment and data integration. Curr Drug Metab 2004;5:389–398.
63. Robertson DG, Reily MD, Sigler RE, Wells DF, Paterson DA, Braden TK. Metabonomics: Evaluation of nuclear magnetic resonance (NMR) and pattern recognition technology for rapid in vivo screening of liver and kidney toxicants. Toxicol Sci 2000;57:326–337.
64. Lindon JC, Keun HC, Ebbels TM, Pearce JM, Holmes E, Nicholson JK. The Consortium for Metabonomic Toxicology (COMET): aims, activities and achievements. Pharmacogenomics 2005;6:691–699.
65. Reily MD, Robertson DG, Delnomdedieu M, Baker JD. High resolution NMR of biological fluids: metabonomics applications in pharmaceutical research and development. Am Pharm Rev 2003;6:105–109.
66. Lindon JC, Holmes E, Bollard ME, Stanley EG, Nicholson JK. Metabonomics technologies and their applications in physiological monitoring, drug safety assessment and disease diagnosis. Biomarkers 2004;9:1–31.
67. van der Greef J, Stroobant P, van der Heijden R. The role of analytical sciences medical systems biology. Curr Opin Chem Biol 2004;8:559–565.
68. Moolenaar SH, Engelke UFH, Wevers RA. Proton nuclear magnetic resonance spectroscopy of body fluids in the field of inborn errors of metabolism. Ann Clin Biochem 2003;40:16–24.
69. Coen M, O'Sullivan M, Bubb WA, Kuchel PW, Sorrell T. Proton nuclear magnetic resonance-based metabonomics for rapid diagnosis of meningitis and ventriculitis Clin Infect Dis 2005;41:1582–1590.
70. Dunne VG, Bhattachayya S, Besser M, Rae C, Griffin JL. Metabolites from cerebrospinal fluid in aneurysmal subarachnoid haemorrhage correlate with vasospasm and clinical outcome: a pattern-recognition ^{1}H- NMR study. NMR Biomed 2005;18:24–33.
71. Swanson MG, Vigneron DB, Tabatabai ZL, Males RG, Schmitt L, Carroll PR, et al. Proton HR-MAS spectroscopy and quantitative pathologic analysis of MRI/3D-MRSI-targeted postsurgical prostate tissues. Magn Reson Med 2003:50:944–954.
72. Sitter B, Sonnewald U, Spraul M, Fjosne HE, Gribbestad IS. High-resolution magic angle spinning MRS of breast cancer tissue. NMR Biomed 2002;15: 327–337.

73. Jones GLAH, Sang E, Goddard C, Mortishire-Smith RJ, Sweatman BC, Haselden JN, et al. A functional analysis of mouse models of cardiac disease through metabolic profiling. J Biol Chem 2005;280:7530–7539.
74. Duarte IF, Stanley EG, Holmes E, Lindon JC, Gil AM, Tang HR, et al. Metabolic assessment of human liver transplants from biopsy samples at the donor and recipient stages using high-resolution magic angle spinning ^{1}H-NMR spectroscopy. Anal Chem 2005;77:5570–5578.
75. Sharma U, Chaudhury K, Jagannathan NR, Guha SK. A proton NMR study of the effect of a new intravasal injectable male contraceptive RISUG on seminal plasma metabolites. Reproduction 2001;122:431–436.
76. Clayton TA, Lindon JC, Cloarec O, Antti H, Charuel C, Hanton G, et al. Pharmaco-metabonomic phenotyping and personalized drug treatment. Nature 2006:440:1073–1077.
77. Iles RA, Hind AJ, Chalmers RA. Use of proton nuclear magnetic resonance spectroscopy in detection and study of organic acidurias. Clin Chem 1985;31:1795–1801.
78. Constantinou MA, Papakonstantinou E, Benaki D, Spraul M, Shulpis K, Koupparis MA, Mikros E. Application of nuclear magnetic resonance spectroscopy combined with principal component analysis in detecting inborn errors of metabolism using blood spots:a metabonomic approach. Anal Chim Acta 2004;511:303–312.
79. Engelke UFH, Liebrand-van Sambeek MLF, De Jong JGN, Leroy JG, Morava E, Smeitink JAM, Wevers RA. N-Acetylated metabolites in urine: Proton nuclear magnetic resonance spectroscopic study on patients with inborn errors of metabolism. Clin Chem 2004;50:58–66.
80. Constantinou MA, Papakonstantinou E, Spraul M, Sevastiadou S, Costalos C, Koupparis MA, et al. ^{1}H NMR-based metabonomics for the diagnosis of inborn errors of metabolism in urine. Anal Chim Acta 2005;542:169–177.
81. Moolenaar SH, Engelke UF, Hoenderop SM, Sewell AC, Wagner L, Wevers RA. Handbook of ^{1}H NMR Spectroscopy in Inborn Errors of Metabolism. 1st ed. Heilbronn: SPS Verlagsgesellschaft, 2002.
82. Engelke UFH, Oostendorp M, Wevers RA. NMR spectroscopy of body fluids as a metabolomics approach to inborn errors of metabolism. In: Lindon JC, Nicholson JK, Holmes E, eds. The Handbook of Metabonomics and Metabolomics. Amsterdam: Elsevier. 2007:375–412.
83. Petersen S, Peto V, Scarborough P, Rayner M. Coronary heart disease statistics. London: British Heart Foundation. Available at http://www.heartstats.org.
84. Otvos JD, Jeyarajah EJ, Bennett DW. Quantification of plasma lipoproteins by proton nuclear magnetic resonance spectroscopy. Clin Chem 1991;37: 377–386.
85. Kuller L, Arnold A, Tracy R, Otvos JD, Burke G, Psaty B, et al. Nuclear magnetic resonance spectroscopy of lipoproteins and risk of coronary heart disease in the cardiovascular health study. Arterioscler Thromb Vasc Biol 2002;22:1175–1180.

86. Brindle JT, Antti H, Holmes E, Tranter G, Nicholson JK, Bethell HWL, et al. Rapid and noninvasive diagnosis of the presence and severity of coronary heart disease using ^{1}H-NMR based metabonomics. Nat Med 2002;8:1439–1444.
87. Kirschenlohr HL, Griffin JL, Clarke SC, Rhydwen R, Grace AA, Schofield PM, et al. Proton NMR analysis of plasma is a weak predictor of coronary artery disease. Nat Med 2006;12:705–710.
88. Brindle JT, Nicholson JK, Schofield PM, Grainger DJ, Holmes E. Application of chemometrics to ^{1}H NMR spectroscopic data to investigate a relationship between human serum metabolic profiles and hypertension. Analyst 2003;128**:**32–36.
89. Swanson MG, Zektzer AS, Tabatabai ZL, Simko J, Jarso S, Keshari KR, et al. Quantitative analysis of prostate metabolites using ^{1}H HR-MAS spectroscopy. Magn Reson Med 2006;55:1257–1264.
90. Griffin JL, Lehtimäki KK, Valonen PK, Gröhn OHJ, Kettunen MI, Ylä-Herttuala S, et al. Assignment of ^{1}H nuclear magnetic resonance visible polyunsaturated fatty acids in BT4C gliomas undergoing ganciclovir-thymidine kinase genetherapy-induced programmed cell death. Cancer Res 2003;63: 3195–3201.
91. Cheng LL, Lean CL, Bogdanova A, Wright SC Jr, Ackerman JL, Brady TJ, Garrido L. Enhanced resolution of proton NMR spectra of malignant lymph nodes using magic angle spinning. Magn Reson Med 1996;36:653–658.
92. Griffiths JR, McSheehy PMJ, Robinson SP, Troy H, Chung YL, Leek RD, et al. Metabolic changes detected by *in vivo* magnetic resonance studies of HEPA-1 wild-type tumors and tumors deficient in hypoxia-inducible factor-1β (HIF-1β):evidence of an anabolic role for the HIF-1 pathway. Cancer Res 2002;62:688–695.
93. Howe FA, Barton SJ, Cudlip SA, Stubbs M, Saunders DE, Murphy M, et al. Metabolic profiles of human brain tumors using quantitative *in vivo* ^{1}H magnetic resonance spectroscopy. Magn Reson Med 2003;49:223–232.

17

Unraveling Glycerophospholipidomes by Lipidomics

Kim Ekroos

Summary

In recent years, proteomics has driven developments of mass spectrometric approaches. Simultaneously, the research community has regained a major interest in lipids, with mass spectrometry unambiguously facilitating the development of lipidomics. Quantitative determination of molecular lipids is essential for addressing the role of lipids in cellular membranes and in metabolic dysfunctions, potentially leading to a disease state. Lipidomics has therefore evoked great interest in academic research laboratories and in the pharmaceutical industry, particularly in biomarker and drug discovery. A high-throughput oriented lipidomics methodology enabling quantitative analysis of the glycerophospholipidome in an automated fashion is described in this chapter. The methodology explicitly shows enormous potential and promises to play a key role in cell biology, molecular medicine, and drug discovery.

Key Words: glycerophospholipid; lipid profiler; lipidomics; mass spectrometry; mouse liver; precursor ion scanning

1. INTRODUCTION

Glycerophospholipids are the main constituents of cellular membranes *(1)*, consisting of a polar head-group with a phosphate moiety and two fatty acids that are attached to the glycerol backbone. Many structural variants are therefore possible within each class of lipids as the head group can be combined with a large pool of fatty acids that vary in both chain length and degree of saturation. This yields a multitude of similarly built molecules having a large diversity of physical properties, which allows the living cell to regulate the intrinsic heterogeneity of membranes in a dynamic fashion *(2)*.

Dietary lipid profile is a profound regulator, and the diverse metabolic effects of changing the lipid composition of the diet are well documented *(3–5)*. In addition, endogenous lipogenesis plays an important role in the *de novo* production of lipids, the first of which are saturated. Both

From: *Methods in Pharmacology and Toxicology: Biomarker Methods in Drug Discovery and Development*
Edited by: F. Wang © Humana Press, Totowa, NJ

dietary and endogenously produced lipids undergo a range of modifications via elongase and desaturase enzymes, the activities of which, in turn, have been linked to profound metabolic alterations both positive and negative ***(6–8)***. Thus, perturbations in glycerophospholipid metabolism can be closely linked to both cellular and metabolic dysfunctions that have impact in the development of diseases, such as obesity ***(9,10)***, type 2 diabetes ***(11)***, and atherosclerosis ***(12)***. Because perturbations are likely to occur at the molecular lipid level, it becomes essential to comprehensively resolve the glycerophospholipidome quantitatively. In the context of lipid-related diseases, this will undoubtedly improve the biomarker discovery and unambiguously gain new insight into the biological mechanisms facilitating improved drug discovery.

Electrospray ionization–mass spectrometry is a sensitive and specific tool for characterizing glycerophospholipids (GPLs) in complex lipid extracts ***(13–16)***. Development of hybrid quadrupole time-of-flight (QqTOF) mass spectrometers accompanied with nanoelectrospray ionization ***(17)*** has enabled characterization of multiple molecular GPLs from microliter sample concentrations in a single mass spectrometric analysis ***(18)***. In addition, by applying lipid class specific synthetic internal standards, hundreds of molecular GPLs could directly be quantitatively determined, from sample sizes in the range of several hundred cells ***(19)*** or microgram wet-weight tissue homogenates ***(10)***. In concert with robotic autosampling and computer-assisted data interpretation, sample throughput of less than 30 min per sample can be achieved ***(19)***, thus defining itself as a high-throughput lipidomics methodology and making it applicable to applications in cell biology, molecular medicine, and drug discovery.

In this chapter, I will focus on how the lipidomics methodology is applied in GPL biomarker studies using liver tissue samples of mice fed normal chow-diet, mice given an atherosclerosis-inducible diet ***(20,21)***, and mice exposed to lipopolysaccharide (LPS) for inducing inflammation ***(22)*** as examples.

2. MATERIALS

2.1. Chemicals and Solvents

1. Chloroform (Merck, Darmstadt, Germany) and methanol (Rathburn Chemicals Ltd, Walkerburn, Scotland) are of LiChrosolve and HPLC grade, respectively.
2. Glacial acetic acid and 99.99% ammonium acetate (Merck). Prepare a 100 mM ammonium acetate stock solution in methanol and store a at –20°C.
3. Two-milliliter safe-lock Eppendorf tubes and 96-well twin-tec PCR plates (Eppendorf AG, Hamburg, Germany).

4. Two hundred microliter and 1000 μL ART filter tips (Molecular BioProducts, San Diego, CA).
5. Ten microliter and 100 μL Hamilton gastight syringes (Hamilton, Reno, NV).
6. Thermo-sealing foil (Eppendorf AG).

2.2. Lipid Standards

Internal standard mixture containing 20 μM each of 1,2-diheptadecanoyl-*sn*-glycerol-3-phosphocholine (PC 17:0/17:0), 1,2-diheptadecanoyl-*sn*-glycerol-3-phosphoethanolamine (PE 17:0/17:0), 1,2-diheptadecanoyl-*sn*-glycerol-3-phosphoserine (PS 17:0/17:0), 1,2-diheptadecanoyl-*sn*-glycerol-3-phosphoglycerol (PG 17:0/17:0), 1,2-diheptadecanoyl-*sn*-glycerol-3-phosphate (PA 17:0/17:0), 1-heptadecanoyl-2-hydroxy-*sn*-glycerol-3-phosphocholine (LPC 17:0), and *N*-heptadecanoyl-D-*erythro*-sphingosylphosphorylcholine (SM 17:0) (Avanti Polar Lipids, Inc., Alabaster, AL) is prepared in chloroform:methanol 1:2 (v/v) and stored in Fiolax-amber glass vials (Schott AG, Mainz, Germany) in 1-mL aliquots at –20°C.

2.3. Quadrupole Time-of-Flight Mass Spectrometry

QSTAR XL quadrupole time-of-flight mass spectrometry (MDS Sciex, Concord, ON, Canada) was equipped with a robotic chip-based nanoflow ion source (TriVersa NanoMate; Advion BioSciences Ltd, Ithaca, NJ).

2.4. Software

1. QSTAR XL MS is controlled by Analyst QS 1.1.
2. TriVersa NanoMate is controlled by ChipSoft 7.1.1.
3. Lipid Profiler 1.0.95 (MDS Sciex, Concord, ON, Canada).
4. Simca-P (Umetrics, Umeå, Sweden).
5. Microsoft Excel (Microsoft, Redmond, WA).

2.5. Lipid Profiler Software Settings

It is highly recommended to use Lipid Profiler software according to the manufacturer's instructions (see Ref. ***(19)*** for more details). Lipid Profiler software settings, *Adjust Settings*, used for multiple precursor ion scanning (MPIS) analysis data processing follows below.

In *Spectrum* Window

1. *Analyst Peak Finding Threshold* should be set to 1% (this is set in Analyst QS software).
2. *TOF PIS* and *Deisotope* is activated.

3. *Q1 Mass Tolerance*, *Minimum % Intensity*, *Minimum S/N*, and *Flow Injection* are set to 0.3, 0.1, 10, and 50, respectively.

In *Lipid Details* Window

4. *FA Scan* is activated in *Identify Species*.
5. *Analysis type* is set for "Glycerophospholipids" and all listed GPL classes are activated.
6. *Total Double Bonds* are set to 8.
7. *Lyso species* and *Odd Chain Fatty Acids/Ether GPL* are activated.
8. *Find Internal Standard Peaks* is activated. Internal standards to be used are set in the *Internal Standard Details* window.

3. METHODS

Ultimately, it is important to determine alterations in the GPL profiles quantitatively and in an efficient and reproducible way. This can be accomplished through automated high-throughput lipidomics. Molecular GPLs are selectively monitored by Fatty Acid Scanning (FAS) and Head Group Scanning (HGS) on a QqTOF mass spectrometer, based on multiple precursor ion scanning (MPIS), by accurately selecting collision-induced dissociation (CID) generated acyl and head-group specific fragment ions *(18,23)*. This is automated by robotic autosampling. In general, 40 to 50 mass spectra are generated from a single sample (depending on the number of selected anions) that are automatically processed using the Lipid Profiler software for molecular lipid identification and quantification. Alterations in the GPL profiles are rapidly determined and visualized using a combination of multi- and univariate data tools.

The lipidomics working process is illustrated in **Fig. 1**, possessing the following elements described below: (i) total lipid extraction, (ii) integrated autosampling nanoelectrospray ion source (TriVersa NanoMate), (iii) quantitative analysis of molecular glycerophospholipids by fatty acid scanning and head-group scanning on a QSTAR XL mass spectrometer *(18,19,23)*, (iv) computer-assisted identification and quantification of detected molecular lipids using Lipid Profiler software *(19)*, and (v) uni- and multivariate data analysis setups.

3.1. Total Lipid Extraction

1. Freshly prepared or frozen (–80°C) liver tissue homogenates (**Note 1**) are adjusted to 10 μg total protein/μL (approximately 30 μg wet tissue/μL).
2. Seventy-five microliters is transferred to a clean 2-mL safe-lock Eppendorf tube and 500 μL methanol is added. The sample is rigorously mixed, using an Eppendorf Thermomixer (at 1400 rpm), at ambient temperature for 10 min, followed by addition of 1000 μL chloroform and another 10 min of rigorous

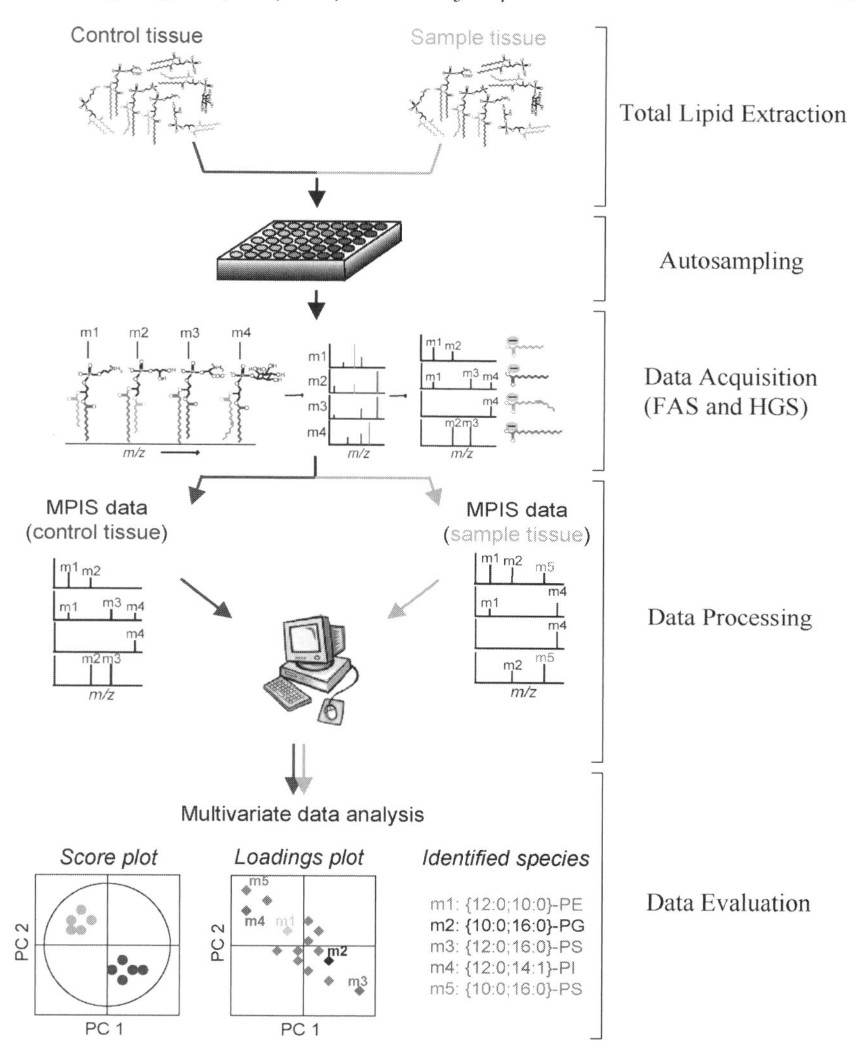

Fig. 1. Schematic overview of the lipidomics working process.

mixing (**Note 2**). At this stage 50 μL of the 20 μM of the GPL standard mixture is added to the mixture (**Note 3**).

3. The sample is centrifuged 5 min at 14,000 × *g* at room temperature, and the clarified supernatant is transferred to a new tube.
4. Three hundred microliters of 20 mM acetic acid is added to the clarified supernatant and the sample is again vigorously mixed for 10 min.
5. Complete phase separation is achieved by centrifugation at 500 × *g* for 5 min at 20°C.

6. The lower phase (A) (organic containing GPLs) is transferred to a new 2-mL safe-lock Eppendorf tube. Rinsing the pipetting tips once in chloroform:methanol (1:2, v/v) before aspirating the lower phase minimizes contamination with the upper phase (organic phase must not contain any water).
7. The upper phase is washed briefly with 500 µL chloroform and subjected to phase separation as described above.
8. The lower phase (B) is transferred and pooled with the lower phase (A).
9. The sample is dried under a gentle stream of nitrogen (or vacuum concentrator) and reconstituted in 200 µL chloroform:methanol (1:2, v/v). Final total lipid extract is either directly subjected to mass spectrometric analysis or stored at –20°C/–80°C (**Note 4**).

3.2. Integrated Autosampling Chip-Based Nanoelectrospray Ion Source TriVersa NanoMate

1. Total lipid extract is diluted 10-fold in chloroform:methanol (1:2, v/v) containing 5.6 mM ammonium acetate (final ammonium acetate concentration is 5 mM) (**Note 5**).
2. Thirty microliters of the diluted total lipid extract is transferred to an Eppendorf twin-tec 96-well plate.
3. After transferring all samples, the 96-well plate is covered with a thermo-sealing foil by heat-sealing (Thermos-Sealer, ABgene, Epsom, UK) for 10 s (**Note 6**). The plate is afterwards centrifuged for 15 min at 1000 × g at room temperature and placed in the TriVersa NanoMate robotic autosampler system.
4. The 96-well plate is kept at 20°C, and prior to sample aspiration, the mandrel is used for piercing the sealing foil. Five-microliter sample aliquots are aspirated in the pipetting tip and delivered to the back plane of the nanoESI chip. Electrospray process, in negative ion mode, is initiated by applying 1.3 kV and 0.3 psi nitrogen head pressure to ensure constant sample flow rates of ~150 nL/min.

3.3. Quantitative Analysis of Molecular Glycerophospholipids by Fatty Acid Scanning and Head Group Scanning on a QSTAR XL Mass Spectrometer

1. QSTAR XL mass spectrometer is calibrated in MS/MS mode using a synthetic lipid standard 1,2-diheptadecanoyl-*sn*-glycerol-3-phosphocholine prior to MPIS experiments *(24)*.
2. FAS and HGS experiments are performed in negative ion mode using a dwell time of 30 ms at a step size of 0.2 Da at unit resolution of the Q1 quadrupole. Collision energy is set to 40 eV [or linearly ramped according to Ejsing et al. *(19)*] and the quadrupole mass range is set from m/z 400 to m/z 900 (including lyso- and diacyl GPLs). Peak enhancement (trapping of fragment ions in the collision cell) is applied according to the manufacturer's instructions *(25)*.
3. The m/z of fragment ions are selected according to **Table 1**.

Table 1
Fatty Acid and Lipid Class–Specific Ions Selected for FAS and HGS Analysis in Negative Ion Mode

Fragment ion	m/z	m/z width	Specific detection of
[Glycerolphosphate–H_2O]$^-$	153.0	0.150	PA/PG/PS/PI
[Cholinephosphate–15]$^-$	168.0	0.150	PC/LPC
FA 10:1	169.1	0.150	Glycerophospholipid species
FA 10:0	171.1	0.150	Glycerophospholipid species
FA 11:1	183.1	0.150	Glycerophospholipid species
FA 11:0	185.2	0.150	Glycerophospholipid species
[Ethanolaminephosphate–H_2O]$^-$	196.0	0.150	PE/LPE
FA 12:1	197.2	0.150	Glycerophospholipid species
FA 12:0	199.2	0.150	Glycerophospholipid species
FA 13:1	211.2	0.150	Glycerophospholipid species
O-14:1	211.2	0.150	Plasmenyl species
FA 13:0	213.2	0.150	Glycerophospholipid species
FA 14:2	223.2	0.150	Glycerophospholipid species
FA 14:1	225.2	0.150	Glycerophospholipid species
FA 14:0	227.2	0.150	Glycerophospholipid species
FA 15:2	237.2	0.150	Glycerophospholipid species
O-16:2	237.2	0.150	Plasmenyl species
FA 15:1	239.2	0.150	Glycerophospholipid species
O-16:1	239.2	0.150	Plasmenyl species
[Inositolsphophate–H_2O]$^-$	241.0	0.150	PI
FA 15:0	241.2	0.150	Glycerophospholipid species

(Continued)

Table 1 (Continued)

Fragment ion	m/z	m/z width	Specific detection of
FA 16:2	251.2	0.150	Glycerophospholipid species
FA 16:1	253.2	0.150	Glycerophospholipid species
FA 16:0	255.2	0.150	Glycerophospholipid species
FA 20:5-CO_2	257.2	0.150	Glycerophospholipid species
FA 20:4-CO_2	259.2	0.150	Glycerophospholipid species
FA 17:2	265.2	0.150	Glycerophospholipid species
O-18:2	265.3	0.150	Plasmenyl species
FA 17:1	267.2	0.150	Glycerophospholipid species
O-18:1	267.3	0.150	Plasmenyl species
FA 17:0	269.3	0.150	Glycerophospholipid species
FA 18:3	277.2	0.150	Glycerophospholipid species
FA 18:2	279.2	0.150	Glycerophospholipid species
FA 18:1	281.3	0.150	Glycerophospholipid species
FA 22:6-CO_2	283.2	0.150	Glycerophospholipid species
FA 18:0	283.3	0.150	Glycerophospholipid species
FA 22:5-CO_2	285.3	0.150	Glycerophospholipid species
FA 22:4-CO_2	287.3	0.150	Glycerophospholipid species
FA 19:2	293.3	0.150	Glycerophospholipid species
O-20:2	293.3	0.150	Glycerophospholipid species
O-20:1	295.3	0.150	Glycerophospholipid species
FA 19:1	295.3	0.150	Glycerophospholipid species
FA 19:0	297.3	0.150	Glycerophospholipid species
FA 20:5	301.2	0.150	Glycerophospholipid species
FA 20:4	303.2	0.150	Glycerophospholipid species
FA 20:3	305.2	0.150	Glycerophospholipid species
FA 20:2	307.3	0.150	Glycerophospholipid species
FA 20:1	309.3	0.150	Glycerophospholipid species
FA 20:0	311.3	0.150	Glycerophospholipid species
FA 22:6	327.2	0.150	Glycerophospholipid species
FA 22:5	329.2	0.150	Glycerophospholipid species
FA 22:4	331.3	0.150	Glycerophospholipid species
FA 22:3	333.3	0.150	Glycerophospholipid species
FA 22:2	335.3	0.150	Glycerophospholipid species
FA 22:1	337.3	0.150	Glycerophospholipid species
FA 22:0	339.3	0.150	Glycerophospholipid species

4. It is important to retrieve sufficient ion statistics per acquired lipid ion. Therefore, MPIS spectra are normally acquired during 20 cycles (20 mass spectra are summed), resulting in a total mass spectrometric analysis time of 25 min per sample.

3.4. Computer-Assisted Identification and Quantification of Molecular Lipids Using Lipid Profiler Software

1. Lipid Profiler software is accessed and operated through the Analyst software.
2. *Process Analyst Data*, in Lipid Profiler software, is applied, and all sample files to be processed are selected (e.g., using *Select All*).
3. Once sample files are processed, select *Confirmed* in *Show Results* (e.g., where at least both fatty acid signals of a GPL have been detected) followed by *View Data*.
4. In *Result View*, select all identified GPLs for quantification.
5. Apply isotope correction factor ***(19)*** for selected GPLs by selecting *Profile Tests*. In the *Target Lipids* window, activate *select all tests* and *set correction factor from lipid names* (this also applies for the used internal standards). In terms of GPL quantification, each identified GPL is to be corrected by an internal standard (e.g., PC species to PC standard), which is set in the same window (**Note 7**). The standard lipid concentrations used is set in the *Sample Data* window.
6. Close *Profile Tests* and return to *Result view*.
7. For absolute quantification, by selecting *Corrected Areas* (or *Intensities*) [for relative quantification use *Corrected Area%* (or *Intensities%*)], the identified GPLs are now isotope and internal standard corrected, and the data can be further exported for further uni- or multivariate data processing.

3.5. Uni- and Multivariate Data Analysis

Multivariate data analysis tools are mainly used to get an overview if samples differ based on their lipid profiles, whereas univariate data analysis tools are used for determining (and visualizing) the lipid quantities. Examples are given below.

1. Exported results are subjected to multivariate data analysis [e.g., principal component analysis (PCA)] ***(26)***. As PCA describes the maximum variation in the data, X (= all quantified GPLs), it is well suited for distinguishing samples based on variations in multiple variables simultaneously as it explains which of the variable(s) cause the differences. Hence, PCA allows the end-user to rapidly visualize and pinpoint alterations in the GPL profile of multiple samples. An example on how PCA can be used to visualize alterations in molecular GPLs, based on absolute amounts, of atherosclerotic, inflammatory, and untreated mouse liver is shown in **Fig. 2**.

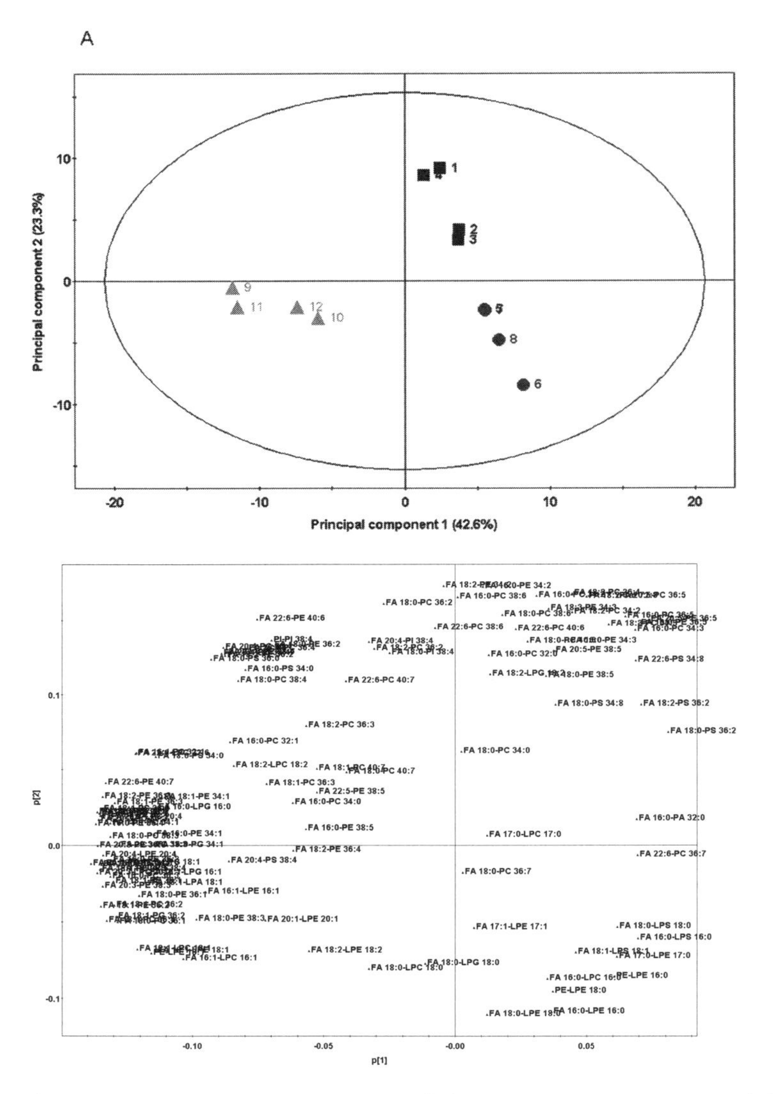

Fig. 2. PCA on mouse liver GPLs. **(A)** Mice fed chow diet (■), an atherosclerosis-inducing diet (△), and chow diet exposed to LPS for acute inflammation induction (•) completely separates into individual classes in the score plot. **(B)** The molecular GPLs causing the separation are presented in the loadings plot. The different treatments lead to selective changes in the GPL profiles. PCA enables rapid pinpointing of the molecular GPLs that alter between the treatments. Absolute amounts of 121 molecular GPLs (variables) and four samples per treatment (n = 4 observations) were used for the PCA. Unit variance scaling was applied.

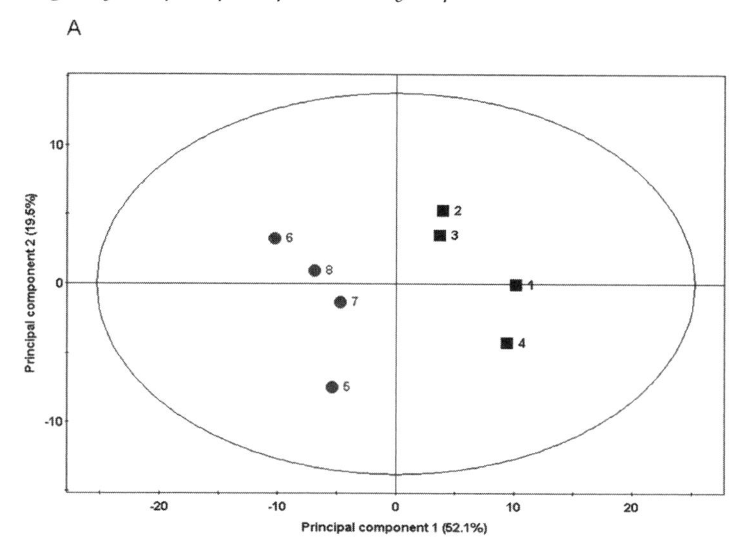

Fig. 3. Most affected liver GPLs upon LPS exposure. **(A)** Mice fed chow diet (■) and chow diet in combination with LPS exposure for acute inflammation induction (•) completely separates into individual classes in the PLS-DA score plot. **(B)** Quantitative alterations of the eight most significant GPLs, retrieved from the PLS-DA generated *Variable of Importance* (*VIP*) list. Absolute amounts of 121 molecular GPLs and four samples per treatment ($n = 4$) were used for PLS-DA. Unit variance scaling was applied. Mean values and the standard deviation are presented. The p values were calculated by equal variance t-test.

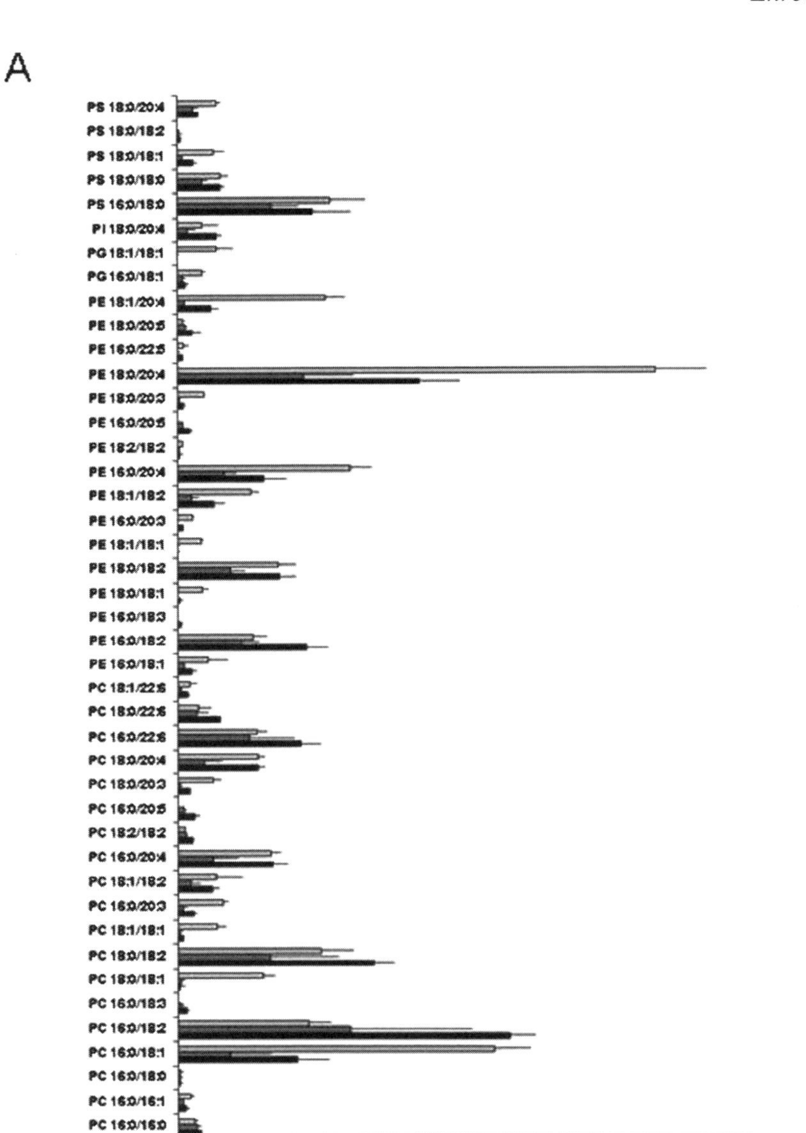

Fig. 4. (Continued)

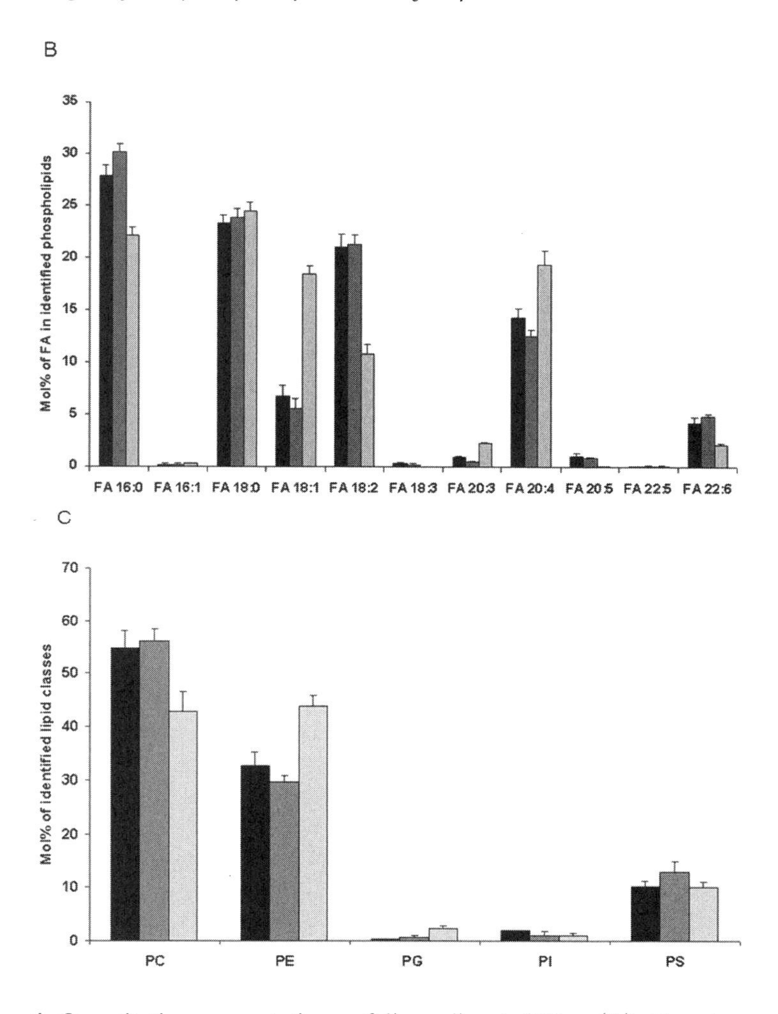

Fig. 4. Quantitative presentations of liver diacyl GPLs. (**A**) Abundance of 43 molecular GPLs of mice fed chow diet (black bars), chow diet and exposed to LPS for inflammation induction (dark-gray bars), and an atherosclerosis-inducing diet (light-gray bars). (**B**) Emulated total fatty acid (FA) profile. The mol% of FA moieties were calculated as the sum of molar concentrations of lipid species in (**A**) containing the respective FA moiety, followed by normalization to the total molar concentration of all FA moieties. The FA concentration corresponding with symmetric lipid species was multiplied by a factor of 2 to account for two identical FA moieties. (**C**) Lipid class profile. The mol% of lipid classes was calculated as the sum of the mol% of lipid species (in **panel A**) of the respective lipid class. Four samples per treatment (n = 4) were analyzed. Mean values and the standard deviation are presented.

2. As partial least squares–discriminant analysis (PLS-DA) explains the most important variables in X *(26)*, it can be used to pinpoint molecular GPLs that differ the most between samples, thus revealing the most significant GPLs. An example on how PLS-DA and univariate data tools can be used to extract and visualize the molecular GPLs that are the most affected, in absolute amount, upon acute inflammation in mouse liver is shown in **Fig. 3A, B**.
3. To get an overview of the absolute amounts of molecular GPLs and how much individual lipid has altered upon treatment, it is suitable to use univariate data tools. The absolute amounts of molecular GPLs, in atherosclerotic, inflammatory, and untreated mouse liver, are plotted and shown as an example in **Fig. 4A**. From the exported GPL data, an emulated total or lipid class–based fatty acid profile can be established (typical obtained by gas chromatography), directly describing alterations in the fatty acid content. An example of the total fatty acid content (mol%) of GPLs in atherosclerotic, inflammatory, and untreated mouse liver is shown in **Fig. 4B**. In similar ways, lipid class profiles are obtainable, which is exemplified in **Fig. 4C**.

4. NOTES

1. GPLs are inherently labile in tissue homogenates mainly due to the presence of active enzymes, water, and oxygen. To obtain reliable and reproducible results, it is therefore important that the tissue homogenate (this can also be snap frozen in liquid nitrogen) is directly extracted in organic solvent (e.g., chloroform and methanol). The final GPL extract should further be evaporated under a gentle stream of nitrogen and reconstituted in fresh organic solvent prior to analysis or otherwise stored at –80°C in an oxygen free atmosphere.
2. Total lipid extraction method according to Folch et al. *(27)* enables high recovery yields of GPLs from tissues samples in a robust manner.
3. Mass spectrometric response of a molecular GPL, in QqTOF instruments, is rather dependent on lipid class than the actual fatty acid composition (e.g., fatty acid chain length and saturation degree) *(19)*. Therefore, inclusion of internal GPL standards of respective lipid class, prior to lipid extraction, allows absolute or relative quantification of molecular GPLs.
4. Long-term sample storage should include low amounts (0.1%) of antioxidants as butylated hydroxytoluene (BHT).
5. Robotic chip-based nanoelectrospray mass spectrometry using the TriVersa NanoMate enables sample infusion flow rates of approximately 100 to 150 nL/min, and because of the high analysis sensitivity, a quantitative description of the GPL complement can be achieved from 1 to 3 μL sample sizes. To maintain this sensitivity, it is therefore important to reduce the chemical noise (e.g., of contaminants) by working with organic solvent–resistant materials and solvents of high purity.
6. To obtain reliable and reproducible results, it is important to use closed containers and covered 96-well plates to minimize sample evaporation, stored at –20°C/–80°C prior to analysis.

7. Because LPC 17:0 is the only commercial available synthetic heptadecanoyl lyso lipid, all lyso glycerophospholipids are quantified by the LPC 17:0 standard. Notably, quantification can be done either relatively or absolutely.

REFERENCES

1. Fahy E, Subramaniam S, Brown HA, et al. A comprehensive classification system for lipids. J Lipid Res 2005;46(5):839–862.
2. Simons K, Vaz WLC. Model systems, lipid rafts, and cell membranes. Annu Rev Biophys Biomol Struct 2004;33(1):269–295.
3. Pan DA, Hulbert AJ, Storlien LH. Dietary fats, membrane phospholipids and obesity. J Nutr 1994;124(9):1555–1565.
4. Hulbert AJ, Turner N, Storlien LH, Else PL. Dietary fats and membrane function: implications for metabolism and disease. Biol Rev Cambridge Philos Soc 2005;80(1):155–169.
5. Lichtenstein AH, Schwab US. Relationship of dietary fat to glucose metabolism. Atherosclerosis 2000;150(2):243.
6. Busch AK, Gurisik E, Cordery DV, et al. Increased fatty acid desaturation and enhanced expression of stearoyl coenzyme A desaturase protects pancreatic {beta}-cells from lipoapoptosis. Diabetes 2005;54(10):2917–2924.
7. Gutierrez-Juarez R, Pocai A, Mulas C, et al. Critical role of stearoyl-CoA desaturase-1 (SCD1) in the onset of diet-induced hepatic insulin resistance. J Clin Invest 2006;116(6):1686–1695.
8. Sampath H, Ntambi JM. Stearoyl-coenzyme A desaturase 1, sterol regulatory element binding protein-1c and peroxisome proliferator-activated receptor-alpha: independent and interactive roles in the regulation of lipid metabolism. Curr Opin Clin Nutr Metabol Care 2006;9(2):84–88.
9. Min Y, Lowy C, Ghebremeskel K, Thomas B, Bitsanis D, Crawford MA. Fetal erythrocyte membrane lipids modification: preliminary observation of an early sign of compromised insulin sensitivity in offspring of gestational diabetic women. Diabetic Med 2005;22(7):914–920.
10. Linden D, William-Olsson L, Ahnmark A, et al. Liver-directed overexpression of mitochondrial glycerol-3-phosphate acyltransferase results in hepatic steatosis, increased triacylglycerol secretion and reduced fatty acid oxidation. FASEB J 2006;20(3):434–443.
11. Yang J, Xu G, Hong Q, et al. Discrimination of type 2 diabetic patients from healthy controls by using metabonomics method based on their serum fatty acid profiles. J Chromatogr B 2004;813(1–2):58.
12. Tsimikas S, Kiechl S, Willeit J, et al. Oxidized phospholipids predict the presence and progression of carotid and femoral atherosclerosis and symptomatic cardiovascular disease: five-year prospective results from the Bruneck Study. J Am Coll Cardiol 2006;47(11):2228.
13. Han X, Gross RW. Electrospray ionization mass spectroscopic analysis of human erythrocyte plasma membrane phospholipids. Proc Natl Acad Sci USA 1994;91(22):10635–10639.

14. Han X, Gross RW. Structural determination of picomole amounts of phospholipids via electrospray ionization tandem mass spectrometry. J Am Soc Mass Spectrom 1995;6(12):1210.
15. Kerwin J, Tuininga A, Ericsson L. Identification of molecular species of glycerophospholipids and sphingomyelin using electrospray mass spectrometry. J Lipid Res 1994;35(6):1102–1114.
16. Brugger B, Erben G, Sandhoff R, Wieland FT, Lehmann WD. Quantitative analysis of biological membrane lipids at the low picomole level by nano-electrospray ionization tandem mass spectrometry. Proc Natl Acad Sci USA 1997;94(6):2339–2344.
17. Wilm M, Mann M. Analytical properties of the nanoelectrospray ion source. Anal Chem 1996;68(1):1–8.
18. Ekroos K, Chernushevich IV, Simons K, Shevchenko A. Quantitative profiling of phospholipids by multiple precursor ion scanning on a hybrid quadrupole time-of-flight mass spectrometer. Anal Chem 2002;74(5):941–949.
19. Ejsing CS, Duchoslav E, Sampaio J, et al. Automated identification and quantification of glycerophospholipid molecular species by multiple precursor ion scanning. Anal Chem 2006;78(17):6202–6214.
20. Tangirala R, Rubin E, Palinski W. Quantitation of atherosclerosis in murine models: correlation between lesions in the aortic origin and in the entire aorta, and differences in the extent of lesions between sexes in LDL receptor-deficient and apolipoprotein E-deficient mice. J Lipid Res 1995;36(11):2320–2328.
21. Williams T, Chambers J, Roberts L, Henderson R, Overton J. Diet-induced obesity and cardiovascular regulation in C57BL/6J mice. Clin Exp Pharmacol Physiol 2003;30(10):769–778.
22. Leitinger N. Oxidized phospholipids as modulators of inflammation in atherosclerosis. Curr Opin Lipidol 2003;14(5):421–430.
23. Ekroos K, Ejsing CS, Bahr U, Karas M, Simons K, Shevchenko A. Charting molecular composition of phosphatidylcholines by fatty acid scanning and ion trap MS3 fragmentation. J Lipid Res 2003;44(11):2181–2192.
24. Ekroos K, Shevchenko A. Simple two-point calibration of hybrid quadrupole time-of-flight instruments using a synthetic lipid standard. Rapid Commun Mass Spectrom 2002;16(12):1254–1255.
25. Chernushevich IV. Duty cycle improvement for a quadrupole-time-of-flight mass spectrometer and its use for precursor ion scans. Eur J Mass Spectrom 2000;6:471–479.
26. Eriksson L, Johansson E, Kettaneh-Wold N, Wold S. Multi- and Megavariate Data Analysis; Principles and Applications. Umetrics AB, Umeå, Sweden 2001, ISBN 91-973730-1-X.
27. Folch J, Lees M, Stanley GHS. A simple method for the isolation and purification of total lipides from animal tissues. J Biol Chem 1957;226(1):497–509.

Index

A

ACTG, *see* Automatic Correspondence of Tags and Genes (ACTG)
Angiogenesis, 11
Antibody microarrays, 273, 274, 282
Anticlotting mechanism, 10
Assay qualification, 2, 11, 12, 17, 21, 22
  angiogenic cytokines, multiplexed assay, 11–12
    Amgen, antiangiogenic therapeutics, 11
    angiogenesis, 11
    fibroblast growth factor, basic (bFGF), 11
    Meso Scale Discovery (MSD), 11
    plasma preparation, 11–14
    platelet-poor plasma, 12
    VEGF/PlGF recovery, effect of NaCl, 17
  ex vivo stimulation of blood, 17–23
    cytokine/lipopolysaccharide (LPS), 18
    endotoxin contamination, 20
    tests, two donors, 21–22
  gene expression
    anticoagulants, effect of, 18–19
    endotoxin, effect of, 20
Automatic Correspondence of Tags and Genes (ACTG), 92

B

Basic fibroblast growth factor (bFGF), 12, 14
Bead processing
  bead assay, 61
  dilution of bead sets/cRNA targets, 60
  hybridization/washing/staining, 60–61
Biofluids/tissue, analysis of
  serum/plasma, 344
  tissue, 344–345
  urine, 344
Bioinformatics, 67, 88, 212, 233, 295
  SAGE, 91–92
    ACTG, 92
    Genie/Map, 91–92
    MGC, 91
Biomarkers
  definition/application, 29
  discovery, scheme for (flow chart), 292
  predictive toxicology, definition, 31
  sample collection/handling, early-phase drug development
    assay qualification, 11–23
  toxicity with gene expression profiling
    cytochrome P450 1A1 (CYP1A1), induction, 29
    diagnostic/predictive, development/use of, 29–31

DrugMatrix (Iconix Biosciences) signatures, 30
FGSIs, 29
gene expression signatures, 28–29
global gene expression changes, 30
immunohistochemistry, 29
Notch signaling, 29
predictive gene expression-based biomarkers, development/use of, 31–33
toxicogenomics, 28–29
in vitro gene expression–based biomarkers, development of, 33–37
genotoxicity, for, 36
predictive signatures for *in vitro* use, identity of, 35
Taqman Low Density Array/ArrayPlate platforms, 37
in vitro toxicogenomics paradigm, 35
See also Gene expression-based biomarkers
Bio-Rad Equalizer Beads, 256
Bio-Rad Pattern Analysis Software, 262
Bio-Rad's Serum fractionation kit, 256
Biotinylated targets, preparation of
cDNA synthesis (first/second strand)/cleanup, 57–58
cRNA synthesis/cleanup/ fragmentation, 58–59
Bradford assay, 214
See also Protein

C

CAGE, *see* Cap analysis of gene expression (CAGE)
Cap analysis of gene expression (CAGE), 90
Chromatographic arrays
anion exchange (ProteinChip Q10 arrays), 253–254
cation exchange (ProteinChip cm10 arrays), 254
immobilized metal affinity capture (ProteinChip imac30 arrays), 254
normal phase (ProteinChip NP20 arrays), 253
preactivated surfaces (RS100 and PS20 ProteinChip arrays), 255
reverse phase (ProteinChip H4 and H50 arrays), 254–255
specialty arrays (ProteinChip gold/SEND-ID/PG20 arrays), 255
CID, *see* Collision-induced dissociation (CID)
Collision-induced dissociation (CID), 247, 372
Consortium for Metabonomic Toxicology (COMET), 356
Coronary heart disease (CHD), 359
Correlation spectroscopy (COSY), 346
COSY, *see* Correlation spectroscopy (COSY)
Cy dye, 190, 192, 194

D

2-DE, *see* Two-dimensional polyacrylamide gel electrophoresis (2-DE)
2-DE, plasma biomarker discovery, 171–185
gel image analysis, 182
HPLC, 173
MARS, 173
materials
equipment, 173–184
reagents and solutions, 174
methods
equilibration, 179

IEF
image acquisition and data analysis, 184–185
in-gel rehydration, 177–178
low-abundant proteins, plasma fractionation/concentration, 174–176
protein pattern, *see* Protein pattern visualization
protein quantification, 177
SDS-PAGE
See also Universal Protein Precipitating Agent (UPPA)
Derivatization, 323–326
Diagnostic biomarkers, development/use of, 29–31
See also Biomarkers
Difference in-gel electrophoresis (DIGE), 189–208
DIGE, *see* Difference in-gel electrophoresis (DIGE)
DIGE, high-resolution protein biomarker research tool, 189–208
first-dimension IEF, 196–197
image acquisition
range (bit depth), 201–202
resolution, 200–201
scan content, 202–203
image analysis, software principles/workflows
image warping and spot matching, 204–205
spot delineation/outline, 205–207
minimal labeling strategy, 190
preparative gels for spot picking, 207
sample, preparation
labeling, 192–196
protease inhibitor cocktails, 192
purification and assay, 191–192
second-dimension SDS-PAGE (2-DE), 197–199
DNA methylation, 120, 124, 128, 136
Drug discovery, 27, 28, 31, 33, 34, 37, 38, 43, 51, 172, 190, 292, 355, 370
Dulbecco's Modified Eagle's Medium (DMEM), 234

E

EDM, *see* Expression Difference Mapping (EDM)
Electrospray ionization, 370
ELISA, *see* Enzyme-linked immunosorbent assay (ELISA)
Enzyme-linked immunosorbent assay (ELISA), 10, 13, 14, 21, 274–287
ExacTag analysis software, protein identification/quantitation, 241–242
Expression Difference Mapping (EDM), 261

F

FGSIs, *see* Functional γ-secretase inhibitors (FGSIs)
Flame ionization detection (FID), 318, 321–322
Fractionation, 173, 174, 176, 256, 261
Functional γ-secretase inhibitors (FGSIs), 29

G

Gas chromatography (GC), 293, 304, 318, 319, 321, 342, 382
Gas chromatography-mass spectrometry, comprehensive two-dimensional (GCxGC-MS), 332
Gas chromatography-mass spectrometry (GC/MS), 317, 319, 332
analytical technology/sample preparation, developments in, 332–335
GCxGC-MS, 332–333

HGA, 334–335
sample, analysis of, 333–334
SPME, 335
data analysis, 329–332
data processing, 329–330
laser capture microdissection (LCM), 334
metabolite identification/ biomarker qualification, 331–332
statistical analysis, 330–331
derivatization, 323–326
advantages, 324
reagents, 324–326
experimental methods, 326–329
GC-MS analysis, 328–329
sample preparation (plasma/urine/tissue), 326
limitations/advantages
assay development and applicability, 336–337
metabolite identification, 335, 336
metabolome, coverage of, 335
sensitivity, 336
technology, *see* GC-MS technology
GC/MS, *see* Gas chromatography-mass spectrometry (GC/MS)
GC-MS-based metabolomics, 317–337
GC-MS technology
columns, 321
detectors, 321–323
FID, 321
Fourier transform ion cyclotron resonance mass spectrometry (FT-ICR-MS (or) FT-MS), 322
mass spectrometry-based detectors, 322
orbitrap, 323
single quadrupole, 322
time-of-flight, 322
ionization modes for GC-MS experiments, 323
chemical ionization, 323
electron impact ionization, 323
sample introduction, 320–321
programmed temperature vaporizer, 320–321
split/splitless interface, 320–321
GCxGC-MS, *see* Gas chromatography-mass spectrometry, comprehensive two-dimensional (GCxGC-MS)
Gene expression-based biomarkers
drug safety, of, 27–44
attrition, causes, 27
toxicity with gene expression, *see* Biomarkers
hepatotoxicity, of, 37–39
bile duct hyperplasia, 38
nephrotoxicity, of, 39–40
toxicity in blood, of, 40–43
globin reduction, 41–42
human peripheral blood mononuclear cells (PBMCs), possibility of, 42–43
scientific/technical challenges, 40–43
Gene expression profiles, PAXgene preparations, 8–9
Gene expression signatures as biomarkers, 28–29
See also Biomarkers
Gene expression signatures profiling, fluorescent microspheres, 50–62
bead-based secondary screening platform, 52
materials, 52
methods
bead, *see* Bead processing
biomarker signature, selection of, 52–54

biotinylated targets, preparation of, *see* Biotinylated targets, preparation of
data analysis, 61
linking probes to luminex xMAP beads, 54–55
oligonucleotide probe selection, perfect match/mismatch, 54
RNA isolation, *see* RNA isolation
tissue treatment, 55–56
notes, 61–62
Glycero-phosphocholine (GPC), for liver, 357
Glycerophospholipids, unraveling by lipidomics
materials, 370–372
lipid profiler software settings, 371–372
lipid standards, 371
methods, 372–382
chip-based nanoelectrospray ion source TriVersa NanoMate, integrated autosampling, 374
lipid extraction, 372–374
Glycerophospholipids (GPLs), 370
GPLs, *see* Glycerophospholipids (GPLs)

H

HCA, *see* Hierarchical cluster analysis (HCA)
Head Group Scanning (HGS), 372
Headspace gas analysis (HGA), 334–335
Heteronuclear single quantum coherence (HSQC), 346, 349, 350
HGA, *see* Headspace gas analysis (HGA)
HGS, *see* Head Group Scanning (HGS)
Hierarchical cluster analysis (HCA), 352–353
High-performance liquid chromatography (HPLC), 173
High-performance liquid chromatography-mass spectrometry (HPLC-MS), 293, 294, 296, 298, 305
HPLC, *see* High-performance liquid chromatography (HPLC)
HPLC-MS, *see* High-performance liquid chromatography-mass spectrometry (HPLC-MS)
HSQC, *see* Heteronuclear single quantum coherence (HSQC)

I

ICAT, *see* Isotope coded affinity tag (ICAT)
IEF, *see* Isoelectric focusing (IEF)
IEM, *see* Inborn errors of metabolism (IEM)
Immunohistochemistry, 29, 144, 154
Inborn errors of metabolism (IEM), 358
International Prognostic Index (IPI), 79
Isobaric tagging for relative and absolute quantitation (iTRAQ), 190, 212, 232
Isoelectric focusing (IEF), 171, 173, 178–179, 196–197
load IPG strips onto multiphor II, 178
multiphor II unit, to prepare, 178
Isotope coded affinity tag (ICAT), 212, 232, 237
iTRAQ, *see* Isobaric tagging for relative and absolute quantitation (iTRAQ)

J

J-resolved (JRES) spectroscopy in metabolomics, 347
See also NMR methods, metabolomics

K

KnowItAll, 350, 354

L

Label-free mass spectrometry
  cisplatin resistance biomarkers, human ovarian cancer cells
    biological/technical replicate, 222
    cell lines and study design, 222–223
    data interpretation, 225–227
    data normalization/ assessment, 223
    ranking/prioritization of data, 223–225
  protein quantification, 212–213
Label-free protein quantification method, peak intensity-based
  Bradford assay, 214
  mass spectrometric analysis, 215–216
  protein classification/pathway analysis, 220
  protein extraction/reduction/ alkylation/digestion, 213–215
  protein identification, prioritization of, 217
  protein identification (SEQUEST and X!Tandem algorithms), 216–217
  protein quantification, 217–219
  quality assurance/control, 219
  quantile normalization, 218
  statistical analysis, 219–220
  Triple-Play mode, 215
  See also Protein
Laser capture microdissection (LCM), 334
LCM, *see* Laser capture microdissection (LCM)
LC-MS, *see* Liquid chromatography-tandem mass spectrometry (LC-MS)
LC-MS-based metabonomics in biomarker discovery
  data processing/analysis, 301–303
    filtering methods, 301
    partial least squares discriminant analysis (PLSDA), 302
  metabolite identification, 303–305
  sampling/sample preparation (data collection), 296–300
LC-MS metabonomics
  application, 305–309
  methodology, *see* LC-MS-based metabonomics in biomarker discovery
Lipidomics, 372, 373
Lipid profiler, 369–370
  software setting, 371–372
Liquid chromatography (LC), 212
Liquid chromatography-tandem mass spectrometry (LC-MS), 233, 236, 241, 294, 299, 301, 302, 304, 305, 335, 336
Liquid chromatography-tandem mass spectrometry (LC-MS/MS), 233

M

Magnetic resonance spectroscopy (MRS), 360
MALDI-MS, *see* Matrix-assisted laser desorption ionization mass spectrometry (MALDI-MS)
Mammalian Gene Collection (MGC), 91
MARS, *see* Multiple Affinity Removal System (MARS)
Mass tagging, 232, 233, 236, 244, 246, 247
Matrix-assisted laser desorption ionization mass spectrometry (MALDI-MS), 5
Meso Scale Discovery (MSD), 11
Metabolic profiling, 294, 295, 300, 302, 303, 326, 342, 343, 348, 356, 357, 359, 361
Metabolite analysis, 317
Metabolomics, definition, 291, 319

MGC, *see* Mammalian Gene Collection (MGC)
Microarrays, 28, 35, 51, 52, 74, 77, 161, 273–274, 282
MicroSAGE, protocol for
day 1
kinase linkers, 93
total RNA isolation, 93
day 2
blunt ending released tags, 99–100
cleavage of cDNA with anchoring enzyme (NlaIII), 96
isolation of polyA + RNA and cDNA synthesis, 93–96
ligating linkers to cDNA, 97
ligating tags to form ditags (Part 1), 100
release tags using tagging enzyme (BsmF1) of cDNA, 98–99
day 3
ditags, isolation of, 103–104
ligating tags to form ditags (Part 2), 100
PCR amplification of ditags, 101–102
day 4
ditags, purification of, 104–106
day 5
cloning concatamers, 108–109
ligation of ditags to form concatamers, 106–108
day 6
electroporation, ligation products, 109–110
day 7
colony PCR, 110–111
Molecular lipids
identification/quantification using lipid profiler software, 377
quantification of, 377
Monocyte chemoat-tractant protein-1 (MCP-1), 10
Mouse liver, 378, 380–381
MPIS, *see* Multiple precursor ion scanning (MPIS)
MRS, *see* Magnetic resonance spectroscopy (MRS)
MS based label-free protein quantification technology (flow chart), 214, 227
Multiple Affinity Removal System (MARS), 172–173
Multiple precursor ion scanning (MPIS), 371, 372, 374, 377
Multiplex protein quantification technologies, 232

N

NanoDrop instrument, 6
National Institutes of Health (NIH) Biomarkers Definitions Working Group, 29
NMR-based metabolomics, biomarker discovery
applications, 355–361
cancer, 360–361
cardiovascular disease, 359–360
clinical biomarkers, 356–358
IEM, 358
toxicology, 355–356
biofluids/tissue, *see* Biofluids/tissue, analysis of
data preprocessing, 350–351
HCA, 343
methods, *see* NMR methods, metabolomics
partial least squares (PLS), 343
PCA, 343
statistical analysis
hierarchical cluster analysis, 352–353
methods, improvising NMR spectra, 354–355
orthogonal signal correction (OSC), 353–354

partial least squares, 353
PCA, 352
NMR methods, metabolomics
chemical derivatization, 348–350
CPMG experiment, 345–346
2D J-resolved spectroscopy, 347
2D NMR methods, 346
COSY, 346
HSQC, 346
TOCSY, 346
standard 1D NMR methods, 345
PRESAT, 345
TOCSY, biofluid analysis, 347–348
NMR spectroscopy, *see* Nuclear magnetic resonance (NMR) spectroscopy
Nuclear magnetic resonance (NMR) spectroscopy, 293

P

Partial least squares (PLS), 343
PCA, *see* Principal component analysis (PCA)
PCR, *see* Polymerase chain reaction (PCR)
PD effects, evaluation, 2
Peripheral blood mononuclear cells (PBMCs), 42–43
Pharmacodynamic (PD) biomarkers, 1, 2, 4, 11
Pharmacokinetics (PK), 1
Plasma protein, 172, 173, 182, 296, 326
PLS, *see* Partial least squares (PLS)
Polymerase chain reaction (PCR), 120
Precursor ion scanning, 372
PRESAT, *see* Presaturation (PRESAT)
Presaturation (PRESAT), 345
Principal component analysis (PCA), 293, 343, 377
Protein
classification and pathway analysis, 220
extraction/reduction/alkylation/ digestion, 213–215
identification, prioritization of, 217
protein identification (SEQUEST and X!Tandem algorithms), 216–217
quantification
analysis of variance (ANOVA) statistical model, 219
group effect, 219
HIGH category protein, 218
log transformation, 218–219
quantile normalization, 218
See also Label-free protein quantification method, peak intensity-based
ProteinChip, 253–267
Protein pattern visualization
silver staining, 183–184
investigator silver stain kit, 182
staining solutions, 182–183
Proteomics, 205, 211, 212, 232, 233, 251, 273, 292, 337, 342
Pyrosequencing, 119–136
analysis of results, 134–136
batch runs using the PSQhs96A, 133–134
entering an SNP run, 131
entering assay details, 130–131
individual plate run for PSQhs96/PSQhs96A, 131–133

Q

QSTAR XL mass spectrometer, 372, 374–377
Quadrupole Time-of-Flight Mass Spectrometry, 371
Quantile normalization, 218
Quantitative proteomics, 211

R

RBM, *see* Rules Based Medicine (RBM)
Real-time PCR, applications
  biomarker validation after microarray analysis, phases
    candidate biomarkers, technical validation of, 74–76
  gene copy number detection, 83–84
    drug metabolizing enzyme (DME) genes, 83
  microRNA expression profiling, 80–82
    drug potency, correlation with, 82
    stem-loop reverse transcription (RT), 81
    TaqMan miRNA assays, schematic description of, 81
Real-time PCR gene expression assays
  biomarker discovery and validation, 63–84
    applications, *see* Real-time PCR, applications
    principles, *see* TaqMan real-time PCR assays, principles of
Reverse transcription-polymerase chain reaction (RT-PCR), 41, 65, 70, 71, 87
RNA expression profiling, 51
RNA isolation
  cleanup using Qiagen RNEasy mini columns/extraction, 56–57
Rosetta Resolver Error Model, 7
RT-PCR, *see* Reverse transcription-polymerase chain reaction (RT-PCR)
Rules Based Medicine (RBM), 10

S

SAGE, *see* Serial analysis of gene expression (SAGE)
SAGE, analysis, 88–91
  Cap analysis of gene expression (CAGE), 90
  methodological versions, comparison, 90
SAGE analysis, phenotype single nucleotide polymorphism identification, 87–111
  bioinformatics, *see* Bioinformatics, SAGE
  Expressed sequence tags (ESTs), 88
  human transcriptome, characterization, 87
  Massively Parallel Signature Sequencing (MPSS), 88
  protocol for microSAGE, *see* MicroSAGE, protocol for
  RT-PCR, 87
  SAGE, 88–91
  single nucleotide polymorphisms (SNPs), 88
  SNPS, influence on interpretation of sage/MPSS experiments, 92
    GLGI-MPSS methodology, 92
Sample types and preanalytical properties
  gene expression profiles, PAXgene preparations, 8–9
  protein analyte measurement from blood, 10–11
    RBM, 10
  RNA/DNA collection from blood, 5–10
    cellular RNA, 6
    free nucleic acids, measurement of, 6
    PAXgene literature, 7
    PAXgene tubes, 7
    resistance mechanisms, 4
    Rosetta Resolver Error Model, 7
  sample selection
    coagulation, initiation of, 5
    laser scanning cytometry, slides, 4

MALDI MS, 5
resistance mechanisms, 4
surrogate tissues, 4
Sandwich ELISA, 274, 275, 282
Sandwich ELISA microarray, reliable/reproducible assays for high throughput screens
antibody microarray, types, 274
capture antibody, 274
chip incubation protocol, 283–287
chip methodology
blocking/storage of slides, 281–282
capture antibody, 278–279
pin cleaning, 277–278
printer/setup, 276–277
slide preparation, 279–280
materials, 275
general supplies, 275–276
instruments, 275
samples, 283
standards, 282–283
SDS-PAGE, *see* Sodium dodecyl sulfate-polyacrylamide gel electrophoresis (SDS-PAGE)
SELDI, *see* Surface-enhanced laser desorption ionization (SELDI)
SELDI technology, identification of protein biomarkers, 251–267
chromatographic arrays, *see* Chromatographic arrays
data acquisition, 258–259
focus mass, 260
laser intensity, 259
mass attenuation, 260
mass range, 259
pulse number, 260
spot partition, 260
data analysis, 261–263
EDM analysis, 261
Mann-Whitney test, 261
ProteinChip data manager, 261–262
direct sequencing applications, 265–266
peptide mapping protocols, 263–265
in-gel digestion, 263–264
passive elution/protease digestion in solution, 264–265
protease digestion on-chip, 265
protein-protein interaction assays, 266–267
sample preparation, 255–256
Bio-Rad Equalizer Beads, 256
SELDI, process, 257–258
See also Surface-enhanced laser desorption ionization (SELDI)
SELDI-assisted purification, 263
SEQUEST and X!Tandem algorithms, 216–217
Serial analysis of gene expression (SAGE), 88–91
Serum glutamic oxaloacetic transaminase (SGOT), 10
SIMCA, *see* Soft independent modeling of class analogy (SIMCA)
Single nucleotide polymorphisms (SNPs), 67, 88, 91, 119, 124, 130
SLR, *see* Stepwise linear regression (SLR)
SNPs, *see* Single nucleotide polymorphisms (SNPs)
SNPs and DNA methylation analysis, pyrosequencing methods, 119–136
materials
agarose gel electrophoresis, 121
bisulfate conversion of genomic DNA and SssI methylase treatment, 120–121
DNA template, 120
polymerase chain reaction, 121
pyrosequencing, 121–122
methods
bisulfate conversion of DNA, 122–124

PCR for pyrosequencing, 127–128
PCR optimization, 125–127
PCR primer design/pseudogenes, 124–125
PCR processing for pyrosequencing, 129–130
pyrosequencing, *see* Pyrosequencing
pyrosequencing primer design, 125

Sodium dodecyl sulfate-polyacrylamide gel electrophoresis (SDS-PAGE), 171, 174, 179, 185, 190, 197, 199, 235–237, 239, 256, 263
gels and buffers, preparation of, 180
load IPG strips and run, 178

Soft independent modeling of class analogy (SIMCA), 361

Solid-Phase Microextraction (SPME), 335

SPME, *see* Solid-Phase Microextraction (SPME)

Stable isotope labeling, 232, 294

Statistical heterospectroscopy (SHY), 355
See also NMR-based metabolomics, biomarker discovery

Stepwise linear regression (SLR), 343

Surface-enhanced laser desorption ionization mass spectrometry (SELDI-MS), 5, 253

Surface-enhanced laser desorption ionization (SELDI)
binding step, 258
chip surface selection, 257
energy-absorbing molecules, 258
preequilibration step, 257
SELDI chip processing, 257

T

TaqMan-based real-time PCR gene expression assays, 64

TaqMan Low Density Arrays (TLDAs), 79

TaqMan real-time PCR assays, principles of
design strategies, 66–70
minor groove binder (MGB), 69
nonfluorescent quencher (NFQ), 69
instrumentation, 70–71
charge-coupled device (CCD), 70–71
5' nuclease chemistry (using TaqMan probes), 66
Förster-type energy transfer, 66
protocols, PCR reaction, 71–74
mix/reaction plate, preparation, 72
plate/individual tube, running, 73
results, analysis, 73
storage, 73
thermal cycler conditions, 71
reaction, 67
technology, 65–66
transcription-polymerase chain reaction (RT-PCR), 65

TOCSY, *see* Total correlation spectroscopy (TOCSY)

Top-down proteomic analysis, multiplexed isobaric mass tagging strategy, 231–248
liquid chromatography-tandem mass spectrometry (LC-MS/MS), 233
materials
cell culture/serum/reagents, 234
ExacTag reagents, 234–235
in-gel digestion, 234, 236
liquid chromatography/mass spectrometry, 236

protein separation by sodium dodecylsulfate-polyacrylamide gel electrophoresis (SDS-PAGE)/gel staining, 235–236
methods
ExacTag labeling/desalting, 238–239
in-gel digestion, 239–241
mass spectrometry acquisition, 241
protein identification/quantitation, mascot and ExacTag analysis software, 241–242
protein separation, SDS-PAGE/Coomassie blue staining, 239
sample preparation, 237–238
top-down multiplexed protein labeling, workflow, 237
results
ExacTag-labeled peptide, spectrum, 242
nucleolar proteome changes after actinomycin D treatment, 242–246
Total correlation spectroscopy (TOCSY), 336
Toxicogenomics, 28–29
Toxicology, NMR, 355–356
Transcriptomics, 292, 318, 337
Triple-Play mode/method, 215
Tumor necrosis factor-α (TNF-α), 10
Two-dimensional gel electrophoresis, 189, 212, 232
Two-dimensional polyacrylamide gel electrophoresis (2-DE), 171–174, 179, 183, 184, 197, 212, 232

U

Ultra-performance liquid chromatography (UPLC), 294
Uni/multivariate data analysis, 377–382
Universal Protein Precipitating Agent (UPPA), 177
UPLC, *see* Ultra-performance liquid chromatography (UPLC)
UPPA, *see* Universal Protein Precipitating Agent (UPPA)

V

Variable of Importance (VIP), 379
Vascular endothelial growth factor (VEGF), 10
VEGF, *see* Vascular endothelial growth factor (VEGF)
VIP, *see* Variable of Importance (VIP)

Printed in the United States of America.